Plant Metabolomics

Xiaoquan Qi · Xiaoya Chen
Yulan Wang
Editors

Plant Metabolomics

Methods and Applications

Chemical Industry Press

Springer

Editors
Xiaoquan Qi
Institute of Botany
Chinese Academy of Sciences
Beijing
China

Xiaoya Chen
Shanghai Institute for Biological Sciences
Chinese Academy of Sciences
Shanghai
China

Yulan Wang
Wuhan Institute of Physics and Mathematics
Chinese Academy of Sciences
Wuhan, Hubei
China

ISBN 978-94-017-9290-5 ISBN 978-94-017-9291-2 (eBook)
DOI 10.1007/978-94-017-9291-2

Library of Congress Control Number: 2014951658

Springer Dordrecht Heidelberg New York London

Jointly published with Chemical Industry Press, Beijing
ISBN: 978-7-122-21436-2, Chemical Industry Press, Beijing

Printed on acid-free paper

Springer is part of Springer Science+Business Media (www.springer.com)

Preface

Life sciences progress quickly with each passing day. The improvement of genomics and related analytical techniques greatly promoted the rapid development of transcriptomics, proteomics, metabolomics, and phenomics. Thus, the means of system integration can be employed to reveal life phenomenon at multiple levels. The above research thoughts and methods gave birth to systems biology. Metabolomics is an important part of systems biology. Metabolites are closest to phenotype, thus the change in metabolites can more directly reveal the function of genes. And metabolic markers have important application values in the early diagnosis of diseases. There are a wide range of plant species in nature. Different groups of plant species synthesize different special compounds. It is estimated that there are 0.2–1 million kinds of metabolites synthesized by plants. The structural and physicochemical properties vary widely, making plant metabolomics research more challenging. Since the year 2002 when the first International Plant Metabolomics Conference was held in Wageningen, analytical techniques and methods of plant metabolomics have been developing rapidly and applied in several areas, such as plant scientific research, biotechnology safety assessment, crop breeding, etc., and play important roles in the study of gene function and the analysis of metabolic pathway and metabolic network regulation. Plant metabolomics research in China started around 2005, and currently has a good development trend. This book written in cooperation by researchers active in plant metabolomics in China, not only introduces the latest advances in plant metabolomics and analyzes the development trend in the next few years, but also demonstrates new studies of authors in their respective scientific projects, reflecting the current study level of China very well.

This book includes three parts introducing and demonstrating plant metabolomics. The first part includes an overview of plant metabolomics and the principles, methods, issues, considerations, and developments of metabolite analytical technologies, which mainly include mass spectrometry and nuclear magnetic resonance; the second part includes metabolomics data analysis, metabolites determination, metabolomics database and metabolic network study; the third part includes detailed application examples in plant metabolomics, which mostly are the current research achievements in recent years. We strive to be realistic and practical in this book, and hope that this book can promote the rapid development of plant metabolomics in China. Immense thanks to the authors of each chapter for taking the time from busy research and teaching tasks. Many authors of this book received the funding of "Metabolism and Regulation of Special Crop Nutrients (2007CB108800)" and the funding of "The Formation Mechanism and Control Approaches of Harmful Substances in Animal Products (2009CB118800)" from "973" plans of Ministry of Science of China. This book received the funding of "Metabolism and Regulation of Special Crop Nutrients" project office, too. Warm thanks to Prof. Chun-Ming Liu from the Institute of Botany of Chinese Academy of Sciences for his concern and help in the publishing of this book. Readers are welcome to criticize and correct errors and shortcomings in this book.

November 2010

Xiaoquan Qi
Xiaoya Chen
Yulan Wang

Contents

Chapter 1
Overview

Xiaoya Chen, Xiaoquan Qi and Li-Xin Duan

Metabolism reflects all the (bio) chemical changes during life activities, and metabolic activity is the essential characteristics and material basis of life. The central dogma of molecular biology believes that life information flows from deoxyribonucleic acids (DNAs) to messenger ribonucleic acids (mRNAs), then to proteins, and then to metabolites catalyzed by enzymes (mostly protein enzymes). And finally, these products converge and interact to produce a wide variety of biological phenotypes. DNA, as the carrier of life information, plays a crucial role in above-mentioned process. Genomics that comprehensively analyzing the constitution and function of DNAs in various species is the earliest 'omics.' Genomics research has greatly accelerated the development of life sciences. The success of genomics has also promoted the development of many other 'omics,' such as transcriptomics, proteomics, metabolomics, and phenomics (Fig. 1.1). Therewith systems biology that integrating above-mentioned 'omics' means to multi-levelly and comprehensively reveal biological phenomena come into being.

Metabolomics or metabonomics aims to study all the small molecular metabolites and their dynamic changes in an organism or a tissue or even a single cell (Oliver et al. 1998; Fiehn 2002). As early as in 300 AD, the ancient Greeks realized that they can predict disease through observing changes of body fluids or tissues. The idea is consistent with metabolomics in disease diagnosis (Nicholson and Lindon 2008). As an interdisciplinary combination of organic chemistry, analytical chemistry, chemometrics, informatics, genomics and transcriptomics, metabolomics has penetrated into all aspects of life sciences research. Metabolomics is a very important node of systems biology and the nearest node to phenotype. Therefore,

X. Chen (✉)
Institute of Plant Physiology and Ecology, Shanghai Institutes for Biological Sciences, Chinese Academy of Sciences, Shanghai 200032, China
e-mail: xychen@sibs.ac.cn

X. Qi · L.-X. Duan
Institute of Botany, Chinese Academy of Sciences, Beijing 100093, China
e-mail: xqi@ibcas.ac.cn

L.-X. Duan
e-mail: nlizn@ibcas.ac.cn

X. Qi et al. (eds.), *Plant Metabolomics*, DOI 10.1007/978-94-017-9291-2_1

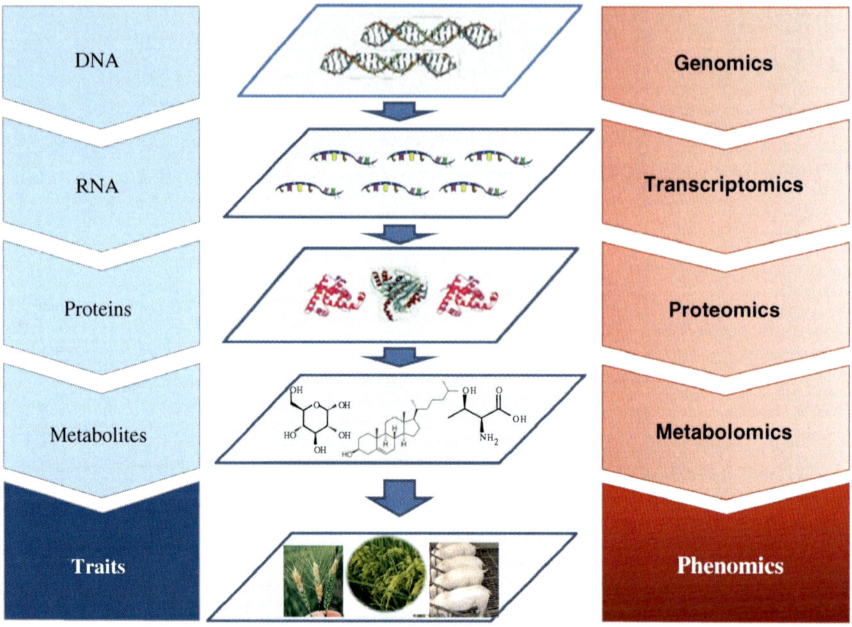

Fig. 1.1 Metabolomics is a branch of systems biology. The multilevel and systematic study is expected to reveal the molecular and metabolic basis for the formation of plant traits

metabolomics research is able to reveal gene function more holistically, thereby providing scientific basis to the application of biotechnology.

Plant metabolomics is one of important parts of metabolomics research. It is estimated that there are about 0.2–1 million metabolites produced by about 0.3 million known plant species in the world (unknown plant species are not included yet) (Dixon and Strack 2003). Of note is that the structures of secondary metabolites are widely different. In terms of the current (or after a period of time) level of instrument analysis, there is no any analytic method that is able to detect all the metabolites, which makes plant metabolomics research more challenging. Other chapters in this book will detail the analysis techniques of plant metabolites, data analysis methods, and application examples of metabolomics. This chapter summarizes the main developments and challenges that plant metabolomics face.

1.1 The Development of Metabolite Analysis Techniques

1.1.1 The Automation of Sample Preparation

Sampling, metabolite extraction, and pretreatment (derivatization) are the three critical steps of sample preparation in metabolomics and the premise for obtaining reliable data. In order to rapidly and efficiently sample and extract metabolites and keep the metabolites with good uniformity and stability, generally add the extraction solution rapidly after the plant tissue is quickly freeze in liquid nitrogen and ground into powder. The commonly used extraction solutions are methanol–chloroform–water, methanol–isopropanol–water, and methanol–water–formic acid. The extract must be dried and derivatized prior to be analyzed by gas chromatography–mass spectrometry (GC–MS). A two-step derivatization method is commonly used: The first step is adding methoxyamine pyridine solvent to reduce the cyclization of reductive sugar and protect carbonyl group; the second step is trimethyl silanization reaction to reduce the boiling point of analyte. Commonly used silanization reagents are bis (trimethylsilyl) trifluoroacetamide (BSTFA) and N-Methyl-N- (trimethylsilyl) trifluoroacetamide (MSTFA). The silanization effect of BSTFA and MSTFA is similar, but as the boiling point of MSTFA is lower than BSTFA, the chromatographic peak times of silanization reagents and by-products is earlier, and thus, the influence to metabolite analysis is less. Prior to be analyzed by liquid chromatography–mass spectrometry (LC–MS), the extract needs to be filtrated to remove insolubles to prevent blocking the separation column. Some efficient sampling, extraction, and derivatization methods have been developed to realize the globality, reproducibility, and high throughput that are required for metabolomics analysis. For example, (Weckweth et al. 2004) based on methanol–chloroform–water (2.5:1:1, v/v/v) extraction solution simultaneously extracted metabolites, proteins, and RNAs from 30 to 100 mg Arabidopsis (*Arabidopsis Thaliana*) fresh leaves, which can be subjected to metabolomics, proteomics, and transcriptomics analysis, respectively, providing convenience for systems biology research.

The automated sampling, extraction, and pretreatment technology come into being to prevent large deviation caused from the tedious and complicated sample preparation process (Nikolau and Wurtele 2007). The online derivatization and autoinjection system armed with the multifunctional autosampler decreases error that is caused from the tedious manual derivatization and the time difference of derivatization. The mechanization and automation of sample preparation technology is the tendency, because it can minimize the experimental error, thus making data more stable and reproductive.

1.1.2 Plant Single-Cell Separation Technology and High-Resolution Imaging Mass Spectrometry Techniques

At present, the main sources of plant metabolomics samples are plant organs, tissues, and suspension cultured cell lines which contain plant cells with different types and in different developmental stages and are treated with different environmental and experimental stimuli or treatment. The types and quantities in different functional cells are different. Therefore, metabolomics samples are different among themselves (Saito and Matsuda 2010). Furthermore, metabolites can be transported through vascular bundle between plant tissues. For example, methionine glucosinolate mainly accumulates in Arabidopsis seeds and buds, but the biosynthesis-related genes are found expressing in internode vascular bundle (Nour-Eldin and Halkier 2009). Although traditional methods of sampling and extraction in spite of the difference of developmental stages and cell types are quick and easy in sample preparation, to a greater degree, it weakens the ability of metabolomics in plant uncovering life activities.

The single-cell separation technology and high-resolution imaging MS are the trend of plant metabolomics development. With the combination of precise cell and tissue separation technology and ultra-high sensitive detection technology, researchers can study the intracellular metabolism and the intercellular transportation of metabolites. Recently, the imaging MS is used to detect the spatial distribution of metabolites, which uses a continuous laser to scan the surface of the plant tissue. The excitation of the laser makes metabolites of the surface of the plant tissue ionized and detected by mass spectrometer. And the distribution of metabolites in tissue or single cell can be observed by means of mass spectrum intensity. An example of such a technology is the use of the colloidal graphite-assisted laser desorption ionization mass spectrometry (MS) imaging technology to detect specific accumulation of flavonoids in flowers and petals (Saito and Matsuda 2010).

1.1.3 The Comprehensive Two-Dimensional Gas Chromatography High-Resolution Time-of-Flight Mass Spectrometry Coupling Technique (GC × GC-TOF/MS)

The separation and detection of metabolites are two core components of plant metabolomics analysis technology. The separation technique consists of a variety of chromatographic separation methods, such as gas chromatography (GC), liquid chromatography (LC), and capillary electrophoresis (CE). And detection technique consists of MS and nuclear magnetic resonance (NMR). The efficient combination of above two techniques is able to meet the fundamental needs of plat metabolomics analysis. Metabolomics aims at comprehensively, qualitatively, and quantitatively analyzing all the metabolites of samples. The problem is that there are about hundreds to thousands of metabolites in a single plant cell and different tissues, different cell types, different subcellular organelles and cells biosynthesize and accumulate different metabolites. In the meanwhile, the type and content of metabolites that are biosynthesized and accumulated are influenced by the changes of developmental stages and the difference of growth environment. Metabolites extracted from plant tissue materials have a huge number of types (at least thousands of types), complicated structures, various analogues, and extremely wide range of variation in content (it is estimated that the range of content variation is about 10^7) (Hegeman 2010). To comprehensively, qualitatively, and quantitatively analyze metabolites, the separation and detection instruments must have such features as good stability, good qualitative ability to compounds, high resolution and sensitivity, fast detection speed, and wide dynamic range of detection.

MS is such a technique that by detecting the size and abundance of charge-to-mass ratio of molecular ions or fragment ions after the ionization of interested components analyzes interested components qualitatively and quantitatively. MS is classified into various types depending on the type of mass analyzer. Compared with other analysis techniques, MS has higher sensitivity, faster detection speed, and wider dynamic range and can be combined with other techniques, such as GC and LC, thereby improving the analysis ability to complex matrices (Ekman et al. 2009). The time-of-flight (TOF) MS technique is superior to the quadrupole MS detector that was commonly used in earlier plant metabolomics analysis. The TOF detector analyzes the charge-to-mass ratio of different ions according to the difference of flight time of charged ions in a vacuum flight tube with extremely high sensitivity and fast scan speed (the data acquisition rate can achieve up to 500 full spectrograms per second). Thereby, it is beneficial for the fast analysis and improving the effect of spectrogram deconvolution. The dynamic range of mass detection can be over 10^5. With the high-resolution (a general capillary column has about more than 1 million theoretical plates) and good stability of GC, when GC is coupled with TOF/MS detector, the required qualification of metabolomics will be achieved. Generally, a standard voltage for electron impact ionization (−70 eV) is employed, the mass spectrum for a compound is usual stable and can be used as the structure characteristics of the compound. Generic compound libraries, such as National Institute of Standards and Technology Mass Spectral Database, have been established, which greatly ease the difficulties of compound qualitative analysis in the plant metabolomics research. Weckwerth and colleagues (Weckwerth et al. 2004) quantified about 1,000 compounds using GC time-of-flight mass spectrometer (GC-TOF/MS) and discriminated sugar biosynthesis isomerase mutant by metabolic network analysis. Wagner and colleagues qualitatively analyzed metabolites by the retention indices and mass spectrum of GC-TOF/MS (Wagner et al. 2003) and established a plant metabolome database, i.e., the Golm Metabolome Database (Kopka et al. 2005).

The maturely developed comprehensive two-dimensional GC (GC × GC) further strengthens the ability to detect complex metabolites. GC × GC is such a technique that connecting in series two chromatographic columns which with different stationary phases and are independent of each other. Each component separated from the first column is trapped and focused by modulator and then enter into the second column in pulse mode. The second column is very short, so separation is quick. Then, the separated component is subjected to MS scan in a speed of up to 500 spectrograms per second to get the two-dimensional GC data. Comprehensive two-dimensional GC combined with time-of-flight MS with high separation capacity and sensitivity is one of the most powerful separation tool widely used in the separation and analysis of complex systems such as metabolomics (Wang et al. 2010). Recently, Zoex Company introduced a instrument which combine comprehensive two-dimensional GC with high-resolution time-of-flight mass spectrometer detector, GC × GC-HiResTOF/MS. The resolution of TOF/MS is 4,000–7,000. The mass precision can be in three decimal places. The precise mass number can be used to speculate molecular formula. The highly precise mass of fragment ion peaks makes the deconvolution of overlap peaks more accurate and easier. As a result, the qualitative ability of MS to compounds is greatly enhanced. It can be predicted that this type of equipment will be widely applied in plant metabolomics research and will be a member of the plant metabolomics technology mainstream.

Components that are suitable for GC analysis are those that can be easily vaporized, which have low polarity and low boiling point. The examples of such components are various volatile compounds, or those have low boiling point after derivatization, such as amino acids, organic acids, sugars, and alcohols. GC can detect some primary metabolites of plant extract, but GC–MS alone cannot fully reveal the changes of all the plant metabolites.

1.1.4 Ultra-High-Performance Liquid Chromatography with Tandem Quadrupole Time-of-Flight Mass Spectrometry

Compared with GC, LC is not influenced by the volatility and thermal stability of the sample, pretreatment of sample is very simple, and injection can be directly done after filtering. Therefore, LC combined with MS can effectively analyze the abundant plant secondary metabolites, including various terpenoids, alkaloids, flavonoids, and glucosinolates. LC–MS-based analytical equipments have been progressed to be the indispensable analytical equipment for metabolomics.

There are a variety of MS types combined with LC, such as quadrupole mass spectrometry (Q/MS), tandem triple quadrupole mass spectrometry (QQQ/MS), ion trap mass spectrometry (IT/MS), time-of-flight mass spectrometry (TOF/MS), tandem quadrupole time-of-flight mass spectrometry (Q-TOF/MS), tandem ion trap time-of-flight mass spectrometry (IT-TOF/MS), fourier transform ion cyclotron resonance mass spectrometry (FT-ICR/MS), and linear trap quadrupole orbitrap mass spectrometer (LTQ-Orbitrap/MS). There are also a variety of ion sources that can be used for LC-MS, such as electrospray ionization (ESI), atmospheric pressure chemical ionization (APCI), matrix-assisted laser desorption ionization (MALDI), and atmospheric pressure photo ionization (APPI). And there are a lot of scan modes, such as selected ion monitoring (SIM), selected reaction monitoring (SRM), multiple reaction monitoring (MRM), full scan and tandem mass spectrometry MS/MS, or multistage MS MS^n scanning. Among many MS types, the high-resolution quadrupole time-of-flight mass spectrometer (Q-TOF/MS) is able to meet the plant metabolomics research requirement to the greatest extent. The present Q-TOF has a scanning speed of about 20 spectrograms per second. The latest triple TOF has a scanning speed of 100 spectrograms per second and a resolution of over 40 thousands and a wide dynamic range of over 10^5. LC-Q-TOF/MS has become the widely used analytical equipment for plant metabolomics research and has been successfully used in tomato metabolomics research (Moco et al. 2006). Diode array

(PDA) detector was connected behind LC, and ESI source was used for systematic analysis of metabolites with moderate polarity in tomato. With the combination of retention time, accurate mass, UV spectra, and double MS, a tomato metabolite database MoTo DB was built. Similarly, a comprehensive analysis of metabolites in Arabidopsis root and leaf was done with the use of Waters CapLC combined with Q-TOF/MS (von Roepenack-Lahaye et al. 2004). LC-Q-TOF/MS has been successfully used for the analysis of metabolite changes in vegetative growth of 14 ecotypic accessions and 160 recombinant inbred lines of *A. thaliana* and found that 75 % of the mass peaks are stably heritable and are mapped to Arabidopsis genome by means of metabolite quantitative trait locus (QTL) (Keurentjes et al. 2006; Fu et al. 2007).

Chromatography linear trap quadrupole mass spectrometer (LC-LTQ-MS/MS^2) can finish 3 full spectrum scans per second or 3 double MS scans per second and provide quantitative and qualitative information in the mean time (Evans et al. 2009).

LTQ-Orbitrap-MS and Fourier transform ion cyclotron resonance mass spectrometry (FT-ICR-MS) have ultra-high resolution of up to 60,000–2,500,000 and the MS^n capacity of up to 10. From the above two techniques, the precise molecular weight of the compound can be obtained for the prediction of molecular formula, the qualitative analysis of compounds, and the establishment of compound MS database. However, in order to obtain ultra-high resolution, a longer scanning time is required.

The development of ultra-high performance liquid chromatography (UPLC or UHPLC) technology is the icing on the cake for the metabolomics analytic technology. The UPLC makes use of a chromatographic column with the particle size of packing <2.0 μm and has overcome the traditional HPLC pressure limit. Column pressure can be increased to 15,000 psi. Therefore, the column efficiency is enhanced, peak widths are narrower, chromatographic resolution is increased, and the analysis time is shorter. It is very ideal for UPLC to combine with Q-TOF/MS which has a high scanning speed for the high-throughput analysis in plant metabolomics research. Recently, there is a breakthrough for the chromatographic packings technology. The hydrophilic interaction liquid chromatography (HILIC) employs a kind of polar stationary phase (such as silica gel and amino-bonded silica gel) and water and polar organic solvent as the mobile phase. It is particularly suitable for the separation of strong polar and strong hydrophilic small molecules. It is a supplement for the reverse chromatography (Tolsticov and Fiehn 2002; Cubbon et al. 2010).

In short, LC–MS has simple sample extraction requirements, easy to implement high-throughput and automation, can detect most of the plant metabolites, and is bound to play a greater role in plant metabolomics research. There have been various combinations of LC with MS, and still, there will be newcomers in the future development. The ultra-high-performance LC coupled with high-resolution tandem quadrupole time-of-flight MS technology will be the mainstream platform of plant metabolomics analysis.

1.1.5 Analysis of Other Special-Purpose Technology

(1) **Capillary Electrophoresis Mass Spectrometry (CE-MS) Technique**. CE technology is a new separation technology that is developed in the early 1980s, which based on the difference of mobility and distribution behavior between components to be separated. It has such features as high speed, high efficiency, high resolution, good reproducibility, and easy to automation, etc. The main advantage of CE-MS is the ability to detect ionic compounds, such as phosphorylated sugars, nucleotides, organic acids, and amino acids. Researchers have detected 200 metabolites from Arabidopsis and have identified 70–100 compounds from the 200 metabolites using CE-MS (Ohkama-Ohtsu et al. 2008).

(2) **Nuclear Magnetic Resonance (NMR) Technique**. NMR technology is a non-biased, universal analytical technique, with simple pretreatment requirement and a variety of detecting method. NMR includes liquid high-resolution NMR, high-resolution magic angle spinning NMR (HR-MAS), and in vivo magnetic resonance spectroscopy (MRS) technology. NMR methods also have their limitations, for example, its detecting sensitivity is low and the dynamic detecting range is limited, which makes it difficult to detect components with great difference in content in the same sample at the same time (Zhu et al. 2006). Recently, combined with LC separation, solid phase extraction (SPE) enrichment, full deuterated solvent elution, and online LC-UV-SPE-NMR-MS have been used in the structure identification of plant metabolites (Exarchou et al. 2003; Lin et al. 2008).

(3) **Fourier Transform-InfraRed (FT-IR) Technique**. FT-IR is based on the mechanism: The infrared ray gives rise to the vibrations of chemical bonds in the molecule or rotational energy level transitions, which lead to the production of absorption spectrum. The infrared spectrum of a plant sample is the superposition of the infrared spectra of all compounds therein, therefore having the fingerprint characteristics. FT-IR is capable of screening metabolic mutants from a large population, as it can conduct fast and high-throughput scanning of more than 1,000 samples per day without destroying them (Allwood et al. 2008). The disadvantage is that it is difficult to identify and discriminate metabolites with similar structure types.

There is a big chemical diversity of plant metabolites. The content of some components is minimal and the dynamic range is wide. The biosynthesis and accumulation of metabolites are vulnerable to the external environment. Metabolomics cannot predict the structure of metabolites from genomic information just as proteomics and transcriptomics do. Currently, the panorama qualitative and quantitative analysis cannot be done by one single mean, but using a variety of analytical tools that can complement each other to monitor and track changes of plant metabolites as many as possible.

1.2 Current Development and Challenges of Metabolomics Data Analysis

1.2.1 The Experimental Design and Standardization in Metabolomics Research

Just like transcriptomics, plant metabolomics research faces the needs and challenges in experimental design and standardization. On the one hand, different metabolites are biosynthesized and accumulated in different developmental stages, different tissues, different organs, and different cell types, and the content of them is extremely vulnerable to the growth environment. On the other hand, with a variety of instruments and analysis conditions being applied in metabolite analysis, a large number of non-comparable data can be easily produced.

The strict experimental design is the first step to achieve the success of metabolomics experiments. The experimental design requires to: ① set substantially identical plant growth and environmental conditions. And if the completely identical conditions cannot be achieved in each experiment, do ensure the identical growth and environmental conditions of different treatment or materials within the same experiment; ② set the experimental replications, typically 4–6 replications, which will further eliminate the environmental and experimental operation error and obtain statistically significant data. In order to control and monitor errors coming from sample extraction, pre-treatment and instrument analysis generally require to: ① set blank control: blank control is only free of the sample to be analyzed. It can detect the purity of the organic solvent, the miscellaneous peak of derivatization reagent, the plasticizers and other foreign contaminants from plastic tubes and pipette tips; ② set quality control samples: the quality control sample is a mixture of different types of standard materials. It can also be the mixture of a small part of each test sample, which contains all the types of compounds to analyze test samples. The systemic drift and deviation, the reduced response to metabolites and other unknown changes that are easily caused by the large number of test samples in one metabolomics experiment and the too long instruments running time, column pollution, reduced column efficiency, and the pollution of inlet as well as the aging of the mass detector can affect the detection of metabolism profile. Therefore, the quality control sample plays a very important role in the detection and correction of metabolomics data; ③ set the internal standard: by adding a known amount of an internal standard substance in the plant during the extraction process, errors produced in the process of extraction and analysis can be detected; ④ add the standard substance for the retention time index: The said standard substance is generally n-alkane or homologues of saturated fatty acid methyl ester, which can be used to calculate the retention index (RI), qualitatively analyze metabolites, and correct the drift of retention time.

Metabolomics develops rapidly, and the number of papers in various areas of metabolomics increases rapidly. Metabolomics mostly studies nontargeted unknown components. The data analytic methods, the metabolites qualification standard, and experimental report formats from different researchers may not be the same and hence hinder the exchange of data, the peer evaluation, and the reproduction of experimental results. Therefore, it is necessary to standardize metabolomics experiments and reports. John C. Lindon of the British Empire University of London took the lead and established the Metabolomics Standards Initiative (MSI) and proposed the standard metabolic reporting structure (SMRS) for metabolomics or metabonomics standardization (Lindon et al. 2005). Jenkins et al. proposed the architecture for metabolomics (ArMet) for plant metabolomics report (Jenkins et al. 2004). ArMet focus on the organization of metabolomics data, while SMRS puts forward more detailed description of data, including which parameter or data are necessary (Fiehn et al. 2007). Three papers that are published in the journal Metabolomics proposed, respectively, the MSI organization (Fiehn et al. 2007), the minimum reporting standards for chemical analysis (Sumner et al. 2007), and the minimum reporting standards for data analysis (Goodacre et al. 2007) in metabolomics. The minimum reporting standards for chemical analysis detailed the methods and technical parameters used in chemical analysis and proposed a lot of new standards and guiding principles (such as the structural determination level of metabolites and naming guidance of metabolites). The minimum reporting standards for data analysis describe in detail methods of univariate statistics, multivariate statistics, and informatics and defined many terms, such as deconvolution and pretreatment.

1.2.2 Metabolic Pathways and Metabolite Database

The establishment of metabolites database is conductive to the connection between metabolomics and other systems biology branches. At present, there are over 100 online databases (Tohge and Fernie 2009), in which the one with good comprehensiveness and containing metabolic pathways database is the Kyoto Encyclopedia of Genes and Genomes (KEGG) (http://www.genome.jp/kegg/). KEGG allows query of metabolic pathways, which include carbohydrates pathway, nucleotides pathway, amino acids pathway, and secondary metabolites pathway. KEGG provides not only all possible metabolic pathways, but also comprehensive annotations of enzymes involved in each step of the catalytic reaction, amino acid sequences, and the link to PDB database. KEGG is a comprehensive bioinformatics database for genome annotation.

MetaCyc (http://metacyc.org/) is a sub-database of BioCyc, which is a database of metabolic pathways and enzymes. It expounds metabolic pathways of more than 1,600 organism species and contains metabolic pathways, reactions, enzymes, and substrates that obtained from a large number of documents and online resources, including more than 1,200 pathways, 5,500 enzymes, 5,100 genes, and 7,700 metabolites. PlantCyc (http://www.plantcyc.org/) database and query system for plant were also established, including 12 species of Arabidopsis, poplar, rice, sorghum, and other species.

In addition to the large commercial databases on metabolites identification, such as NIST and WILEY, databases specifically for metabolomics that are established by experimental groups or research centers also come into being.

The Golm Metabolome Database (GMD) (http://gmd.mpimp-golm.mpg.de/) contains GC-MS and GC-TOF-MS mass spectrogram database of metabolites after derivatization. The current GMD database contains more than 2,000 evaluated mass spectrograms provided by both quadrupole and TOF mass spectrometry technical platform, including 1,089 non-redundant mass spectral tags (MSTs) and 360 identified MSTs. Additionally, GMD database also includes mass spectra, retention time and retention index, greatly improving the identification of structurally similar compounds. GMD can be queried free.

The METLIN metabolite database (http://137.131.20.83/metabo_search_alt2.php) (SCRIPPS) that is established by the Biological Mass Spectrometry Center of SCRIPPS Institute includes more than 23,000 endogenous and exogenous metabolites, small molecule drugs and drug metabolites, small peptides (about 8,000 dipeptides and tripeptides) and the like. METLIN contains corresponding LC-MS, MS/MS, and FTMS mass spectrometry data of each compound and can be freely retrieved by the mass, chemical formula, and structure of a compound online.

Madison-Qingdao Metabolomics Consortium Database (MMCD), http://mmcd.nmrfam.wisc.edu/) database is developed and maintained by the National Magnetic Resonance Facility at Madison of Wisconsin-Madison University and Qingdao Institute of Bioenergy and Bioprocess Technology, Chinese Academy of Sciences. It contains data of over 20,000 small molecule metabolites that are gathered from electronic databases and the scientific literatures and provides search engines of texts, chemical structures, NMR data, MS data, etc.

KNApSAcK (http://kanaya.naist.jp/KNApSAcK/) covers the majority of connections between plant species and metabolites, which including information on more than 40,000 metabolites and 8,000 plant species and can be easily queried by users to obtain reported information of a certain plant species.

MassBank (http://www.massbank.jp/) mass spectrogram database is jointly established by many universities and research institutions in Japan. It mainly collects high-resolution mass spectrogram produced by various mass spectrometer, such as the ESI-Q-TOF-MS/MS, ESI-QqQ-MS/MS, ESI-IT-$(MS)^n$ (Ion Trap, IT), GC-EI-TOF-MS, and LC-ESI-TOF-MS. The reference spectrum comprises the information of multistage MS. So far, 24,993 mass spectrograms of more than 12,000 primary metabolites and secondary metabolites have been obtained in positive and negative ion modes. Users can input text format of mass spectrogram to search and compare to the three-dimensional mass spectrogram.

But, so far, the number of metabolites in the plant metabolites database is not enough, and most of the metabolic pathways of most of plant species are not complete, which is the main challenges and opportunities faced by the plant metabolomics research.

1.2.3 The Structural Identification of Metabolite Is the Bottleneck of Metabolomics Development

The determination of metabolites is the focus and difficult point of metabolomics research. The minimum reporting standards for chemical analysis bases on the degree of determination divided metabolites into four categories, i.e., identified compounds, putatively annotated compounds, putatively characterized compound classes, and unknown compounds (Sumner et al. 2007).

The main approach of metabolite identification using GC–MS is by software deconvolution to obtain pure mass peak and search in the existing database. The more higher the similarity of an interested component to the compound in the database, the more possible it can be regard as the same compound. In recent years, the RI is used to assist determination, which to some extent can differentiate isomers with similar mass spectragram. Aligning with a standard is the most accurate determination method. For LC-MS, there is no standard spectrogram database, and thus, the determination is mainly done by accurate mass and MS^n spectral tree. Therefore, the application of high-resolution mass spectrometry instruments to determine the molecular weight of the compound is essential. In addition, the online HPLC-DAD-SPE-NMR-MS can be combined with high-resolution mass spectrometry and NMR technology to accelerate the determination and analysis of metabolite structure.

The first challenge of metabolomics bioinformatics is the comprehensive annotation of metabolites and the dissection of metabolic pathways. At present, the rebuilding of metabolic pathways of a plant species can be done by ways of hunting-related functional genes in genomic information and EST data as the KEGG and PlantCyc do. On the other hand, the previously reported metabolic pathways-related literature data can be used not only as the reference for the identification of metabolites, but also to infer metabolic pathways; The second challenge is the collection of structure identification data of natural product, including multistage mass spectrometry data; the third challenge is to develop an algorithm for the similarity comparison and search of multistage mass spectrometry; the fourth challenge is the de novo resolution of mass spectrogram, which start the automated structure identification from mass spectrogram and predict the mass spectrogram of compounds with known structure (Saito and Matsuda 2010).

1.2.4 The Metabolomics Data Annotation Facing the Challenge

The main research content of the metabolomics data annotation and the main goal of metabolomics research are to obtain a large number of qualitative and quantitative data of metabolites, classify these disorder data according to plant metabolic pathways or metabolic network and the functions involved, and then to get data with biological significance.

Plants synthesize a large number of compounds through metabolic pathways. Metabolites in the same pathway interrelated to each other to complete one or more biological functions. At present, only a small part of metabolic pathways have been dissected, which have been collected by KEGG and PlantCyc. On the one hand, the metabolic basis of certain biological functions can be systematically revealed by the annotation of differential metabolites that are collected from metabolomic analysis into metabolic pathways; on the other hand, some metabolic pathways can be predicted according to the association between metabolomics data and biological function, so as to accelerate the dissection of plant metabolic pathways. For example, after the type and content changes of tomato fruit metabolites were analyzed using GC–MS and annotated into metabolic pathways, the relationship of metabolic pathways and main agronomic traits was established (Schauer et al. 2006). Plants initiate certain metabolic pathways and produce certain disease-resistant metabolites immediately after the pathogen invasion. The concern to above-said metabolic pathways is beneficial for the dissection of disease resistance mechanism in terms of metabolites. For example, metabolomics studies have shown that the expression of salicylic acid and the regulation of metabolic pathways are closely associated with plant disease resistance (Kachroo and Kachroo 2009; Thomma et al. 1998). Fatma Kaplan et al. used GC–MS to study the changes of metabolic profile caused by low- and high-temperature stimuli in Arabidopsis and found some temperature stress-related metabolites (Kaplan et al. 2004). And then, the above-said data were co-analyzed with transcriptomics data to reveal the mechanism of temperature stress (Kaplan et al. 2007). The metabolomics data annotation will help resolving the function of new genes. For example, Jander et al. 2004 identified a threonine aldolase gene using LC–MS to high-throughput screen Arabidopsis mutation population. In addition, the annotation of metabolic pathways can be used for the determination of the activity of the enzyme involving in the metabolic pathway for the metabolic flux analysis.

The biological metabolism is a complete system, in which various metabolic pathways are interrelated and constitute a complex metabolic network. Combined with genomics, transcriptomics and metabolomics information, the annotation or reconstruction of metabolic network can be done. Patrick May et al. reconstructed the metabolic network of *Chlamydomonas reinhardtii* by the analysis of 1,069 proteins using LC–MS and 159 metabolites using GC × GC-TOF/MS and the integrating of genomic information (May et al. 2008). The correlation between metabolites can be established through correlation analysis of metabolic pathways. For example, some associated RNAs, metabolites, and closely correlated metabolic pathways have been identified in the ripening process of tomato by the study and statistical analysis of the content changes of primary metabolites and secondary metabolites and the combination with transcriptomics data (Ciarrari et al. 2006).

1.3 Plant Metabolomics Applications

Since the term metabolomics was proposed (Oliver et al. 1998), metabolomics has developed rapidly in aspects of analytical technique, data analysis, and application, got widely attention in the field of human disease study and disease diagnosis, and has broad application prospects (Nicholson and Lindon 2008). Plant metabolomics has been gradually applied to the basic biology study, such as the study of gene function, the dissection of metabolic pathway, and the regulation mechanism of metabolic network (Fiehn et al. 2000; Naoumkina et al. 2010; Keurentjes et al. 2006; Fernie et al. 2004); plant metabolomics also started to be applied to the field of plant breeding, such as crop yield and nutritional components (Schauer et al. 2006; Tarpley et al. 2005).

1.3.1 Plant Metabolomics Applications in Basic Biology

As an essential part of systems biology, plant metabolomics plays an increasingly important role in uncovering the basic vital movement and mechanism of life. Information on gene expression patterns and metabolites accumulation can be obtained by the analysis of expression profiles and metabolome. The comprehensive analysis of gene expression patterns, metabolites accumulation, and genomic data provides an effective way for the analysis of gene function. It is generally believed that certain genes, proteins, or metabolites involving in a biological process coexist in the same control system, with the relationship of coordinated regulation and co-expression. Therefore, if an unknown gene co-express with known genes, it can be assumed that the unknown gene may involve in the same biological process with the co-expressed known genes. The co-occur principle can be further applied in the co-accumulation relationship. If a metabolic pathway was modified by gene mutations or environmental changes, the modification process can be shown through changes of the metabolic profile. The analysis of gene expression profiles and metabolic profile can predict which gene(s) may be involved in this modification process. Candidate genes can be obtained by correlation analysis between metabolites and genes, and the function of the candidate genes can be studied by reverse genetics or reverse biochemical methods (Saito and Matsuda 2010). In the study of sulfur hunger, the change of metabolites and genes related to certain amino acids, lipids, and secondary metabolites such as glucosinolates and flavonoids was found, and 10,000 transcriptomics data (DNA chip) and 1,000 metabolites data (HPLC and CE-FT-MS) were analyzed by batch learning and self-organizing maps, which predicted genes involved in the biosynthesis of glucosinolates, such as genes encoding the sulfotransferase (Hirai et al. 2005), two MYB transcription regulating factor (Hirai et al. 2007), side chain extension-related enzymes (Sawada et al. 2009a), and a putative glucosinolates transporter (Sawada et al. 2009b).

Metabolomics combined with molecular genetics can resolve Arabidopsis metabolic pathways and metabolic network (Keurentjes et al. 2006). Keurentjes et al. using LC-Q-TOF-MS analyzed the full metabolite profile of 14 Arabidopsis accessions derived from different locations around the world and identified 2,475 different mass spectrum peaks, 706 of which are unique for certain accession, only 331 of which are common for the 14 accessions. So the variation of metabolite composition among the 14 accessions is very large. The broad-sense heritability analysis showed that most of the mass peak variations derived from genetic factors. In the 160 recombinant inbred lines (F_{10}) of Landsberg *erecta* (L*er)* and Cape Verde Islands (Cvi), a total of 2,129 mass peaks were detected, 853 of which are recombinant inbred population-specific, indicating that they are generated from genetic recombination. Genetic mapping of quantitative trait loci analysis found that the 1,592 mass peaks (74.8 %) have at least one QTL control ($P < 0.0001$, $q < 0.0002$). The mapped genetic factors at least partially explained the quantity and quality variation of metabolites. The analysis of the distribution of these detected QTL in genome found that they tend to locate in some specific regions of the genome, i.e., the hot spot region.

If a large number of highly relevant mass peaks were located to the same QTL of Arabidopsis, it can be inferred that these mass peaks are regulated by the same key genes. Co-regulated metabolites may be subject to a special regulatory factor control, or a particular step in metabolic pathways may be affected. Kroymann carefully analyzed these intermediates involved in the glucosinolates metabolic pathway of *A. thaliana* and found that all the aliphatic glucosides locate in the *MAM* locus of chromosome 5 and *AOP* locus of chromosome 4. It is known that the *MAM* controls chain extension (Kroymann et al. 2001), and *AOP* controls the modification of the side chain (Kliebenstein et al. 2001). The glycoside metabolic network structure rebuilt by metabolomics analysis is consistent with reported results. Therefore, it is confirmed that the colocation of Arabidopsis glucosinolates is due to a particular step of the metabolic pathway. In addition, the metabolomics analysis can infer the relationship between *AOP* and *MAM.* As shown in Fig. 1.2, metabolites at the *AOP* locus are also mapped at *MAM* locus, the reverse, however, does not apply, and thus, it can be estimated that *AOP* locates at the downstream of *MAM*.

Genetic analysis of metabolites can help resolving unknown metabolic pathways. A series of highly relevant unknown metabolite mass peak (QTL) have been located at 88.6 cM of *A. thaliana* chromosome 1. All of these co-localized metabolites are identified as flavonol glycosides with the analysis of photodiode array (PDA) absorption signal and MS/MS mass spectrum fragments. The additive allelic effect results indicate that the genotype variation of this locus displays an opposite flavonol accumulation mode, i.e., the L*er* allelic genotype accumulates flavonols, while Cvi not. These results indicate that a previously unknown glycosyltransferase exists in L*er*, while not in Cvi. Based on the homologous sequence analysis, two putative glycosyltransferase genes, UGT79B10 and UGT79B11, were found in the QTL locus, providing candidate genes for the functional verification of genes.

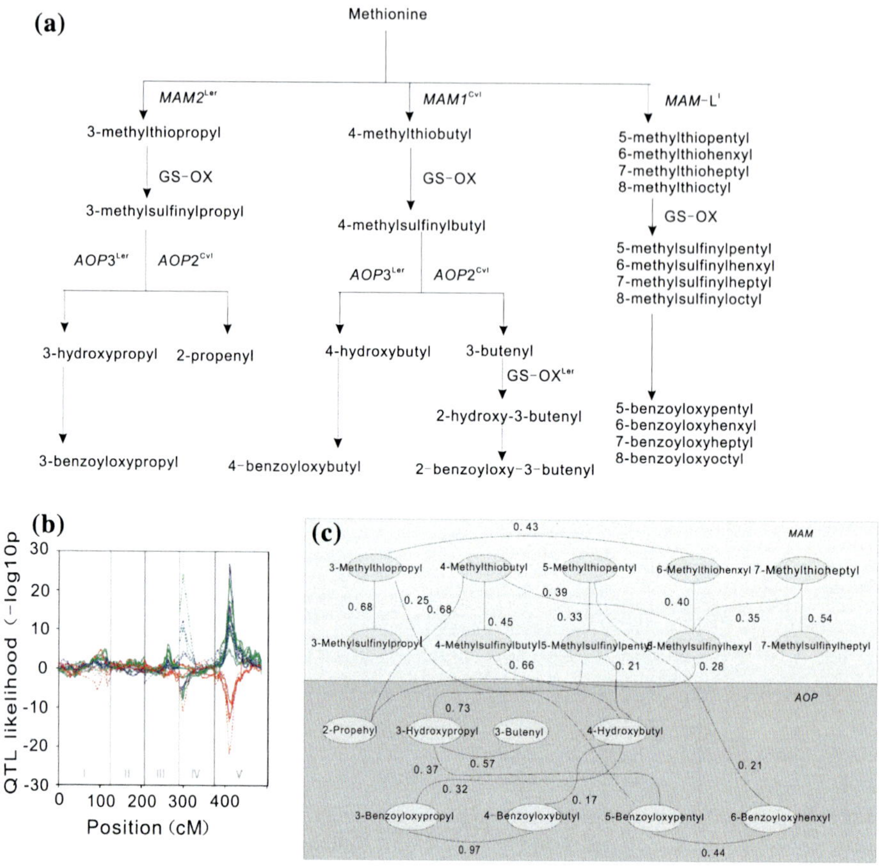

Fig. 1.2 Genetic regulation of aliphatic Glucosinolate accumulation in *A. thaliana*. **a** Scheme of aliphatic Glucosinolate formation. Corresponding loci of enzymatic steps are shown in *bold next to the arrows*. **b** QTL likelihood profiles of aliphatic Glucosinolates detected in the RIL population. The first QTL, at 303.3 cM, is at the *AOP* locus, and the second, at 409.4 cM, is at the *MAM* locus. The sign of the value is related to the additive effect at each marker (+, Cvi; –, Ler). *Solid lines* represent Glucosinolates before side chain modification and *dotted lines* Glucosinolates after side chain modification. Chromosomal borders are indicated by *vertical gray lines*. **c** Second-order genetic correlations between aliphatic Glucosinolates detected in the RIL population. *Upper panel* contains Glucosinolates before side chain modification; *lower panel* contains Glucosinolates after side chain modification. All edges depicted are significant at $\alpha = 0.05$, as determined by permutation. Corresponding correlation values are placed next to edges. In (**b**) and (**c**), colors represent different chain lengths (*red*, 3 C; *blue*, 4 C; *green*, 44 C) (Reprinted by permission from Macmillan Publishers Ltd: [Nature Genetics] (Keurentjes et al. 2006))

1.3.2 Applications of Metabolomics in Crop Breeding and Biotechnology

The safety of transgenic breeding has been the key issue hindering its wide use. Combined with allergic reactions and toxicological reaction experiments, metabolomics can be used for the comprehensive evaluation of the safety of genetically modified (GM) plants (Rischer and Oksman-Caldentey 2006). The metabolic profiles of GM, non-GM, and two cultivated varieties of potato were analyzed by principal component analysis (PCA) and hierarchical clustering analysis (HCA) and show that the most obvious differences in metabolic profiles are between the two potato varieties but not between the wild type and the GM plant (Shepherd et al. 2006).

Metabolomics can be used to study the effect of temperature, moisture, salt, sulfur, phosphorus, heavy metals, and other stress on plant metabolism. Drought resisting has been one of the most concerned issues in plant breeding. When drought occurs, the signal substances secreted by root will be transported to the aboveground part of plant through the xylem which is the channel for transporting water, minerals, and other components in plant. Alvarez et al. using metabolomics studied the changes of metabolites and proteins in maize xylem sap under drought stress (Alvarez et al. 2008) and found content changes in 31 metabolites. They found changes in most of the 31 metabolites such as the phytohormones abscisic acid (ABA) and cytokinin and the presence of high concentrations of the aromatic cytokinin 6-benzylaminopurine (BAP) at the 12 days of drought stress. Several phenylpropanoid compounds (coumaric, caffeic, and ferulic acids) were found in xylem sap. The concentrations of some of these phenylpropanoid compounds changed under drought. In parallel, an analysis of the xylem sap proteome was conducted. They also found a higher abundance of cationic peroxidases, which with the increase in phenylpropanoids may lead to a reduction in lignin biosynthesis in the xylem vessels and could affect the length of leaves and stems.

Plant initiates autoimmune response after the invasion of pathogen, in which metabolite plays a very important role. Once plant recognizes the pathogen, the cell will initiate a series of mobilization activities to activate disease resistance response to resist the invasion of pathogens. This disease resistance response requires a lot of energy, reducing power and carbon skeleton from the primary metabolic pathways to achieve the energy supplement and scheduling (Bolton 2009), and some low molecular weight antibacterial secondary metabolites from the secondary metabolic pathway such as phytoalexins (phenols, isoflavones, terpenes), and some substances that can block pathogen invasion and spread, such as lignin and callose (O'Connell and Panstruga 2006). On the other hand, after invading to the plant, pathogen will usually interfere with the normal metabolism of plant in order to meet their nutritional needs (Solomon et al. 2003; Swarbrick et al. 2006; Divon and Fluhr 2007). By the qualitative or quantitative analysis of metabolites in organism under certain conditions, metabolomics can detect changes in metabolites associated with a particular physiological and pathological reaction. In recent years, such method starts to be applied in the study of plant–microbe interaction. By the nontarget metabolite profile analysis of different resistant varieties of wheat and barley, some disease resistance-related mark metabolites have been identified, which play a role as assisting breeding in production practice (Hamzehzarghani et al. 2005; Swarbrick et al. 2006).

The plant–pathogen interaction systems, such as Brachypodium (*Brachypodium distachyon*)–rice blast fungus (*Magnaporthe grisea*) (Allwood et al. 2006; Parker et al. 2009), Arabidopsis (*A. thaliana*)–Arabidopsis nematode (*Heterodera schachtii*) (Hofmann et al. 2010), potato (*Solanum tuberosum* L)–potato late blight fungus (*Phytophthora infestans*) (Abu-Nada et al. 2007), have been studied by metabolomics. The study found that once the pathogen successfully invades into plant, it will seriously interfere with the normal metabolism of plants to meet their own needs of nutrition absorption and usage. Changes of metabolic profiles between resistant and susceptible cultivar after the invasion of pathogen are different. Combined with metabolomics and transcriptomics, Doehlemann et al. expounded that changes in signal transduction and metabolites in corn disease resistance response to tumor smut (*Ustilago maydis* SG200) are controlled by a major gene (Doehlemann et al. 2008). These studies suggest that metabolomics is a very effective method in the study of plant–microbe interaction.

The main traits of the crop, especially nutrition, quality, and other traits, have become the main target of metabolomics studies. Metabolomics is able to distinguish which chemical component determines the taste, such as sweet and sour. Accordingly, it has a good application prospect for breeding in improving nutrition, quality, and food quality. Using ^{1}H-NMR spectrum, combined with multi-dimensional statistical analysis can distinguish wines of different varieties and different regions (Son et al. 2009); by the application of GC–MS to analyze the metabolic profile of green tea, the quality of green tea is able to be evaluated (Pongsuwan et al. 2008); and the use of FT-IR to distinguish between different regions of olive oil ingredients (Rischer and Oksman-Caldentey 2006; Galtier et al. 2007).

The improvement of the nutritional quality and taste of the tomato can be achieved by changing the composition and content of the metabolites. The following section will be a more detailed introduction to literatures of metabolomics research in the molecular mechanisms of plant phenotype and metabolite content of tomato fruit. The study used 76 introgression lines containing chromosome segments of a wild tomato species (*Solanum pennellii*) in the genetic background of a cultivar M82 (*Solanum lycopersicum*) as the material and used high-throughput GC-MS to obtain metabolomics data of tomato fruit, combined with tomato whole plant phenotype data, and to explore the molecular genetic basis of metabolite composition formation of tomato fruit (Schauer et al. 2006). With the use of variance analysis, a total of 889 quantitative fruit metabolic loci and 326 loci that modify yield-associated traits were identified.

To further analyze the correlation among fruit metabolites, change, plant growth and development, and the important agronomic traits, correlation analysis of metabolites and 83 traits such as total fruit yield, harvest index and Brix was done, and a total of 280 positive correlations and 22 negative correlations were obtained with the significant level $P = 0.0001$ (Fig. 1.3). Module analysis and correlation

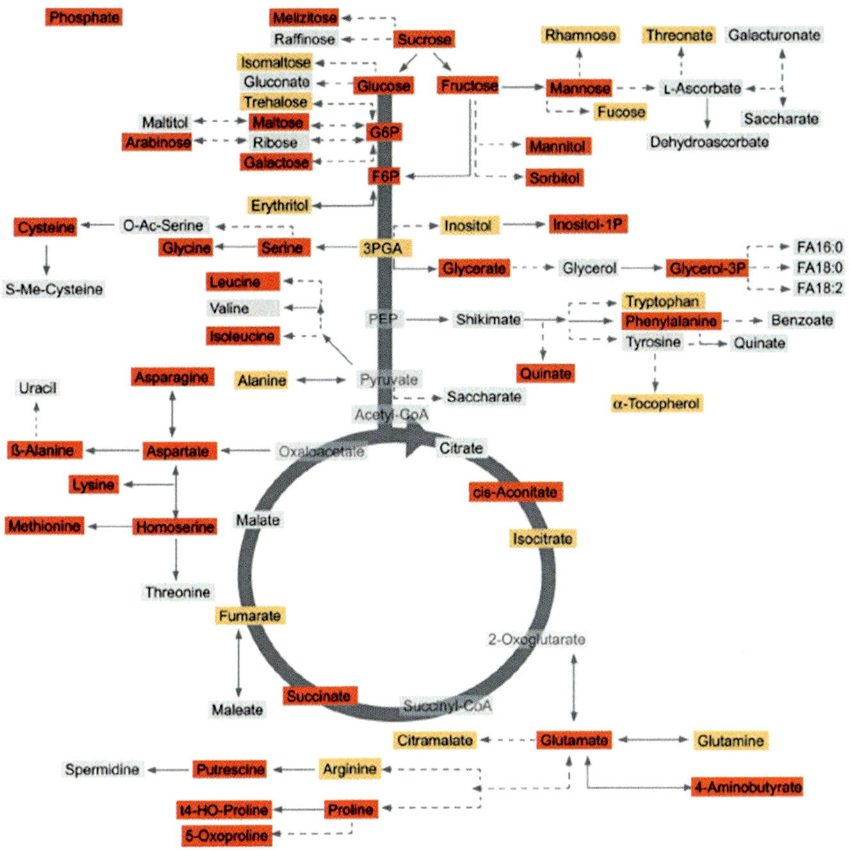

Fig. 1.3 The diagram of phenotype-associated and phenotype-independent metabolites in tomato fruit. Metabolites marked in *red* and *orange* were determined to be highly phenotype associated ($p \geq 0.005$) and phenotype associated ($p \geq 0.05$), respectively; those marked in *gray* were determined to be phenotype-independent because they did not correlate to any of the true phenotypic traits at either the strict or the permissive level. Traits colored *pale gray* were not analyzed in this study (Reprinted by permission from Macmillan Publishers Ltd: [Nature Biotechnology] (Schauer et al. 2006))

analysis show that many of the metabolic loci are associated with at least one QTL that modify whole plant phenotype traits. According to the degree of correlation between metabolic and whole plant phenotypic traits, the association was classified into three categories: whole plant phenotype associated, whole plant phenotype independent, and the intermediate. The proportion of the three types is 50, 27, and 23 %, respectively. In addition, correlation analysis of metabolic QTL and whole plant phenotypic QTL showed that 46 % of the 889 metabolic QTL are shared with whole plant phenotypic QTL.

Data from many introgression lines containing bin 6F (chromosome segment number) indicate that bin 6F contains 7 metabolic QTL associated with whole plant phenotype traits and 1 QTL associated with decreased harvest index and increased Brix. Previous studies have shown that bin 6F contains the *SELF-PRUNING* (*SP*) gene which regulates vegetative to reproductive switching of plant (Pnueli et al. 1998). Therefore, it is not difficult to understand why the harvest index-associated QTL locates in bin 6F. In order to further assess the causality among *SP* gene, harvest index and Brix, the phenotype and metabolite content of three near-isogenic lines (*sp* recessive mutants in Gardener and VFNT varieties and M82) and wild-type control were directly compared. The data show that the harvest index of the *SP* line was much lower than that of the *sp* recessive mutant, whereas the Brix value was always higher, indicating that there is a strong negative association between these traits. For Brix determination, the content of soluble solids, i.e., the soluble metabolites, was detected. From above experiments, it can be considered that the changes of plant phenotype can also affect the changes of fruit metabolites. Results from the F_2 population of M82 that segregated for a recessive male sterile mutation also proved that plant morphology contributes strongly to determining the metabolic composition of fruits at harvest time. The above findings provide target traits for efficient selection and laid the foundation for efficient breeding.

In short, plant metabolomics is developing rapidly and is playing an increasingly important role in revealing the molecular mechanisms of plant growth and development and adversity adaptation. At present, there are a lot of technical bottleneck and challenges in terms of metabolite analysis technique, compound structure identification, and data analysis, providing research objectives for those engaged in plant metabolomics research. This chapter and other chapters only enumerate several application examples of plant metabolomics; however, with the improvement of plant metabolomics techniques and methods, it is bound to be applied to every field of botany by the majority of plant science researchers.

References

Abu-Nada Y, Kushalappa EA, Marshall WD, Al-Mughrabi K, Murphy A. Murphy temporal dynamics of pathogenesis-related metabolites and their plausible pathways of induction in potato leaves following inoculation with *Phytophthora infestans*. Eur J Plant Pathol. 2007;118:375–91.

Allwood JW, Ellis DI, Heald JK, Goodacre R, Mur LAJ. Metabolomic approaches reveal that phosphatidic and phosphatidyl glycerol phospholipids are major discriminatory non-polar metabolites in responses by *Brachypodium distachyon* to challenge by *Magnaporthe grisea*. Plant J. 2006;46:351–68.

Allwood JW, Ellis DI, Goodacre R. Metabolomic technologies and their application to the study of plants and plant–host interactions. Physiol Plant. 2008;132:117–35.

Alvarez S, Marsh EL, Schroeder SG, Schachtman DP. Metabolomic and proteomic changes in the xylem sap of maize under drought. Plant Cell Environ. 2008;31:325–40.

Bolton MD. Primary metabolism and plant defense—fuel for the fire. Mol Plant Microbe Interact. 2009;22:487–97.

Carrari F, Baxter C, Usadel B, Urbanczyk-Wochniak E, Zanor M, Nunes-Nesi A, Nikiforova V, Centero D, Ratzka A, Pauly M, Sweetlove L, Fernie A. Integrated analysis of metabolite and transcript levels reveals the metabolic shifts that underlie tomato fruit development and highlight regulatory aspects of metabolic network behavior. Plant Physiol. 2006;142:1380–96.

Cubbon S, Antonio C, Wilson J, Thomas-Oates J. Metabolomic applications of HILIC–LC–MS. Mass Spectrom Rev. 2010;29:671–84.

Divon HH, Fluhr R. Nutrition acquisition strategies during fungal infection of plants. FEMS Microbiol Lett. 2007;266:65–74.

Dixon RA, Strack D. Phytochemistry meets genome analysis, and beyond. Phytochemistry. 2003;62:815–6.

Doehlemann G, Wahl R, Horst RJ, Voll L, Usadel B, Poree F, Stitt M, Pons-Kuehnemann J, Sonnewald U, Kahmann R, Kämper J. Reprogramming a maize plant: transcriptional and metabolic changes induced by the fungal biotroph Ustilago maydis. Plant J. 2008;56:181–95.

Ekman R, Silberring J, Westman-Brinkmalm A, Kraj A. Mass spectrometry: instrumentation interpretation and applications. New Jersey: Wiley; 2009.

Evans AM, DeHaven CD, Barrett T, Mitchell M, Milgram E. Integrated, nontargeted ultrahigh performance liquid chromatography/electrospray ionization tandem mass spectrometry platform for the identification and relative quantification of the small-molecule complement of biological systems. Anal Chem. 2009;81:6656–67.

Exarchou V, Godejohann M, van Beek TA, Gerothanassis IP, Vervoort J. LC-UV-solid-phase extraction-NMR-MS combined with a cryogenic flow probe and its application to the identification of compounds present in greek oregano. Anal Chem. 2003;75:6288–94.

Fernie A, Trethewey R, Krotzky A, Willmitzer L. Innovation—Metabolite profiling: from diagnostics to systems biology. Nat Rev Mol Cell Biol. 2004;5:763–9.

Fiehn O. Metabolomics—the link between genotypes and phenotypes. Plant Mol Biol. 2002;48:155–71.

Fiehn O, Kopka J, Dormann P, Altmann T, Trethewey RN, Willmitzer L. Metabolite profiling for plant functional genomics. Nat Biotech. 2000;18:1157–61.

Fiehn O, Robertson D, Griffin J, van der Werf M, Nikolau B, Morrison N, Sumner L, Goodacre R, Hardy N, Taylor C, Fostel J, Kristal B, Kaddurah-Daouk R, Mendes P, van Ommen B, Lindon J, Sansone S-A. The metabolomics standards initiative (MSI). Metabolomics. 2007;3:175–8.

Fu J, Swertz MA, Keurentjes JJB, Jansen RC. MetaNetwork: a computational protocol for the genetic study of metabolic networks. Nat Protoc. 2007;2:685–94.

Galtier O, Dupuya N, Dr′eau YL, Ollivier D, Pinatel C, Kister J, Artaud J. Geographic origins and compositions of virgin olive oils determinated by chemometric analysis of NIR spectra. Anal. Chim. Acta. 2007;595:136–44.

Goodacre R, Broadhurst D, Smilde AK, Kristal BS, Baker JD, Beger R, Bessant C, Connor S, Calmani G, Craig A, Ebbels T, Kell DB, Manetti C, Newton J, Paternostro G, Somorjai R, Sjostrom M, Trygg J, Wulfert F. Proposed minimum reporting standards for data analysis in metabolomics. Metabolomics. 2007;3:231–41.

Hamzehzarghani H, Kushalappa AC, Dion Y, Rioux S, Comeau A, Yaylayan V, Marshall WD, Mather DE. Metabolic profiling and factor analysis to discriminate quantitative resistance in wheat cultivars against fusarium head blight. Physiol Mol Plant Pathol. 2005;66:119–1113.

Hegeman AD. Plant metabolomics—meeting the analytical challenges of comprehensive metabolite analysis. Brief Funct Genomics. 2010;9:139–48.

Hirai MY, Klein M, Fujikawa Y, Yano M, Goodenowe DB, Yamazaki Y, Kanaya S, Nakamura Y, Kitayama M, Suzuki H, Sakurai N, Shibata D, Tokuhisa J, Reichelt M, Gershenzon J, Papenbrock J, Saito K. Elucidation of gene-to-gene and metabolite-to-gene networks in Arabidopsis by integration of metabolomics and transcriptomics. J Biol Chem. 2005;280:25590–5.

Hirai MY, Sugiyama K, Sawada Y, Tohge T, Obayashi T, Suzuki A, Araki R, Sakurai N, Suzuki H, Aoki K, Goda H, Nishizawa OI, Shibata D, Saito K. Omics-based identification of Arabidopsis Myb transcription factors regulating aliphatic glucosinolate biosynthesis. Proc Natl Acad Sci USA. 2007;104:6478–83.

Hofmann JL, El Ashry AEN, Anwar S, Erban A, Kopka J, Grundler F. Metabolic profiling reveals local and systemic responses of host plants to nematode parasitism. Plant J. 2010;62:1058–71.

Jander G, Norris SR, Joshi V, Fraga M, Rugg A, Yu S, Li L, Last RL. Application of a high-throughput HPLC-MS/MS assay to Arabidopsis mutant screening; evidence that threonine aldolase plays a role in seed nutritional quality. Plant J. 2004;39:465–75.

Jenkins H, Hardy N, Beckmann M, Draper J, Smith AR, Taylor J, Fiehn O, Goodacre R, Bino RJ, Hall R, Kopka J, Lane GA, Lange BM, Liu JR, Mendes P, Nikolau BJ, Oliver SG, Paton NW, Rhee S, Roessner-Tunali U, Saito K, Smedsgaard J, Sumner LW, Wang T, Walsh S, Wurtele ES, Kell DB. A proposed framework for the description of plant metabolomics experiments and their results. Nat Biotech. 2004;22:1601–6.

Kachroo A, Kachroo P. Fatty acid-derived signals in plant defense. Ann Rev Phyto. 2009;47:153–76.

Kaplan F, Kopka J, Haskell DW, Zhao W, Schiller KC, Gatzke N, Sung DY, Guy CL. Exploring the temperature-stress metabolome of Arabidopsis. Plant Physiol. 2004;136:4159–68.

Kaplan F, Kopka J, Sung DY, Zhao W, Popp M, Porat R, Guy CL. Transcript and metabolite profiling during cold acclimation of Arabidopsis reveals an intricate relationship of cold-regulated gene expression with modifications in metabolite content. Plant J. 2007;50:967–81.

Keurentjes JJB, Fu J, de Vos CHR, Lommen A, Hall RD, Bino RJ, van der Plas LHW, Jansen RC, Vreugdenhil D, Koornneef M. The genetics of plant metabolism. Nat Genet. 2006;38:842–9.

Kliebenstein DJ, Lambrix VM, Reichelt M, Gershenzon J, Mitchell Olds T. Gene duplication in the diversification of secondary metabolism: tandem 2-oxoglutarate–dependent dioxygenases control glucosinolate biosynthesis in Arabidopsis. Plant Cell. 2001;13:681–93.

Kopka J, Schauer N, Krueger S, Birkemeyer C, Usadel B, Bergmuller E, Dormann P, Weckwerth W, Gibon Y, Stitt M, Willmitzer L, Fernie AR, Steinhauser D. GMD@CSB.DB: the golm metabolome database. Bioinformatics. 2005;21:1635–8.

Kroymann J. A gene controlling variation in Arabidopsis glucosinolate composition is part of the methionine chain elongation pathway. Plant Physiol. 2001;127:1077–88.

Lin Y, Schiavo S, Orjala J, Vouros P, Kautz R. Microscale LC-MS-NMR platform applied to the identification of active cyanobacterial metabolites. Anal Chem. 2008;80:8045–54.

Lindon JC, Nicholson JK, Holmes E, Keun HC, Craig A, Pearce JTM, Bruce SJ, Hardy N, Sansone S-A, Antti H, Jonsson P, Daykin C, Navarange M, Beger RD, Verheij ER, Amberg A, Baunsgaard D, Cantor GH, Lehman-McKeeman L, Earll M, Wold v, Johansson E, Haselden JN, Kramer K, Thomas C, Lindberg J, Schuppe-Koistinen I, Wilson ID, Reily MD, Robertson DG, Senn H, Krotzky A, Kochhar S, Powell J, Ouderaa FVD, Plumb R, Schaefer H, Spraul M. Summary recommendations for standardization and reporting of metabolic analyses. Nat Biotech. (2005);23:833–8.

May P, Wienkoop S, Kempa S, Usadel B, Christian N, Rupprecht J, Weiss J, Recuenco-Munoz L, Ebenhoh O, Weckwerth W, Walther D. Metabolomics- and Proteomics-assisted genome annotation and analysis of the draft metabolic network of chlamydomonas reinhardtii. Genetics. 2008;179:157–66.

Moco S, Bino RJ, Vorst O, Verhoeven HA, de Groot J, van Beek TA, Vervoort J, de Vos CHR. A liquid chromatography-mass spectrometry-based metabolome database for tomato. Plant Physiol. 2006;141:1205–18.

Naoumkina MA, Modolo LV, Huhman DV, Urbanczyk-Wochniak E, Tang Y, Sumner LW, Dixon RA. Genomic and coexpression analyses predict multiple genes involved in triterpene saponin biosynthesis in *M. truncatula*. Plant Cell. 2010;22:850–66.

Nicholson JK, Lindon JC. Metabonomics Nature. 2008;455:1053–6.

Nikolau B, Wurtele E. Concepts in plant metabolomics. Dordrecht, The Netherlands: Springer; 2007. p. 11–5.

Nour-Eldin HH, Halkier BA. Piecing together the transport pathway of aliphatic glucosinolates. Phytochem Rev. 2009;8:53–7.

O'Connell RJ, Panstruga R. Tête à tête inside a plant cell: establishing compatibility between plants and biotrophic fungi and oomycetes. New Phytol. 2006;171:699–718.

Ohkama-Ohtsu N, Oikawa A, Zhao P, Xiang C, Saito K, Oliver DJ. A {gamma}-glutamyl transpeptidase-independent pathway of glutathione catabolism to glutamate via 5-oxoproline in Arabidopsis. Plant Physiol. 2008;148:1603–13.

Oliver SG, Winson MK, Kell DB, Baganz F. Systematic functional analysis of the yeast genome. Trends Biotechnol. 1998;16:373–8.

Parker D, Beckmann M, Zubair H, Enot DP, Caracuel-Rios Z, Overy DP, Snowdon S, Talbot NJ, Draper J. Metabolomic analysis reveals a common pattern of metabolic re-programming during invasion of three host plant species by *M. grisea*. Plant J. 2009;59:723–37.

Pnueli L, Carmel-Goren L, Hareven D, Gutfinger T, Alvarez J, Ganal M, Zamir D, Lifschitz E. The SELF-PRUNING gene of tomato regulates vegetative to reproductive switching of sympodial meristems and is the ortholog of CEN and TFL1. Development. 1998;125:1979–89.

Pongsuwan W, Bamba T, Yonetani T, Kobayashi A, Fukusaki E. Quality prediction of Japanese green tea using pyrolyzer coupled GC/MS based metabolic fingerprinting. J Agric Food Chem. 2008;56:744–50.

Rischer H, Oksman-Caldentey KM. Unintended effects in genetically modified crops: revealed by metabolomics? Trends Biotechnol. 2006;24:102–4.

Saito K, Matsuda F. Metabolomics for functional genomics, systems biology, and biotechnology. Annu Rev Plant Biol. 2010;61:463–89.

Sawada Y, Kuwahara A, Nagano M, Narisawa T, Sakata A, Saito K, Hirai MY. Omics-based approaches to methionine side-chain elongation in Arabidopsis: characterization of the genes encoding methylthioalkylmalate isomerase and methylthioalkylmalate dehydrogenase. Plant Cell Physiol. 2009a;50:1181–90.

Sawada Y, Toyooka K, Kuwahara A, Sakata A, Nagano M, Saito K, Hirai MY. Arabidopsis bile acid: sodium symporter family protein 5 is involved in methionine-derived glucosinolate biosynthesis. Plant Cell Physiol. 2009b;50:1579–86.

Schauer N, Semel Y, Roessner U, Gur A, Balbo I, Carrari F, Pleban T, Perez-Melis A, Bruedigam C, Kopka J, Willmitzer L, Zamir D, Fernie AR. Comprehensive metabolic profiling and phenotyping of interspecific introgression lines for tomato improvement. Nat Biotech. 2006;24:447–54.

Shepherd LV, McNicol JW, Razzo R, Taylor MA, Davies HV. Assessing the potential for unintended effects in genetically modified potatoes perturbed in metabolic and developmental processes. Targeted analysis of key nutrients and anti-nutrients. Transgenic Res. 2006;15:409–25.

Solomon PS, Tan K-C, Oliver RP. The nutrient supply of pathogenic fungi: a fertile field for study. Mol Plant Pathol. 2003;4:203–10.

Son HS, Hwang GS, Kim KM, Ahn HJ, Park WM, Van Den Berg F, Hong YS, Lee CH. Metabolomic studies on geographical grapes and their wines using ^{1}H NMR analysis coupled with multivariate statistics. J Agric Food Chem. 2009;57:1481–90.

Sumner LW, Amberg A, Barrett D, Beale MH, Beger R, Daykin CA, Fan TWM, Fiehn O, Goodacre R, Griffin JL, Hankemeier T, Hardy N, Harnly J, Higashi R, Kopka J, Lane AN, Lindon JC, Marriott P, Nicholls AW, Reily MD, Thaden JJ, Viant MR. Proposed minimum reporting standards for chemical analysis. Metabolomics. 2007;3:211–21.

Swarbrick P, Schulze-Lefert P, Scholes J. Metabolic consequences of susceptibility and resistance (race-specific and broad spectrum) in barley leaves challenged with powdery mildew. Plant Cell Environ. 2006;29:1061–76.

Tarpley L, Duran AL, Kebrom TH, Sumner LW. Biomarker metabolites capturing the metabolite variance present in a rice plant developmental period. BMC Plant Biol. 2005;8:1–12.

Thomma B, Eggermont K, Penninckx I. Separate jasmonate-dependent and salicylate-dependent defense-response pathways in Arabidopsis are essential for resistance to distinct microbial pathogens. Proc Natl Acad Sci USA. 1998;95:15107–11.

Tohge T, Fernie AR. Web-based resources for mass spectrometry based metabolomics: a user's guide. Phytochemistry. 2009;70:450–6.

Tolstikov VV, Fiehn O. Analysis of highly polar compounds of plant origin: combination of hydrophilic interaction chromatography and electrospray ion trap mass spectrometry. Anal Biochem. 2002;301:298–307.

Von Roepenack-Lahaye E, Degenkolb T, Zerjeski M, Franz M, Roth U, Wessjohann L, Schmidt J, Scheel D, Clemens S. Profiling of Arabidopsis secondary metabolites by capillary liquid chromatography coupled to electrospray ionization quadrupole time-of-flight mass spectrometry. Plant Physiol. 2004;134:548–59.

Wagner C, Sefkow M, Kopka J. Construction and application of a mass spectral and retention time index database generated from plant GC/EI-TOF-MS metabolite profiles. Phytochemistry. 2003;62:887–900.

Wang B, Fang A, Heim J, Bogdanov B, Pugh S, Libardoni M, Zhang X. DISCO: distance and spectrum correlation optimization alignment for two-dimensional gas chromatography time-of-flight mass spectrometry based metabolomics. Anal Chem. 2010;82:5069–81.

Weckwerth W, Wenzel K, Fiehn O. Process for the integrated extraction, identification and quantification of metabolites, proteins and RNA to reveal their co-regulation in biochemical networks. Proteomics. 2004;4:78–83.

Zhu H, Tang HR, Zhang X, Liu ML. NMR based metabolomics. Chemistry. 2006;69:1–9.

Chapter 2
Gas Chromatography Mass Spectrometry Coupling Techniques

Zhen Xue, Li-Xin Duan and Xiaoquan Qi

Relative to other metabolomics analysis techniques, gas chromatography mass spectrometry (GC/MS) is one of the earliest applied analysis techniques in metabolomics. The first paper on metabolomics (metabolic profiling) is derived from the application of GC/MS analysis in urine and tissue extracts (Dalgliesh et al. 1966). With the arrival of omics era and the proposing of metabolomics concept, people began to try using a variety of analytical techniques to obtain metabolomics data. These techniques include chromatography, capillary electrophoresis, mass spectrometry, nuclear magnetic resonance (NMR), infrared spectroscopy, and electrical chemical methods, etc. GC/MS and NMR are the main technologies applied in the early development of metabolomics; in later stage, high-resolution liquid chromatography mass spectrometry (LC-MS) with fast scanning capability is widely used in metabolomics analysis; in recent years, people began trying to integrate a variety of analytical techniques in order to bring into play the advantage of a variety of methods and to make up for the lack of a single analysis. However, there is no one technology can perform quantitative and qualitative analysis for all the endogenous metabolites in the biological sample. GC/MS is the most mature chromatography mass spectrometry coupling technology, suitable for the analysis of metabolites with low polarity, low boiling point, or volatile after being derivatized. GC/MS has been one of the main analytical platforms in plant metabolomics due to the high resolution, high sensitivity, good reproducibility, a large number of standard metabolite spectra libraries, and the relative low cost. This chapter introduces the GC/MS technology principle, plant metabolite extraction and pretreatment technologies, main problems of GC/MS in plant metabolomics analysis, notes, as well as the latest developments.

Z. Xue (✉) · L.-X. Duan · X. Qi
Institute of Botany, Chinese Academy of Sciences, Beijing 100093, China
e-mail: xuezhen@ibcas.ac.cn

L.-X. Duan
e-mail: nlizn@ibcas.ac.cn

X. Qi
e-mail: xqi@ibcas.ac.cn

X. Qi et al. (eds.), *Plant Metabolomics*, DOI 10.1007/978-94-017-9291-2_2

2.1 GC/MS Principles and Key Technologies

GC can well separate complex mixtures, and MS can detect these compounds. The combination of the two has a more favorite place, for example, both GC and MS can run in the gaseous state; thus, they can be connected directly, and the interface is very simple. Simply speaking, the performance of GC/MS is stable, and the reproducibility is good.

GC plays a role in separation and introduces target substances into MS system by directly injecting analytes into chromatographic column or introducing analytes into chromatographic column after injecting and heating. The chromatographic column is heated thermostatically or program-controlled. Each component is separated by the difference of thermodynamic properties (the difference of boiling points and the difference of selective absorption in the stationary phase) and the different distribution in stationary phase and mobile phase (carrier gas). MS is in fact a detector, mainly including ionization source, mass analyzer, and electron multiplier tubes. Target substances enter into MS through GC and ionized into gaseous ions in the ionization source and then enter into mass analyzer. Ions with different mass-to-charge ratio are sequentially separated and reach the electron multiplier, generating electrical signal, in order to give the 3D information of the target substances, making qualitative analysis more accurate by using ion fragment information. Figure 2.1 shows a schematic diagram of the main parts of GC/MS.

The following are the key technologies of GC/MS (Villas-boas et al. 2007).

2.1.1 Key Technologies of GC in GC/MS

2.1.1.1 Gas Supply System and Mobile Phase

Helium is the commonly used mobile phase in GC/MS, which is provided by compressed gas cylinder, and the flow rate is controlled by pressure and flow

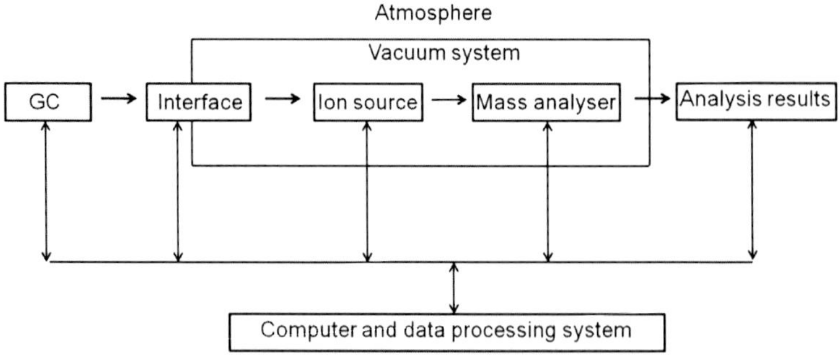

Fig. 2.1 The schematic diagram of the constitution of GC/MS

regulators. GC/MS analysis requires stable airflow, so the gas supply system is a critical component in GC; most of the advanced flow rate control systems are very stable and capable of providing stable and accurate flow rate and pressure control. Gas quality is also an important aspect guaranteeing GC/MS analysis. Gas quality includes gas purity and gas supply. Poor gas quality will cause the emergence of "ghost peaks" in chromatogram. In addition, gas residues, especially oxygen and water, will damage chromatographic column (polar substances in the column are very sensitive to oxygen), and oxygen will reduce the lifetime of lamp filament, and the presence of hydrocarbons will increase signal background. Thus, it is inevitable to use high-purity carrier gas and gas purification apparatus to remove trace oxygen and water in carrier gas. Gas purity should reach 99.9995 %, the purer the better.

2.1.1.2 Chromatographic Columns and Oven

GC columns include packed column and capillary column. The packed column is to coat stationary liquid to carrier particles with uniform size and pack the coated carrier into metal, glass, or plastic tube. Capillary column is to coat stationary phase to the inner wall of the capillary tube, and the capillary tube has no packing. Generally, GC/MS-specific column with MS identification is chose in GC/MS. So far, GC column is classified by the difference of mobile phases: The apolar methyl silicone column is the most used stationary phase; the methyl silicone containing 5 % phenyl groups is the medium polar column; the cyano-propyl methyl silicone stationary phase makes the polarity of the column greater; and the column with the largest polarity is the carbowax-containing stationary phase. These stationary phases are chemically bond into the inner wall of the column, cross-linked to increase its stability. All the columns have their own temperature tolerance ranges; generally speaking, the greater the polarity of the column, the worse the high temperature tolerance.

The diameter of the capillary column and the thickness of the stationary phase determine the β value (the distribution ratio of substance between the gas phase and the stationary phase), that is, the amount of substances distributed in the gas phase and the stationary phase. β value is a core parameter in selecting column. Low β value will increase the retention ability of analytes in the column (the equivalent of more analytes retained in the stationary phase), while reducing the number of plates. Thus, column with thick-film stationary phase (low β value) is typically used for the analysis of volatile compounds, and thin-film column is beneficial for the analysis of less volatile compounds with high boiling point.

Chromatographic peak width is proportional to the square root of the column length, while the column length and the number of theoretical plates have an important relationship. Retention time is proportional to the retaining time of the substance in the stationary phase, that is, the longer the substance retained in the column, the wider the peak width. According to the principle that separation efficiency (theoretical plate number N) is proportional to the square root of column length, if column length increases four times, for example, the resolution will be doubly increased.

Since the distribution efficiency of substances between two phases is strongly depend on the gas temperature, the control of GC temperature is also very important. Suitable temperature program can effectively improve the separation efficiency of substances in the analysis process. At the same time, temperature program can also be used to optimize analysis time.

2.1.1.3 GC Injection System

Injection system is the most critical part of GC, with the aim to convert liquid samples into gaseous state and focus them in the starting end of the column. In metabolomics analysis, imperfections of injection system will cause the incomplete vaporization of sample and the sample cannot timely enter into the column, resulting in peak broadening. Split/splitless injection commonly used injection modes in metabolomics, and also the focus is discussed below.

Split/splitless injection is based on the technique that the sample is rapidly evaporated in a small heated chamber, and the volatile substances are transferred into the column under the action of carrier gas. In the packed column, the flow rate of the carrier gas is high (30–50 ml/min), which is very easy to transfer the sample into the column rapidly and efficiently. However, the introduction of capillary column reduces the flow rate (typically 1–2 ml/min), making the need to change prior techniques to adapt to the characteristics of the capillary column. Initially, it is resolved by venting a part of the sample out of the sample inlet with high flow rate, but this will result in the loss of sample, reducing the sensitivity—this is known as split injection. A later development was to close the split vent during injection, while focusing the analyte in the column rapidly—splitless injection. Figure 2.2 shows the schematic diagram of the split/splitless injection system.

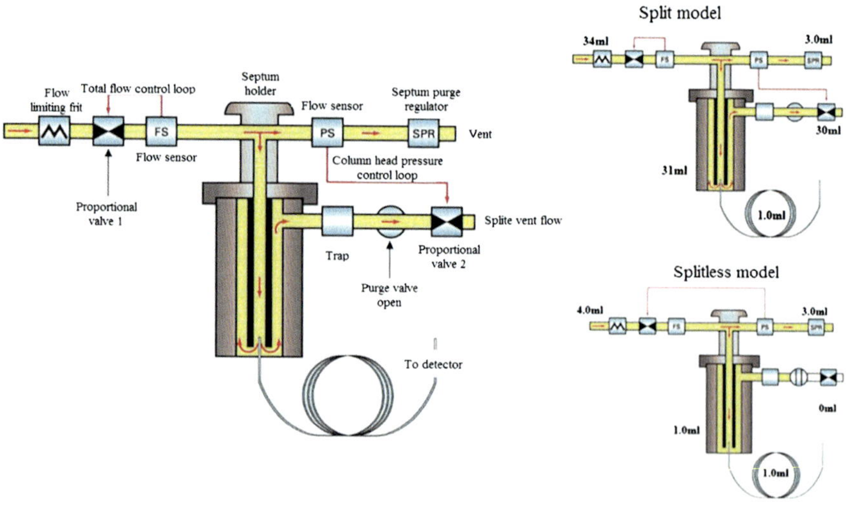

Fig. 2.2 The schematic diagram of the split/splitless injection system

The liner, typically a glass tube, is installed in the heated block of GC inlet. As the evaporation chamber of sample, it has a lot of design patterns, such as with/without packing, various deactivation methods, insertion modes, and sizes. A large volume liner (wide bore) is often used for splitless injection and a smaller liner (narrow bore) for split injection. The liner diameter is usually 2–4 mm, the length is usually 8–10 cm, and the volume of wide-bore liner is about 1 ml.

Injection is to instantly and completely transfer the sample into GC column, which begins from the penetration of syringe into the septum/seal. When the plunger is pushed down, the sample is injected into the hot glass liner, where solvents and analytes are ideally flash-evaporated. Evaporation is a rather complex process. Incomplete evaporation is the main problem it may be encountered. Incomplete evaporation will cause the involatile matrix, droplets, and involatile substances hit and get deposited on the glass of the liner and then slowly released by thermal degradation, seriously affecting the chromatographic separation and even causing the appearing of "ghost peak." Another problem is that the gas flows through the liner too fast and enters into the inlet of the column before the droplet is completely evaporated, or enters into the inlet of the column by spraying means. Finally, a noteworthy problem is the sample overfilling the injector. 1 μl solvent evaporates to produce 0.5–1 ml gas, thus completely filling a normal wide-bore liner. For split injection, the gas flows through the injector quickly and the evaporated solvent is quickly removed, so it is not prone to overfilling problem, usually by reducing the split ratio to increase the injection volume. But in case of splitless injection where the flow rate through the injector is very low, a larger volume of solvent vapors may overfill the injector. For example, overfilling the liner will lead to cross-contaminations, high variability, and high background.

Split injection is mainly for high-concentration samples. In the split injection, the majority of the carrier gas is split from the bottom of the liner and be vented away. The flow amount into the column is mainly adjusted by split vent. When split injection is carried out, the flow rate of the carrier gas is relatively high, and there is a long distance between injection syringe and the column inlet (injector bottom), allowing more time for the evaporation of the sample. Typically, a narrow bore liner can give effective heat transfer, ensuring the maximum concentration of sample vapors. Although split injection can give sharper peaks, most of the sample will be lost (it will loss 97 % of the whole sample), and nonlinear split will cause quantification distortion.

Splitless injection is used to increase the amount of sample directly into the column by closing the split vent during the injection to make all the gas flow arrive the column through the liner. Since the flow rate of the column is limited in the range of several milliliters per minute, so that it takes quite a while for all the samples entering into the column, typically in the range of 30–90 s, which will cause the expansion of original spectrum very easily. Hence, measures must be taken to focus the sample before it enters into the column, in order to get a better chromatographic separation. In simple terms, the injection time has to be short compared with the peak width in the chromatograms. Solvent recondensation technology can effectively trap sample, concentrating sample at the beginning of the column. Such concentration technology is done by installing a fused silica column which is deactivated but without stationary phase at the beginning of the column. This silica column is about 2–5 m in length. Solvent will be effectively evaporated when passing through this column, thereby concentrating sample molecules to a narrow band. When all the solvent is evaporated, sample molecules will be focused in a narrow injection band with the carrier gas as a small bandwidth and into the column to be separated.

2.1.1.4 Derivatization Technology for GC

GC requires the sample to be efficiently evaporated in the injector, which is very easy for small molecular compounds with low boiling point (lower than 200–300 °C). However, high boiling point compounds need to be chemically derivatized before evaporation. In metabolomics, interested compounds are usually amino acids, sugars, small molecular organic acids, and other polar metabolites, as well as some apolar metabolites, such as fatty acids and sterols, and the majority of the metabolites are in their nonvolatile state. For compounds containing carboxylic, hydroxylic, or amino groups, derivatization can increase their volatility. The following are the main benefits of derivatization for GC/MS analysis (Wang et al. 2001).

① The GC nature of the analytes can be improved. The GC nature of some relatively polar groups, such as hydroxyl and carboxyl, is not good, which has no peak or peak tailing in some common columns. After derivatization, the situation is improved.
② The thermal stability of the analytes can be improved. The thermal stability of some analytes is not good. The analyte will decompose or change in the evaporation or chromatographic process. Derivatization can convert it into stable compound under GC/MS detection condition.
③ The molecular weight of the analyte can be changed. The molecular weight of most of the derivatized analytes is increased, which is beneficial for the separation between analyte and matrix, reducing the impact of background chemical noise.
④ MS behavior of the analyte can be improved. In most cases, derivatized analytes produce more regular and easily be interpreted mass fragments.
⑤ The introduction of halogen atom or electron-withdrawing group makes analytes easily be detected by chemical ionization. In most cases, the detection sensitive can be increased, and the molecular weight of the analyte can be detected.
⑥ Some particular derivatization methods can be used to split some chiral compounds which are difficult to be separated.

Of course, if the derivatization method is applied improperly, some drawbacks will be brought about. For example, some derivatization reagents need to be removed by nitrogen flow before injection, and improper operation will result in sample loss. Incomplete derivatization reaction will reduce the damage sensitivity of detection, and the improper use of derivatization reagent will sometimes make the molecular weight of analyte increase too much, approaching or exceeding the mass range of some small MS detectors. Derivatization will produce some artifacts in the sample and will contain some excess reagent, which will seriously interfere with split/splitless injection, and usually, they are not volatile and deposit in the liner.

Derivatization methods suitable for GC/MS analysis include methylation and silylation. There are also many other derivatization methods, and for details, see relevant books (Drozd 1981; Toyo'oka 1999).

2.1.2 Key Technologies of MS in GC/MS

MS itself can analyze very complex samples. Meanwhile, it is a detector for chromatography, providing very high sensitivity as well as chemical and structural information after connecting chromatography. The development of modern biological MS is more or less driven by the development of metabolomics. MS is a very important analytical means among modern biotechnological analytical methods. Almost all the biotechnological analytical problems can be resolved by MS, from small molecular volatile substances, to complex natural products, as well as proteins and viruses.

The core principle of MS is to determine the mass-to-charge ratio (m/z) of charged compounds. In principle, any charged (or can be charged) substance, which can be transferred into GC at the same time, can be detected by MS. Major development in recent decades is a great expansion of the molecular weight range of MS and the significant improvement of sensitivity. Meanwhile, the mass spectrometer becomes cheaper and easier to be operated. For the instrument constitution of MS, see Fig. 2.3.

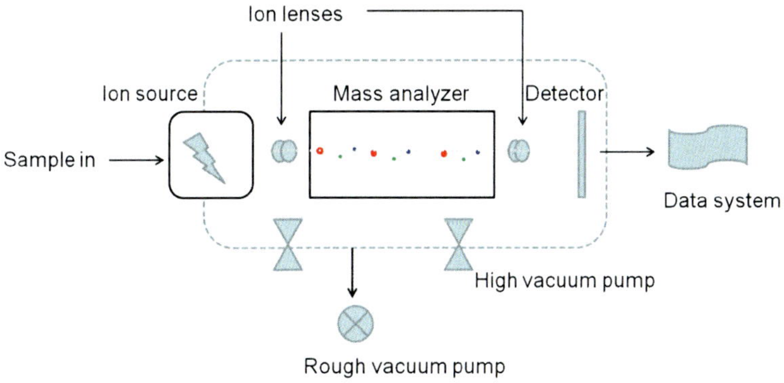

Fig. 2.3 Major constitution of mass spectrometer

2.1.2.1 The Ion Source

GC/MS generally uses open capillary column, the flow rate of carrier gas is low, not to damage the vacuum environment of MS, so that samples can be directly introduced into ion source. The most critical process in ion source is the evaporation, ionization, and transferring of sample into the vacuum system, which depends on the sample type (gas/liquid) and ionization method. These processes can also be carried out in reverse order. For example, the sample is ionized in the solvent, and then, the ions are transferred into gas phase. So far, the most commonly used ionization technique is electron-impact ionization (EI) and electrospray ionization (ESI), the former is mainly coupled with GC, and the later is mainly coupled with LC. EI ion source and GC/MS represent the classic mass spectrometer combination, because the gaseous mobile phase can well flow into MS vacuum. Modern GC/MS system is already very mature technology, is easy to operate, and delivers highly reproducible results.

The lamp filament of EI ion source is usually made of tungsten filament or rhenium filament. Under high vacuum condition, when the current passes through the cathode, the filament temperature can reach up to about 2,000 °C. Hot filament produces electrons, and when the electron energy is higher than the ionization potential of the sample, the sample molecules or atoms are ionized. Ions gain kinetic energy in the electric field and enter into mass analyzer at a certain speed.

In the EI source, the molecules or atoms of analyte lose valence electrons to generate positive ions:

$$M + e^- \rightarrow M^{+\bullet} + 2e^- \tag{2.1}$$

Or capture electrons to generate negative ions:

$$M + e^- \rightarrow M^- \tag{2.2}$$

Generally, the generated positive ions are 10^3 times of negative ions. If not specifically pointed out, conventional MS only studies positive ions. EI energy should be at least equal to the ionization potential of the analyte to make the analyte be ionized to generate positive ions. The ionization potential of elements in the periodic table is 3–25 eV, wherein most of them is less than 15 eV. The ionization potential of organic compounds is 7–15 eV. If EI energy is just equal to the ionization potential of the analyte, all the energy of the electron must be transferred into the analyte to make it be ionized. In fact, the number of molecules or atoms that can obtain all the energy of the electron is limited, and the ionization efficiency is low. Therefore, increasing EI energy is beneficial for increasing the ionization efficiency. To obtain reproducible MS spectra, EI energy is generally 70 eV. But relative high energy will make the remainder energy of the molecular ion greater than the bond energy of some molecules, thereby cracking the molecular ions. Lower ionization voltage can be used to control the number of fragment ions and increase the intensity of molecular ion peaks. The ionization voltage of common MS instruments can be adjusted in the range of 50–100 V.

A major drawback of EI source is that the solid or liquid sample must be gasified before entering into ion source and therefore not suitable for samples difficult to be volatile or with poor thermal stability. Ion source is a component in MS requiring more attention in operation and maintenance, and many ionization parameters play important roles in obtaining results. Especially when introducing the sample into ion source, the use of solvent is at the core position in the ionization process.

Besides the conventional EI ion source for GC/MS, there are chemical ionization (CI) and field ionization (FI) ion sources. The ionization modes of the latter two are relatively mild, which are soft ionization modes and can obtain molecular weight information of compounds. But the spectral reproducibility is not good as that of EI source, so the use is not so widespread.

2.1.2.2 The Mass Analyzer

Mass analyzer separates charged ions according to mass-to-charge ratio to record the mass and abundance of various ions. According to the types of mass analyzer, MS can be classified into magnetic MS, single quadrupole/MS, triple quadrupole tandem MS, TOF/MS, and ion trap (IT)/MS. Magnetic MS and IT MS are rarely used in GC/MS, and metabolomics research is mainly based on GC-single quadrupole/MS and GC-TOF/MS.

Quadrupole mass spectrometer is currently the most commonly used mass spectrometer. This instrument is characterized by small size, simple structure, low cost, and good performance. Especially for general purposes, it has advantages in cost and performance. The mass analyzer of quadrupole mass spectrometer consists of four parallel electrodes where a radio frequency (RF) voltage supply is connected to adjacent electrodes creating an alternating electric field between the electrodes. Under certain RF voltage and RF frequency, only those ions with certain m/z can successfully pass and reach detector. The amplitude of other ions will constantly increase until they hit the electrode pole and neutralized by electrons to become neutral particles. Quadrupole is also known as mass filter. For principle, see Fig. 2.4.

TOF analyzer is the simplest analyzer, with high scanning speed, very high ion collection efficiency, wide mass range, a resolution of up to 40,000, and dynamic range of up to 10^5. TOF is very suitable for analyzing very complex metabolomics samples. It can quickly scan the whole spectrum and obtain qualitative information.

The detection principle of TOF technology is that the ion beam is accelerated by high voltage and pushed into the flying tube in pulse, and then "free drift" to reach the detector. Ions with different masses obtain different accelerated speed. Ions with smaller mass have higher speed than ions with larger mass. The arriving time of ion at detector is correlated with ion mass. The principle of TOF detector is relatively simple, and the design art directly affects MS performance. Pusher is used to accelerate particles. In order to get consistent initial kinetic energy and consistent

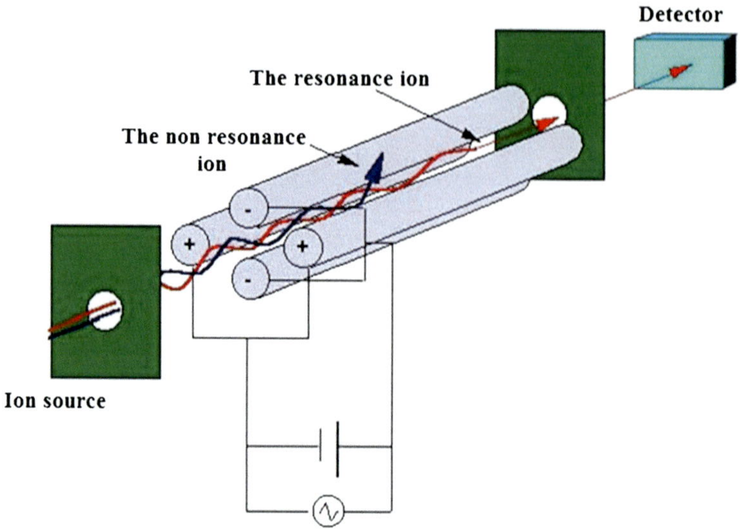

Fig. 2.4 Constitution and analysis principle of quadrupole mass analyzer

initial flight time for charged ions, the ion beam entered into pusher must be very level and narrow to reduce energy diffusion of ions at different directions, which is an important aspect affecting the resolution and accuracy of TOF/MS. The introduction of ion reflector not only increases free flight distance, but also further focuses ions with different kinetic energies. TOF detector requires higher vacuum degree to avoid collisions between ions, and between ions and gas. TOF flight tube requires keeping high stability to reduce subtle changes in mass axis caused by heat expansion and cold contraction, which affects the reproducibility of result and the accuracy of measurement. For typical ion-reflective TOF detector, the flight time of ions with 1,000 Da m/z is not less than 50 μs (50×10^{-6} s). Therefore, the analytic speed of TOF is very fast, requiring the detection speed of ns to ps (10^{-9} to 10^{-11} s). Pusher can reach 20,000 times/s, and the obtained spectrum is the accumulation of multiple ion pushing detection results. To obtain rapid and accurate flight time of ions, the requirement to detector is very high, and TOF-dedicated digital converter (TDC or ADC) is needed. TOF cannot scan ions, cannot store and release ions like ion trap, all the ions are pushed into flight tube one-off, and wait all the ions reach detector. Before next group of ions are pushed into flight tube, the former group of ions cannot be operated. Thus, TOF cannot do selected ion monitoring (SIM) or selected ion recording (SIR), and thus, ion pushing rate affects sensitivity. TOF detector is an ideal match for high-speed GC/MS (advantage in deconvolution) and high-speed HPLC/MS. At present, the quantitative capability of TOF has not yet reached the level of quadrupole, mainly due to the problem previously mentioned for ion pushing in pulse and detector dead time. For principle, see Fig. 2.5.

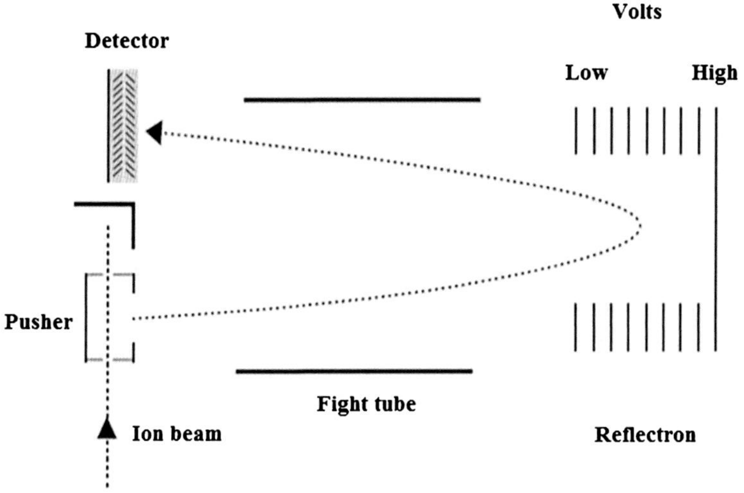

Fig. 2.5 Constitution and analytical principle of TOF mass analyzer

2.1.3 Other Hardware

Besides the above-mentioned components, GC/MS also consists of a pumping system for maintaining required MS vacuum, as well as control electronics and power supplies. The maintenance of high vacuum system is based on two pumping stages to reach a range of 10^{-5} to 10^{-7} hPa, where high-resolution mass analyzer requires higher degree of vacuum. The first stage of pumping system is generally a rotary oil pump backing one or more turbomolecular pumps to achieve minimum pressure for the starting of molecular pump. Generally, these vacuum systems need more care and attention. The second important hardware is the high-voltage power supply system. The stability of MS high-voltage control system is of significance for the resolution, accuracy, and sensitivity of mass spectrometer. Although the high-voltage supply system is very good, they should be maintained over the years, including high-voltage electrical wires and connectors. All modern mass spectrometers are controlled by data analysis system, which not only controls the instrument, but also plays an important role in data analysis. Therefore, data analysis system is known as the fourth leg of mass spectrometer and has equal significance as other accessories.

2.2 Sample Preparation and Analytical Techniques

Small molecular metabolites in the plant will rapidly change when there is a change in external environment. Metabolomics analysis generally requires the sample to be rapidly frozen after collection and stored in the environment below −60 °C until the extraction process begins, to ensure that metabolites are not damaged by enzyme system in the plant. In the extraction process, it should be ensured that metabolites are free from the impact of physical or chemical substances as far as possible. Therefore, it is necessary to keep the ambient temperature constant and avoid using acid or alkali to treat samples. In addition, the use of high-purity solvent for sample pretreatment and the addition of quality control sample in the analytical process are necessary for avoiding the contamination of external samples. The pretreatment of blank sample and target sample in the same batch should be carried out in synchronism. The only difference between the two is that the blank sample does not contain sample materials. In addition, to ensure the reproducibility of biological data, at least six sample replicates are needed.

Prior to sampling, the collection time points should be determined. For example, since the plant leaf is the main part of photosynthesis, the middle time of light cycle should be selected when collecting leaves. For plants in the vegetative stage, collection should be performed before the emergence of the first inflorescence, and the internode at the same part and the same non-aging leaf should be selected. Experience has improved that rapid sampling and rapid quenching are very important. Prior to sample homogenization, all the experimental materials and reagents should be cooled to avoid sample change due to thawing. There are a lot of reports on metabolomics sample pretreatment methods. In addition to referring relevant literatures, experimental methods are also provided by international Web sites maintained by relevant laboratories and institutes. Different laboratories can establish their own methods according to their own needs. Different pretreatment methods may be needed to explore for different experimental materials. Here are just two cases for reference.

Method 1 (Lisec et al. 2006), shown in Fig. 2.6:

Sampling and extraction:

1. 100 mg plant leaf sample is placed in a 2-ml round-bottom tube with screw cap, quickly frozen in liquid nitrogen;
2. Homogenization in a ball mill. Place the sample tube with steel balls in ice-cold ball mill, and grind for 2 min at the vibration frequency of 20 Hz.
   Note: The frozen samples can be stored in −80 °C refrigerator for no more than 3 months;
3. Add samples into 1,400 μl −20 °C precooled methanol, vortex for 10 s;
4. Add 60 μL ribitol (0.2 mg/ml aqueous solution) as an internal standard, vortex for 10 s;
5. Extract using a thermomixer at 70 °C, 950 rpm for 10 min;
6. Centrifuge sample at 11, 000 g for 10 min;
7. Transfer supernatant to another glass vial;
8. Add 750 μl −20 °C precooled chloroform, vortex for 10 s;
9. Add 1,400 μl precooled (−20 °C) dH_2O, vortex for 10 s;
10. Centrifuge sample at 2, 200 g for 15 min;
11. Transfer 150 μl supernatant (polar phase) into a clean 1.5 ml tube;
12. Transfer another 150 μl supernatant as a backup into a clean 1.5 ml tube;

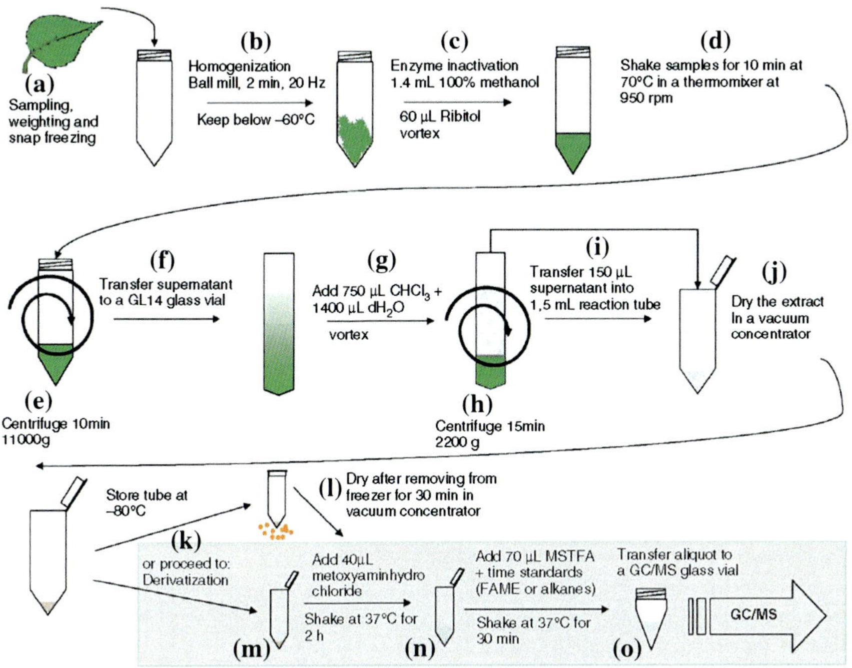

Fig. 2.6 The flowchart of metabolomics sample preparation (Reprinted by permission from Macmillan Publishers Ltd: [Nature Protocols] (Lisec et al. 2006))

13. Drying in a vacuum concentrator at ambient temperature;
14. Fill argon into the sample tube, place all the sample tubes into a sealed plastic bag containing silica-drying agent, and store in −80 °C low-temperature refrigerator;

    Derivatization:
15. Place the sample tubes which are stored at −80 °C in a vacuum concentrator drying for 30 min;
16. Add 40 μl methoxyamination reagent (methoxyamine hydrochloride, 20 mg/ml pyridine solution);
17. Prepare a blank derivatization sample as a control;
18. Shake at 37 °C for 2 h;
19. Prepare silylation reagent MSTFA [N-methyl-N-(trimethylsilyl) trifluoroacetamide], add 20 μl/ml mixture of retention index standard (mixture of saturated fatty acid methyl ester series, FAME (C_8–C_{30}), and dissolve in chloroform at a concentration of 0.4 mg/ml (liquid standard) or 0.8 mg/ml (solid standard));
20. Add the silylation reagent in step 19 into the sample reaction tube;
21. Shake at 37 °C for 30 min;
22. Transfer the derivatized sample to the liner suitable for GC/MS analysis.

Note: The derivatization reagent is extremely toxic, take extra care, wear gloves, and operate under a fume hood.

Key point 1: In the course of derivatization, derivatization reagent is likely to remain in the tube wall or the tube cap of the reaction tube; thus, centrifugation should be performed at each derivatization step;

Key point 2: 15–22 steps are critical. The derivatization reagent in this method is in excess to ensure complete derivatization.

GC/MS instrument model: Agilent G6890 with autosampler, Leco Pegasus IV TOF/MS, MDN-35 capillary column, 30 m length, 0.32 mm inner diameter, 0.25 μm film thickness.

GC injection parameters: injection volume 1 μl, inlet temperature 230 °C, splitless injection mode, helium carrier gas, flow rate at 2 ml/min, autosampler injection. For high-concentration samples, split injection mode is suggested, and the split ratio is set to 1: 25.

GC parameters: MDN-35 capillary column (30 m). GC temperature gradient starts at 80 °C and holds constant temperature for 2 min, followed by ramping at 15 °C/min to 330 °C, which is held for 6 min. Transfer line temperature is set to 250 °C.

MS parameters: Ion source temperature is set to 250 °C, mass scanning range is at m/z 70–600, acquisition rate at 20 spectra/s, MS EI lamp filament starting time at 170 s of chromatographic reagent delay, detector voltage 1,700–1,850 V, MS loss is set to 0, the filament bias current is 70 eV, the instrument is automatically tuned.

Spectral Deconvolution parameters: Deconvolution software ChromaTOF, baseline offset is set to 1 (0.5–1); spectral smoothing for five data points (3–7), peak width 3 s (3–4 s); signal-to-noise ratio (S/N) is 10 (2–15).

Method 2: This method is used for the simultaneously extraction of metabolites, proteins, and RNA (Weckwerth et al. 2004).

Sampling and extraction:

30–100 mg samples of *Arabidopsis* leaves at a developmental stage is harvested.

1. Immediately frozen in liquid nitrogen;
2. Homogenize tissue using the Retsch mill in liquid nitrogen;
3. Add 2 ml precooled (−20 °C) extraction solvent methanol: chloroform:water (5:2:2, v/v/v);
4. Mix vigorously at 4 °C for 30 min to precipitate proteins and DNAs/RNAs, and disassociate metabolites from cell membrane and cell wall components;
5. Centrifuge the sample and transfer supernatant;
6. Add 1 ml precooled (−20 °C) extraction solvent methanol:chloroform (1:1, v/v) for the second extraction;
7. Centrifuge the sample and transfer supernatant;
8. Combine the organic extracts from the two extraction steps;
9. Add 500 μl water to separate the organic phase and aqueous phase. The methanol–water layer mainly contains sugars, amino acids, and small organic molecules. Chloroform layer mainly contains lipids, chlorophyll, and waxes;
10. Add 1 ml extraction buffer (containing 0.05 M Tris, pH 7.6; 0.5 % SDS; 1 % *β*-mercaptoethanol) and 1 ml water-saturated phenol to the residue after extraction;

11. Extraction at 37 °C for 1 h;
12. Centrifuge at 14,000 g and transfer supernatant;
13. Separate the phenol phase from supernatant;
14. Precipitate the phenol-phase proteins with ice-cold acetone at −20 °C overnight;
15. Wash three times with ethanol, and dry at ambient temperature;
16. Precipitate proteins remained in the RNA extraction buffer with 200 μl chloroform;
17. Centrifuge to remove precipitate and separate buffer;
18. Add 40 μl acetic acid and 1 ml ethanol to precipitate RNA at 37 °C for 30 min;
19. Wash with one volume of 3 M sodium acetate for one time, and wash with one volume of 70 % ethanol for two times;
20. The remaining pellet is dissolved in 100 μl RNAse-free water;
21. Detect the amount and purity of RNA by absorbance at 260 nm and gel electrophoresis in agarose;
    Derivatization steps:
22. Evaporate to dry metabolites in the organic phase, and add 50 μl methoxy-amination reagent (methoxyamine hydrochloride, 20 mg/ml pyridine solution);
23. React at 30 °C for 90 min with continuous shaking;
24. Add 80 μl MSTFA;
25. React at 30 °C for 30 min;
26. The derivatized samples are stored at ambient temperature for 120 min before injection.

GC-MS instrument model: GC model: HP 5890, Leco Pegasus IV GC-TOF mass spectrometer, 40 m length, inner diameter 0.25 mm, RTX-5 capillary column with a 10 m precolumn.

GC injection parameters: injection volume 1 μl, inlet temperature 230 °C, splitless injection mode, helium carrier gas, flow rate 1 ml/min, autosampler injection.

GC Parameters: GC temperature gradient starts at 80 °C and holds constant temperature for 2 min, followed by ramping at 15 °C/min to 330 °C, which is held for 6 min.

MS parameters: record 20 spectra per second between m/z 85 and 500. S/N threshold is 20.

2.3 New Technologies and Trends

GC/MS is the most mature chromatography mass spectrometry coupling technology. The one in the spotlight is the comprehensive two-dimensional GC (GC × GC) developed in 1990s, which is a multi-dimensional chromatography. The separation mechanism is to tandem combine two columns with different stationary phases as a two-dimensional chromatography. A modulator is installed between the two columns, playing a role in trapping and retransmitting. Each fraction separated by the first column must firstly enter into the modulator to be focused in a pulse manner

Table 2.1 The comparison between characteristics of one-dimensional GC/MS and comprehensive two-dimensional GC/MS (Reprinted from Kusano et al. (2007), with permission from Elsevier)

	One-dimensional GC/MS	Comprehensive two-dimensional GC/MS
High throughput	High	Medium
Resolution	High	Very high
Sensitivity	Medium	High
Cost	Low	Medium
Deconvolution	Good	Very good
Data file size (in ASCII and CSV format)	Medium (about 200 MB)	Very large (about 1 GB)

before further separated by the second column. Generally, the second column is relatively short, and the analytical time is short. The comprehensive two-dimensional GC has characteristics such as high resolution and high sensitivity, is one of the currently most powerful separation tools, and is widely applied in the separation of petroleum, tobacco, and pharmaceutical and other complex systems. The key component of comprehensive two-dimensional GC is the modulator between the two chromatographic columns, which is required to completely trap and release the first-dimensional fractions in a very short time period. The modulator is a container connecting the two chromatographic columns, such as the two-stage circular tube. Commonly used method is to trap rapidly at low temperature, such as liquid nitrogen or low-temperature cold trap, and then rapidly warm up to release. Currently, there have been a lot of analytical chemists using comprehensive two-dimensional GC/MS in metabolomics research. Due to the limitation in data analysis methods and the control technique of the modulator, this technology has not yet been widely applied. For the high throughput and reproducibility, usually one-dimensional GC/MS is used to obtain full spectrum, and comprehensive two-dimensional GC/MS is used to obtain more detailed peak information (Table 2.1).

2.4 Common Problems and Cautions

2.4.1 Common Problems in GC/MS-Based Metabolomics Research

1. **GC/MS database.** Metabolite qualification has been a difficult problem in metabolomics. Compared with other chromatography and mass spectrometry coupling technologies, GC/MS has a large number of databases, such as NIST database. However, most of the GC/MS peaks still cannot be resolved by existing commercial MS database. The structures of metabolites in plant are complex, and so metabolomics-targeted database is needed. Many databases, such as human metabolomics database, plant metabolomics database, and

species-specific metabolomics database, are being continuously established and improved. Scholars in Max Planck Institute for Molecular Plant Physiology suggest to accumulate metabolomics qualitative data from everyday metabolomics analysis and to establish database based on retention index and MS data. This method is simple and feasible, and the qualitative effect is more accurate than MS (Wagner et al. 2003; Schauer et al. 2005; Strehmel et al., 2008).

2. **Deconvolution**. The aim of deconvolution is to parse overlapping coeluted peaks and obtain the MS peak of single pure substance. Currently, there are only few good deconvolution softwares, free software such as AMDIS, and commercial software such as ChromaTOF and AnalyzerPro. The deconvolution effects of the three softwares were compared with 36 endogenous standard metabolites mixture. The 36 substances were prepared in standard solutions with five different concentrations. The first group is 500 μM, the second group is 350 μM, the third group is 150 μM, the fourth group is 50 μM, one half of the fifth group is 500 μM, and another half of the fifth group is 50 μM. The 36 standard substances were derivatized and produce 51 metabolites and derivatives. AMDIS performed deconvolution and detected peaks of all the metabolites; ChromaTOF did not detect the 8 metabolites and derivatives in the fourth group; most of metabolites and derivatives were not detected by AnalyzerPro; ChromaTOF and AnalyzerPro produced many false-negative results. However, AMDIS detected the most false-positive peaks, obtaining as much as 522–750, while ChromaTOF obtained 78–173 false-positive peaks and AnalyzerPro obtained fewer false-positive peaks. In addition, the number of deconvoluted peaks and the accuracy of deconvoluted spectra are closely correlated with metabolite concentration, i.e., if the concentration is decreased, the number of deconvoluted metabolites is significantly decreased, and the number of metabolites can correctly match the database is decreased, even some peaks cannot be deconvoluted because the S/N is too low (Lu et al. 2008) (Table 2.2).
3. **Multi-peak phenomena.** In addition to the large number of false-positive peaks caused by the algorithms of different deconvolution software, multiple peaks can

Table 2.2 The comparison of deconvolution effects of different deconvolution softwares

	Deconvolution software	The first group	The second group	The third group	The fourth group	The fifth group
Peak number after deconvolution	ChromaTOF	173	161	121	78	162
	AMDIS	720	620	529	522	720
	AnalyzerPro	67	49	38	14	42
Peak number undetected by deconvolution	ChromaTOF	0	0	0	8	0
	AMDIS	0	0	0	0	0
	AnalyzerPro	2	9	17	38	19
Metabolite number correctly matched with deconvolution	ChromaTOF	37	31	28	14	27
	AMDIS	32	30	20	8	26
	AnalyzerPro	28	24	14	5	18

be produced in sample preparation, extraction, derivatization, and analysis process, especially the variation of derivatization has great impact on metabolic profile. The multi-peak phenomena refer to one metabolite producing multiple peaks, which can be caused by sample degradation, by-products formation, and the introduction of exogenous contaminants. The multi-origination phenomena mean one peak has multiple origins (precursors) (Xu et al. 2010).

Reasons causing the multi-peak phenomena in the derivatization process of GC/MS include:

1. **The multi-peak phenomena are generated in the derivatization process,** including ① forming by-products. The silylation derivatization process is to silylate metabolites, but some functional groups, such as aldehyde, amino, carboxyl, ester, ketone group, and phenolic hydroxyl group, may form a plurality of products. In addition, derivatization reagents, organic solvent impurities, and plastic tube contaminants may also cause by-products. All these non-specific products are referred to as artifacts; ② incomplete derivatization. Many compounds contain a plurality of reactive groups available for derivatization, and when the amount or derivatization time of derivatization reagent is not enough, incomplete derivatization will be produced.
2. **The conversion of metabolites structure.** Geometrical isomers of metabolites may lead to the multi-peak phenomena, for example, linear and cyclic D-glucose can be converted with each other in the solvent. Usually, glucose in the solvent shows at least five different tautomeric forms [such as α-D-glucopyranose (62 %), β-D-glucopyranose (38 %), α-D-glucofuranose (trace), β-D-glucofuranose (trace), and linear-D-glucose (0.01 %)], and all of these tautomeric forms maintain a dynamic balance, their contents are impacted by solvent composition, temperature, and pH, and complex chromatograms will be produced after BSTFA derivatization. Again, inositol alcohol has nine different stereoisomers; arginine can be converted into ornithine with MSTFA derivatization (37 °C, 20 min).
3. **Metabolites will be degraded in the extraction, derivatization, and GC/MS analysis process.** Thermal non-stable compounds prone to be thermally degraded, resulting in multi-peak and multi-origination phenomena. Xu et al. (2001) detected two groups of structural and biological-related substances and found that even the thermally stable compounds, such as phosphorylcholine (PC), 1,2-diacetylglycerol-3-phosphate ester (DAG), and hemolysis phosphocholine (LPC), can generate multi-peak phenomena; structurally similar compounds can produce the same peak (multi-origination phenomena); glycerol, phosphoric acid, fatty acids, and some lipid fragments may be the cracking fragments of these structural similar compounds (including free type and bound type).

2.4.2 Notes on GC/MS-Based Metabolomics Research

① Derivatization reagent methoxyamine hydrochloride need temporary preparation, MSTFA need to dry stored in 2–8 °C, avoid absorbing moisture in the air.
② Samples must be randomly injected to eliminate systematic errors.
③ Set control samples and blank samples (e.g., reagent blank, method blank).
④ Low loss injection pad and low loss injection system are important.
⑤ Original GC/MS data are transferred to the server, and long-term data preservation should use DVD backup or transfer to a server mirroring system.
⑥ System background subtraction: Background peaks of plasticizers, phthalates, and silylation reagent and column bleed as well as the water peak of derivatization reagent should be subtracted.

2.5 Prospects

The data processing in GC/MS-based metabolomics has been a bottleneck hindering the development of the platform, such as the routine applications of data modeling tools in bioinformatics and systems biology analysis.

Challenges in the modernization of GC/MS platform are the parallel control techniques of multiple samples, including ① the automation of sample preparation, sample pretreatment, and the high-throughput and reproducible technology after obtaining data; ② the integration of metabolomics data and other omics data, for example, the analysis of samples by using integrated analysis technique platform and combining proteomic data and transcriptomic data; ③ the spectral recognition of trace compounds or signal molecules among a group of metabolites; ④ the combination of full spectrum analysis and flow analysis; ⑤ reproducible quantitative results, clear system nomenclature, and the comparability of data obtained by different laboratories using different GC/MS platforms; and ⑥ the last but not the least, which may be the most important challenge to all the metabolomics analysis, that is, the structural identification of metabolites in complex spectra of metabolites.

References

Dalgliesh CE, Horning EC, Horning MG, Knox KL, Yarger K. A gas-liquid-chromatographic procedure for separating a wide range of metabolites occurring in urine or tissue extracts. Biochem J. 1966;101:792–810.

Drozd J. Chemical derivatization in gas chromatography. J Chromatogr Libr. 1981.

Kusano M, Fukushima A, Kobayashi M, Hayashi N, Jonsson P, Moritz T, Ebana K, Saito K. Application of a metabolomic method combining one-dimensional and two-dimensional gas chromatography-time-of-flight/mass spectrometry to metabolic phenotyping of natural variants in rice. J Chromatogr B. 2007;855:71–9.

Lisec J, Schauer N, Kopka J, Willmitzer L, Fernie AR. Gas chromatography mass spectrometry-based metabolite profiling in plants. Nat Protoc. 2006;1:387–96.

Lu H, Dunn WB, Shen H, Kell DB, Liang Y. Comparative evaluation of software for deconvolution of metabolomics data based on GC-TOF-MS. Trends Anal Chem. 2008;27:215–27.

Schauer N, Steinhauser D, Strelkov S, Schomburg D, Allison G, Moritz T, Lundgren K, Roessner-Tunali U, Forbes MG, Willmitzer L, Ferniea AR, Kopka J. GC-MS libraries for the rapid identification of metabolites in complex biological samples. FEBS Lett. 2005;579:1332–7.

Strehmel N, Hummel J, Erban A, Strassburg K, Kopka J. Retention index thresholds for compound matching in GC-MS metabolite profiling. J Chromatogr B. 2008;871:182–90.

Toyo'oka T. Modern derivatization methods for separation science. New Jersey: Wiley; 1999.

Villas-Boas SG, Roessner U, Hansen MAE, Smedsgaard J, Nielsen J. Metabolomics analysis an introduction. New Jersey: Wiley; 2007.

Wagner C, Sefkow M, Kopka J. Construction and application of a mass spectral and retention time index database generated from plant GC/EI-TOF-MS metabolite profiles. Phytochemistry. 2003;62:887–900.

Wang ZF, Yang SM, Wu MT, Yue WH. Chromatography coupling techniques. Beijing: Chemical Industry Press; 2001.

Weckwerth W, Wenzel K, Fiehn O. Process for the integrated extraction, identification and quantification of metabolites, proteins and RNA to reveal their co-regulation in biochemical networks. Proteomics. 2004;4:78–83.

Xu GW, Ye F, Kong HW, Lu X, Zhao XJ. Technique and advance of comprehensive two-dimensional gas chromatography. Chromatography. 2001;19:132–6.

Xu F, Zou L, Ong CN. Experiment-originated variations, and multi-peak and multi-origination phenomena in derivatization-based GC-MS metabolomics. Trends Anal Chem. 2010;29:269–80.

Chapter 3
LC-MS in Plant Metabolomics

Guo-dong Wang

There are more than 200,000 metabolites in the plant kingdom (here mentioned are compounds with molecular weight <1,000), some of them are primary metabolites necessary for maintaining plant life activities as well as growth and development, and some of them are species-specific secondary metabolites (Dixon and Strack 2003). As the final product in the process of cellular regulation and control, the change of metabolite in kind and quantity is regarded as the final response of plant to the change of gene or environment. The comprehensive qualitative and quantitative analysis of these metabolites dynamically and statically, i.e., the metabolomics study, is an important part of the understanding of plant biological system in the post-genomic era (Fukusaki and Kobayashi 2005; Hagel and Facchini 2008). The data collection platform is indispensable in plant metabolomics research. Currently, common platforms include MS-based and NMR-based platforms, in which the high-performance liquid chromatography–mass spectrometry (HPLC-MS) technology has been more and more widely applied in plant metabolomics research for its powerful separation ability, high throughput, high resolution, and high detection sensibility and can give the qualitative result of the analyzed metabolites. Generally, HPLC-MS should be equipped with auto-sampler, interface necessary for chromatography and MS coupling, ion source, mass analyzer, detector, computer control and data processing system and vacuum system. The following will concentrate on the introduction of LC and MS system.

3.1 LC-MS Fundamentals

3.1.1 HPLC

HPLC technique was developed by combining the theory of gas chromatography and the principles of classic chromatography. Technically, the mobile phase is changed from gas to liquid with high-pressure transportation; the chromatographic

G.-d. Wang (✉)
Institute of Genetics and Developmental Biology, Chinese Academy of Sciences, Beijing 100101, China
e-mail: gdwang@genetics.ac.cn

X. Qi et al. (eds.), *Plant Metabolomics*, DOI 10.1007/978-94-017-9291-2_3

column is packed with materials in small particle size (several micrometers) to make the column efficiency far higher than that of classic chromatography (the plate number per meter can reach tens of to hundreds of thousand); meanwhile, the high-sensitive detector is connected with column, which can continually detect the effluent. At present, the reversed-phase chromatography making use of nonpolar stationary phase as the column packing materials (such as C_{18} and C_8) is about 80 % of the whole HPLC applications. As the currently used chemical separation and analytical techniques, HPLC characteristics are the following:

① High pressure. HPLC uses liquid as the mobile phase (also known as carrier liquid). The liquid suffers big resistance when flowing through the chromatographic column. In order to make the carrier liquid rapidly flow through the chromatographic column, the exerting of high pressure is necessary.
② High speed. The in-column flow rate of the mobile phase is generally 1 ml/min; thus, the separation time is shorter than 1 h, and the speed of ultra-high-pressure liquid chromatography (UPLC) will be even faster.
③ High efficiency. In recent years, many new types of packing materials have been developed for column preparation, making the separation resolution highly improved.
④ High sensitivity. The high-sensitive detector has been widely applied in HPLC, further improving the analysis sensitivity of HPLC. In recent years, with the continuous innovation and development of various techniques, MS detector not only can do quantitative analysis, but also can provide more structural information of analyzed chemicals, receiving more and more tremendously popular.
⑤ Wide application range. Compared with HPLC, GC is limited to analytes that are thermally stable and volatile (or can be made volatile by derivatization) although it has characteristics of good separation ability, high sensitivity, fast analysis speed, easy to be operated, etc. However, the samples for LC analysis only needed to be prepared into solution, and generally, no time-consuming derivatization procedure is required for sample volatility, which commonly used for GC analysis. In principle, organic compounds with high boiling point, bad thermostability, or big relative molecular weight can be separated and analyzed by HPLC.

In general, HPLC system can be divided into preparative and analytical based on the sample loading amount (see Table 3.1): The preparative LC is generally used for purification, while the analytical one is used for the qualitative and quantitative analysis of plant extracts. In recent years, with the rapid development of miniaturization technology and the decrease of column diameter, various (ultra-) micro-quantitative HPLC have been available and applied. The commonly used columns for the (ultra-) micro-quantitative chromatograph are monolithic silica column, whose packing material is not the conventional spherical particles but the monolithic silica structure with micropores, which overcome the drawback of conventional spherical particle packing that difficult to reach high efficiency and high speed

Table 3.1 Different types of HPLC

Types	Diameter of the column	Flow rate
Preparative	2.1 to >200 mm	10 ml/min
Analytical	2.1–4.6 mm	1.0 ml/min
Micro	1.0 mm	200 μl/min
Capillary	300 μm–1 mm	4 μl/min
Nano	25–300 μm	200 nl/min

at the same time (small particle will increase the column efficiency, but will increase column pressure thereby decreasing separation speed). In practical applications, monolithic silica capillary column often shows higher separation efficiency than conventional chromatographic column (Tanaka et al. 2000). Although (ultra-) micro-quantitative chromatography can largely improve instrumental sensitivity and reduce reagent quantity thereby decreasing cost, a bad reproducibility was always observed when using (ultra-) micro-quantitative chromatography, mostly because the mobile phase can not be well-monitored in gradient. Many manufacturers are trying their best to solve this problem.

3.1.2 The Interface Between LC and MS

Compared with the diverse of GC-MS detector, the universality of interface technique for LC-MS is not so good; thus, common commercial LC-MS is equipped with several changeable interfaces to meet the requirement of sample diversity. Three main problems need to be resolved for the interface issue:

① The flow rate of HPLC is relatively large, while MS needs a high vacuum environment to work;
② Enough ions need to be generated from the mobile phase for MS analysis;
③ Impurities in the mobile phase should be removed to prevent possible contamination to MS.

In 1972, Tai'roze et al. first proposed the idea of directly importing the column outlet into MS to resolve the problem on LC-MS interface. Until 1987, the invention of atmospheric pressure ionization (API) interface by Bruins et al. resolved the flow limitation problem. After that, the first commercial LC-MS with API source came into being. API is a general name of MS ionization techniques under atmospheric condition, which currently includes electrospray ionization (ESI) and atmospheric pressure chemical ionization (APCI). The working principles of the two are different, and the interested readers can refer to *Chromatography and Mass Spectrometry Coupling Techniques* (Sheng et al. 2006 and literatures therein). The main difference between APCI and ESI is that ESI can produce multi-charged ions, largely increasing the molecular weight range of compounds to be analyzed and enabling ESI more suitable for the determination and analysis of polar compounds and biological big molecules such as proteins, while APCI mainly produces single charged ions, making it more suitable for the analysis of nonpolar and weak polar small molecular compounds.

3.1.3 Mass Spectrometer

Conventional detectors that can be linked to HPLC are UV detector, differential refraction detector, and fluorescence detector. As for the range of detectable compounds, MS is largely superior to UV detector or fluorescence detector. For example, MS has advantages in analyzing compounds without UV absorption or only have end absorption (saponins, terpenoids, saccharides, etc.). Generally, as long as the compound can be ionized, it can be detected by mass spectrometer. If quantitatively analyzed using UV detector, compound of interest must be separated from all other interfering compounds. Generally, the sensitivity of MS is 1–2 orders of magnitude higher than that of UV detector. If the selected/multiple reaction monitoring (S/MRM) mode is applied, the sensitivity will be even higher. While UV detector and fluorescence detector could only provide functional group information, MS especially MS^n can provide much more information for structural elucidation. Common ion detectors coupled with MS are electron multiplier and the array, ion counter, induced charge detector, and Faraday collector, which will not be described in detail here.

Currently, common mass spectrometers linked to LC are magnetic analyzer, TOF mass spectrometer, quadrupole (Q) mass spectrometer, ion trap (IT) mass spectrometer, and ion cyclotron resonance (ICR) mass spectrometer. The different types of mass spectrometers mentioned above not only can be used separately, but also can be used in combination, forming hybrid mass spectrometer to obtain the best analysis result. At present, the most successful and commonly used hybrid mass spectrometers are the combinations between different mass detectors with TOF-MS, such as Q-TOF. The employment of ion delayed extraction and ion mirror techniques makes the resolution and accuracy of TOF greatly improved. When coupled with quadrupole MS, more structural information of compounds can be obtained, making such a hybrid mass spectrometer be widely applied in plant metabolomics research (see Table 3.2).

Compared with GC-MS which generally use hard ionization techniques such as electron-impact ionization (EI), LC-MS interface commonly uses soft ionization techniques such as API. The abundance of molecular ion which used for analyzing compounds in the MS spectra is relatively high, mainly giving the molecular weight information of compounds. The little information of corresponding fragment peak makes the elucidation of compound structure difficult. Although collision-induced dissociation (CID) and MS^n technologies can be used to obtain structure information, the fragment spectra for the same compound obtained by different analytical instruments are not the same and difficult to be standardized, resulting in the

Table 3.2 Characteristics of different mass spectrometers and their applications in plant metabolomics

Mass spectrometer	Coupled with LC	MS quantitative metabolite profiling	MS^n for structural identification	Mass accuracy	Largest mass detection range
Q	+	+	–	low	2,000
TOF	++	++	–	2–5 ppm	10,000
Q-TOF	++	+	+	2–5 ppm	20,000
QQQ	+	++	+	low	3,500
IT	+	–	++	<1 ppm	4,000
FT-ICR	–	+	++	1 ppm	20,000

lack of "standard mass spectral library" suitable for different LC-MS until now. To date, there is still a long way to go in the establishment of a general-purpose "standard mass spectra library" for LC-MS. In practical, many laboratories engaged in plant metabolomics study have established many "in-house mass spectral libraries" suited to their own use, but cannot be integrated with others. Therefore, in the application of LC-MS, especially when performing compound structure identification, the high-resolution mass spectra and precise molecular weight determination are particularly important.

Currently, single high-resolution mass spectrometers, such as double-focusing magnetic sector and Fourier transform ion cyclotron resonance (FT-ICR) mass spectrometer, could offer the highest levels of resolution and mass precision (typically within 1–2 ppm). These two mass spectrometers alone can reach tandem mass spectrometry (MS/MS) operation, i.e., selectively store ions with a particular charge to mass ratio (m/z), and then directly observe its reaction and give corresponding secondary ion fragment peaks information. But the high price and the continuous improvement of TOF mass spectrometer's resolution and accuracy (currently, there are TOF mass spectrometers with accuracy close to that of FT-ICR on the market) limit its broad application.

Ion trap mass spectrometer (IT-MS) is a dynamic mass spectrometry. There are many similarities as the quadrupole mass spectrometer. In many cases, the difference between quadrupole mass analyzer and ion trap is considered that the former is two-dimensional, while the latter is three-dimensional. The ion trap has many advantages, such as simple structure, cost-effective, high sensitivity, and wide mass detection range (W. Paul, H. Dehmelt, and N. Ramsey won the Nobel Prize in Physics in 1989 for the invention of IT-MS). Previous tandem mass spectrometry is spatially tandem, which is tandem connected by several mass analyzers and hence the price doubled. Now ion trap is tandem connected temporal, and hence, the price is the lowest. The main advantage of ion trap is that it can carry out multistage tandem MS measurement (MS^n, theoretically up to ten stages and usually reach 4–5 stages in practice). This mass analyzer has been widely applied in proteomics research. However, the disadvantage of IT-MS is that it is difficult to do quantitative analysis; thus, its applications in metabolomics are far less than q-TOF mass spectrometer.

It should be noted that the molecular weight calculation of compound is based on the 1/12 of ^{12}C mass as one mass unit. The calculated ion mass value refers to the monoisotope mass of all the elements in this ion, rather than the sum of all the atomic mass numbers of this element in the Periodic Table of the Elements (This value is the average value of the isotopic mass). The introduction of atomic mass in the Periodic Table instead of monoisotopic mass into the molecular weight calculation will result in mass errors.

Compound structural elucidation has been a difficulty troubling LC-MS analysis means. The following points will have some help to compound structural elucidation in practice: ① Collect as much as possible the structure, mass spectral data, and UV data of known chemical components in the sample; ② In analysis, perform LC-MS positive and negative ion switching one stage mass spectrometric detection, with the purpose of determining the molecular weight. It is unreliable to detect and determine the molecular weight of a chemical using only one ion mode, unless the $[M + H]^+$, $[M + Na]^+$, and $[M + K]^+$ peak can be observed simultaneously in positive mode, molecular weight can be determined based on the sequentially difference of 22 (Na–H) and 16 (K–Na); in addition, high-resolution MS is helpful to obtain molecular weight information; ③ combined with the UV spectrum characteristics to determine probable type of the compound (usually, LC-MS is equipped with UV detector); ④ perform multistage mass spectrometric analysis. Possible fragments in the molecule can be determined based on the cracked fragments. Combined with molecular weight, if the compound is known, it can be substantially attributed by now. However, the stereoisomer cannot be attributed. It is difficult to determine the linking position of various kinds of chain and the substitute position of some groups; ⑤ for the compound with new structure and no standards are available, the only way is to purify and enrich using various means and determine the structure with NMR.

3.2 Data Interpretation

After data collection by LC-MS, the original spectra must be subject to noise signal filtering, alignment, feature detection, and normalization to obtain the raw data that can be further processed. Different vendors provide different built-in software for data processing. As for the spectra produced by LC-MS, the peak area integration results are often extracted as raw data, followed by integrating all data for different samples in a uniform format, which is a very crucial step in metabolomics study. The use of high-throughput detection analytical tool can obtain vast amount of data. However, if not be reasonably processed, these interfering data will be detrimental to the study work. Pattern recognition and multi-dimensional statistical methods can be used to extract useful information from the vast amount of data. These methods can conduct data dimensionality reduction, making it easier for visualization and classification. Currently, two commonly used algorithms in data analysis are the pattern finding-based unsupervised method and supervised method. In addition, at each stage of data process and analysis, it must be also careful for the quality control of data and the validity check of model. We only do a brief introduction on their classification in this chapter. For details, refer to Chap. 6 (MS-based metabolomics data processing).

3.2.1 Unsupervised Method

Unsupervised method is used to explore completely unknown characteristics of the data. The original data information is classified in accordance with the characteristics of the sample, and the target data with similar characteristics are classified in the homological class and are visually expressed by appropriate visualization technologies. Common methods used in this field include cluster analysis and principle components analysis (PCA).

3.2.2 Supervised Method

If there is some prior information and assumptions related with the data, the supervised method is more suitable and more effective than unsupervised method. Supervised method establishes class information on the basis of existing knowledge and makes use of the established class information to identify, classify, and predict the unknown data. Because there are train samples available for learning in the model establishment, this method is named as supervised learning. Common methods used in the field include linear discrimination analysis and partial least square discriminant analysis (PLS-DA) (Fukusaki and Kobayashi 2005).

3.2.3 LC-MS-Related Online Database

In the information age, the use of network resources is essential for study, and the establishment of corresponding online database for the multitude of plant metabolites study is also imperative. These metabolomic databases are also conducive to link the relationship between metabolomics and other systems biology branch (Weckwerth and Fiehn 2002). Currently, the most commonly used database is Arabidopsis Web site: TAIR (www.arabidopsis.org). Other commonly used online plant resources include www.maizegdb.org, www.york.ac.uk/res/garnet/garnet.htm, and www.genome.ad.jp/kegg/pathway.html. In addition, there are some online data resources related with metabolism and metabolomics (Yonekura-Sakakibara and Saito 2009). The following is a brief introduction of some commonly used online resources in metabolomics study.

KEGG (www.genome.jp.kegg/) is a comprehensive informatics database on genome decoding, in which the PATHWAY database integrates current knowledge on molecular interaction network, and the COMPOUND/GLYCAN/REACTION database provides knowledge on biochemical complex and reaction.

KNApSAcK (www.kanaya.naist.jp/KNApSAcK) is a Web site covering the relationships between most of plant species and metabolites, including more than 40,000 compounds and 8,000 plant species. Users can conveniently check to know which metabolites in a specific species have been reported.

PlantCyc (www.plantcyc.org) is a database on metabolites, elucidating 643 metabolic pathways in more than 350 plant species (version 3.0) and including metabolic pathways, reactions, enzymes, and substrates obtained from a number of literatures and online resources.

MMP (www.chem.qmul.ac.uk/iubmb/enzyme) describes in detail main metabolic pathways and related key enzymes.

MassBank (www.massbank.jp) is a high-resolution MS data developed and maintained by Japanese society for mass spectrometry. This Web site is updated continuously. By February 2010, it includes about 25,000 high-resolution MS data and provides various mass spectral search means, providing powerful support for the rapid identification of compounds.

3.3 New Trends in LC-MS Development

There are still many challenges in current LC- and LC-MS-based metabolomics study, a part of which lies in the instrument. For example, the resolution of LC is not enough high, it is difficult to cover all chemicals in a run, especially for high complexity sample. Due to the bias of different analytical methods to chemicals, some specific compounds often cannot be detected even in an improper method; another challenge comes from informatics, the structural elucidation of unknown compound is a bottleneck for the rapid development of metabolomics and another challenge is to explore of high-efficient data analysis methods. Bino et al. (2004) proposed that currently, there are two restriction factors limiting the analytical platform in metabolomics research: The first one is how to improve the separation capacity of LC; the second one is how to improve the identification ability to compounds, and so the developmental trend in LC-MS is expanded around these two points.

3.3.1 Ultra-High-Performance Liquid Chromatography

According to Van Deemter chromatography theory, column efficiency is inversely proportional to the size of column packing; thus, it can be considered that instrument resolution can be improved by increasing column efficiency (theoretical plate number *N*), and the use of column packing with particle size less than 2 μm is undoubtedly a good way to increase column efficiency. But as column packing particle size decreases, the system pressure will rise accordingly. In recent years, ultra-high-performance liquid chromatography (UPLC) technology gradually solved the problem of the entire system on the operating pressure. The connection system of the instrument, valve, the equipped column, and so on are able to resist high pressure to achieve commercial standard. Therefore, there is no essential difference in theory between UPLC (the column packing particle size is 1.7 μm) and HPLC (the column packing particle size is typically 3.5 or 5 μm) (Novakova et al. 2006), but UPLC has the following three technical characteristics: ① ultra resolution; ② ultra speed. Under the premise of ensuring the same quality data, UPLC can provide more information per unit time and can reduce the column length without affecting resolution. Thus, small particle size can provide higher analysis speed while save the consumption of mobile-phase solvent; ③ sensitivity. In the past, most works were focused on the detector to improve the sensitivity of LC system, no matter it is an optical detector or mass detector. In fact, the use of UPLC can also increase analytical sensitivity. UPLC can improve column efficiency *N*, so that the peak width *W* becomes narrower, whereas the peak height is increased. At the same time, UPLC uses shorter column (column length *L* is smaller), further increasing the peak height. Therefore, with improving the column efficiency, the system sensitivity is 70 and 40 % higher in the UPLC system using 1.7 μm packing than that using 5 and 3.5 μm packing, respectively. And at the same column efficiency, the UPLC system using 1.7 μm packing can provide sensitivity 3 times and 2 times higher than that of 5 and 3.5 μm packing, respectively. Currently, UPLC and similar products (UPLC is a registered trademark of Waters, and other analytical instrument companies had to use a different name) are more and more widely used in the field of liquid chromatography.

3.3.2 Hydrophilic Interaction Liquid Chromatography

Hydrophilic Interaction Liquid Chromatography (HILIC) is a kind of chromatographic mode using the polar stationary phase (e.g., bare silica gel or amino-silica) and using aqueous mobile phase. Similar to the normal-phase chromatography, the retention times of polar compounds in the HILIC column are longer than that of the nonpolar compounds. And the retention sequence is opposite with that of reversed-phase chromatography. Thus, combined with results of the reversed-phase column, the integrity of the metabolomics information can be ensured (Fig. 3.1). Because a large amount of metabolites such as saccharides and peptides cannot be retained on the conventional reversed-phase column and being washed away out of the column with the solvent front, thus cannot be well analyzed by LC-MS equipped with regular column. Antonio et al. (2008) using HILIC-ESI-MS analyzed the content difference of monosaccharides, glycol, and phosphate sugar in the leaf of wild-type and mutant (*pgm1*) Arabidopsis. These highly polar molecules were all well

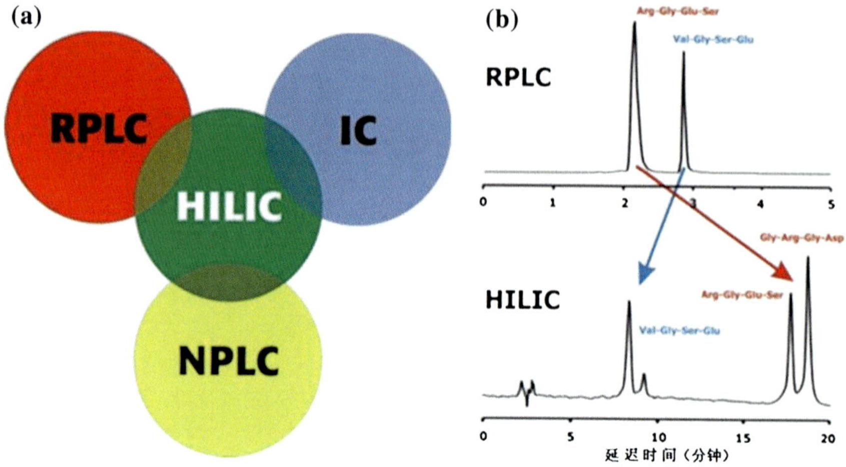

Fig. 3.1 The relationship between HILIC and other LCs and spectral differences. HILIC is a useful complement of other LCs such as reversed-phase LC (*RPLC*), normal-phase LC (*NPLC*), and ion chromatography (*IC*) in analytical capacity (the *left figure*); the *right figure* is a simple schematic of the difference between RPLC and HILIC. The retention sequence of the two is opposite, and the sample being analyzed is short peptides with different polarity

separated by HILIC column. Acetonitrile–water system is often used as the mobile phase in HILIC analysis. Initial conditions include a high proportion of organic solvent; the typical concentration is 95 % organic phase such as acetonitrile and then gradually reduced to the aqueous phase (Cubbon et al. 2009). Therefore, HILIC is also called reversed-phase chromatography. In addition, the same mobile phase system used for HILIC column and reversed-phase column it possible to combine HILIC and reversed-phase chromatography in a two-dimensional chromatography.

3.3.3 Two-Dimensional Liquid Chromatography

The Two-Dimensional Liquid Chromatography (2D-LC) is a high-efficient separation technology developed recently. The resolution and peak capacity of the two-dimensional orthogonal system have been compared with that of one-dimensional LC, and 2D-LC has been a powerful tool in the separation and analysis of complex system and widely used in life sciences, especially in the analysis of small polar molecules (Evans and Jorgenson 2004). 2D-LC is such a separation system that tandem connecting two columns with different separation mechanism and mutually independent. The basic principles are similar to that of conventional two-dimensional thin-layer chromatography, and the difference is that two-dimensional thin-layer chromatography uses different developing solvent in the same separation matrix, while 2D-LC uses both different columns and mobile phases. The sample to be analyzed flows into the interface through the first-dimensional column, being concentrated, enriched or cut, and then switch into the second-dimensional column and detector. 2D-LC often analyze samples using two different separation mechanism, that is, separate the complex mixture to single components based on different characteristics of the sample, including molecular shape, isoelectric point, hydrophilicity, charge, and special intermolecular interaction (affinity), etc. Components, which cannot be completely separated in the one-dimensional separation system, may get well-separated in the two-dimensional system, greatly improving the separation capacity and resolution. Theoretically, the peak capacity of completely orthogonal 2D-LC is the sum of the peak capacity of two kinds of one-dimensional separation system.

Based on the separation purpose, different column systems with different separation mechanisms can be used to build multi-dimensional LC separation system. The ion exchange chromatography (IEC), RPLC, affinity chromatography (AC), size exclusion chromatography (SEC), and normal-phase chromatography (NPC) can all be combined for the special purpose of separation. For the combination of two separation modes, not only the separation selectivity, resolution, peak capacity, column capacity, and analysis speed should be considered, but also the separation of the biological sample, the sample recovery rate, and activity may be also very important. In the practical building process of multi-dimensional separation system, the influence of different factors should be comprehensively considered to select a reasonable separation system.

3.3.4 LC-NMR

In plant metabolomics field, the structural elucidation of unknown compounds has been a very important but very difficult problem. An instrument that can complete the separation, quantitative and qualitative analysis, and structure elucidation of the sample at the same time has been a dream of researchers engaged in metabolomics. As for compound structure analysis, although mass spectrometry, especially high-resolution mass spectrometry, can provide very useful element composition and structure information, in the case of no available chemical standards, it is rarely possible to rely only on the information provided by mass spectrometry to obtain the structure. The structure of unknown compounds will eventually be obtained relying on NMR. Lindon et al. have done the pioneer-work for testing the feasibility of LC, NMR, and MS coupling technique and made some breakthroughs (Lindon et al. 1997), that is, the liquid chromatography and nuclear magnetic resonance coupling technique (LC-NMR), and the working principle can be found in Chap. 4.

According to the different characteristics of the coupling instrument, the current application of LC-NMR detection can be divided into continuous-flow, stopped-flow, collection-analysis, timesharing stopped-flow, and the automatic detection using UV detector to stimulate NMR sampling (Lindon et al. 2000). ① Continuous-flow. This mode was only applicable to ^{1}H NMR technique, not to ^{13}C NMR due to its long signal collection time, which is not suitable for high-throughput omics analysis. In the elution process, the solvent composition is changing continually, making it difficult to suppress the solvent peak (This technique is mainly used for eliminating interference of LC eluent to NMR measurement). Usual solution is to do some preliminary experiments to record the peak at different times, determine the optimal separation conditions, and then acquire the NMR spectrum of the targeted peak; ② stopped-flow. Even if the solution stays in the detection cell for testing, when the component retention time is known, or HPLC-NMR uses a sensitive online detector, this method can be used. In the stopped-flow mode, NMR ^{1}H spectrum or two-dimensional spectrum is detected; ③ collection-analysis. The chemicals separated by LC were collected in advance to each sample cell before the offline NMR analysis; ④ time-sliced stopped-flow. The current mobile phase is suspended at a certain time interval to detect NMR spectrum. This approach is particularly suitable for the component without UV chromophore. The purity of the peaks also can be estimated by HPLC-NMR spectrum; ⑤ UV-detected triggered NMR acquisition. This method is primarily to use software technology. When the component peak is detected by UV, the component is accurately transferred in the detection cell after calculation and activated NMR for sample analysis. Although current LC-NMR [and even LC-NMR-MS/MS (Duarte et al. 2009)] just stay in the exploratory stage in plant metabolomics study, and with some challenging problems, it is still worth foreseeing many applications of this technology in plant metabolomics in near future.

3.4 Common Problems and Notes

3.4.1 LC System

We have pointed out in the foregoing instrument working principle that the separation of the LC is realized in the column. The reversed-phase column is commonly used in plant metabolomics. How to protect the separation column from contamination to extend column lifetime is the key to obtain reliable and rich information content results. There are a few things to note for reader reference:

(1) Adding guard column (pre-column) is an effective way to protect the separation column. Guard column is a short column with packing's nature similar to that of separation column. The guard column is connected in front of the separation column instead of flow filter or coupled with it. The function of the guard column is to prevent chemical waste from separation column port, or else which will gradually accumulate in the separation column head if directly into the separation column and finally makes the column efficiency reduced.
(2) Avoid high-pressure shock to separation column. The column generally can afford high-pressure, but unable to withstand the high-pressure impact of

sudden changes, which will change the column bed volume, affecting the column efficiency.

(3) Reasonably select chromatographic column. We know that the selection of chromatographic separation column is based primarily on two points: one is according to the molecular weight of the component to be tested, and the other one is according to the mobile-phase conditions, including pH value, ionic strength, and solvent polarity. Therefore, we must select appropriate chromatographic column according to the analysis object and the mobile-phase conditions, if necessary, to consult to the company's technical staff.
(4) Rinse chromatographic column on a regular basis. At the end of each analysis, after flushing with buffered saline, the column should be rinsed with a strong solvent. For example, the reversed column can be rinsed by HPLC grade methanol and acetonitrile, on the purpose of removing strongly retained components absorbed in the column. Even the mobile phase is aqueous methanol, the column should be rinsed in this way to recover the column efficiency to a certain extent.
(5) Purifying sample is the most effective way to reduce column contamination. Generally, the cleaner the purified sample, the longer the lifetime of the chromatographic column; on the contrary, if the sample is only extracted without purification, the column efficiency of the chromatographic column will reduce rapidly. There should be a set of effective sample pre-treatment method for the establishment of an ideal chromatographic analytical method. The sample extraction and purification should be paid special attention to minimize impurities content in the treated sample (such as the large molecule protein) to protect the column head and the column.
(6) The extracted and purified sample should be dissolved in the mobile phase. On the one hand to reduce the solvent peak interference in the chromatogram and on the other hand to check the solubility of the sample in the mobile phase. If the sample precipitates in the mobile phase after injection, it will block the sample injector and the column head, and even the sample will decompose, generating strange peaks. If the sample is not soluble in the mobile phase, the dissolve condition should better be changed, such as changing solvents, improving pre-treatment method, or filtering to remove insoluble materials.

3.4.2 The Matrix Effect of the MS System

MS can be understood as a detector, like the commonly used UV and fluorescence detector. Thus, just like UV and fluorescence detector, mass detector also has matrix effect. In the chemical analysis, matrix refers to other components except the analyte in the sample, which often have significant interference on the analysis process of the analyte and affect the accuracy of the analysis result. Mass spectrometry plays a key role in the qualitative and quantitative analysis of the sample, and so if the ion source is contaminated due to matrix effect, the ionization efficiency will be greatly reduced, and this phenomenon is known as ion suppression. The main reason of ion suppression is the failure of the separation stage resulting in the compound co-elution. It will happen in any type of mass spectrometry, so regularly cleaning the ion source is necessary to ensure ionization efficiency (the cleaning cycle can be determined according to the use frequency of the instrument and the complexity of the sample to be analyzed). There are two main methods for evaluating the matrix effect of LC-MS: post-column injection and post-extraction adding method (Rogatsky and Stein 2005).

The matrix effect in LC-MS was firstly discovered in 1993. With increasing the sample complexity, the ESI response value of the target analyte is decreased. The US Food and Drug Administration (FDA) clearly stated in the 2001 *Guidance for Industry: Bioanalytical Method Validation* that the matrix effect should be assessed in the development and validation of LC-MS analysis methods. Matrix effect refers to the combined impact of all components of the sample, which affect the ionization process of the analytes of interest by components co-eluted with them in the chromatographic separation. The co-eluted interfering components can be divided into endogenous impurities and exogenous impurities. Endogenous impurities refer to the organic or inorganic molecules simultaneously extracted in the sample extraction process. If these impurities have relatively high concentration and co-eluted with target compounds from the column and into ion source, the ionization process of the target compounds will be seriously affected. Exogenous impurities are usually ignored, but they can also bring serious matrix effect. These interfering substances are introduced in the sample pre-treatment process. According to reports, they mainly include polymer residues, phthalates, detergents degradation products, ion pair reagent, and ion exchange agents such as organic acid, buffer salts, small SPE column material, and released substances from column stationary phase.

The presence of matrix effect will seriously affect the quantitative accuracy and precision of the analyte. The affecting factors are varied and are difficult to be completely eliminated. In recent years, studies committed to the elimination and compensation of matrix effect are gradually increased, mainly including optimizing MS conditions, optimizing chromatographic separation system, multi-step purification techniques, internal standard quantification, matrix-matched calibration curve quantification, and echo peak technology.

Currently, the most commonly used method of matrix effect elimination is to establish a calibration curve by standard chemicals with known concentration while keep the matrix as unchanging as possible. Solid samples also have strong matrix effect, and the correction is also particularly important. The standard addition method can be used for the impact of complex matrix or matrix with unknown components. In this method, the respond value of sample needs to be measured and recorded. A small amount of standard solution is further added, and then, the response value of the sample is recorded again. Ideally, standard addition should make the analyte concentration increase 1.5–3 times, while the solution added several times should also be consistent. The standard sample volume should be as small as possible to minimize the impact of matrix in the adding process.

3.4.3 Sample Preparation and Precautions

Different from the relatively simple analysis process and strong regularity of sample composition in genetics (four different nucleotides arrangement) and proteomics (about 20 different amino acids arrangement), the atomic composition and arrangement of different metabolites are vastly different, causing the great difference between metabolites in physical and chemical properties (chemical stability, volatility, polarity, solubility). Also, in the same analysis sample, the content of different metabolites is enormously different. For example, it is impossible to detect phytohormones (pmol or less) and primary metabolites such as amino acids and oligosaccharides (mmol) with one analytical platform simultaneously, otherwise the accuracy of the analysis results are bound to be greatly interfered. Different metabolites must be roughly separated according to their physico-chemical properties and contents before being analyzed. Thus, now laboratories engaged in plant metabolomics research have a variety of different analysis platforms to meet the analytical requirements of different components. Take chromatography–mass spectrometry analysis platform, for example, there are GC-MS, LC-MS, and capillary electrophoresis–mass spectrometry (CE-MS). Each analysis platform is equipped with a variety of columns with different packing materials, and each column only has a certain range of application.

As the final product of organisms, in many cases, the difference range of metabolites is larger than that of corresponding transcription and protein levels in organisms, which is an advantage of metabolomics compared with other omics. Meanwhile, the biological metabolic pathways are highly sensitive to the environmental changes, which accordingly causes the great difference of metabolites content between different individual organisms. Studies have shown that even in strictly controlled growth conditions, the biological difference between analysis samples is several times higher than that of other systemic difference (Roessner et al. 2000). Thus, the planting conditions, collection time, and manners must be as consistent as possible. Nevertheless, in the process of sample collection and preparation, it must be considered that the speed of many metabolic reactions in plant cell is very fast (<1 s). Especially, when using scissors to take a certain organ from the whole plant, if the speed is too slow, the wound induction will be induced, and the analysis result cannot reflect the real situation of the plant. Therefore, various enzymatic reactions must be rapidly inhibited when collecting plant materials. The commonly used method is to freeze the material being analyzed into liquid nitrogen and store it in −80 °C low-temperature freezer. Nevertheless, long store time of the plant material in −80 °C freezer (generally not exceeding 4 weeks) will still cause significant differences in analysis results. In addition, the pool collection should better be used when collecting plant materials to eliminate the deviation of analysis results caused by individual difference and reduce biological replicates (usually about 4–6 times). For example, an Arabidopsis seedling is enough to meet the analysis demand, but you need to take at least five seedlings for the sampling and analysis in order to eliminate individual variation. In this respect, the plant suspension cells are a better system, but because it cannot represent various physiological state of the whole plant, the application range is still very narrow. The drawback of multi-organisms sampling is that it will cover a lot of biological information with subtle difference. But if using individuals as analysis object, the replicate number will be increased to dozens, and the workload will increase a lot (Weckwerth et al. 2004).

Sample preparation in plant metabolomics is the main step of introducing system error. The sample preparation of plant metabolites includes tissue sampling, homogenization, extracting, preservation, and sample pretreatment (Fukusaki and Kobayashi 2005). Metabolites are usually, respectively, extracted with water or organic solvent (such as methanol and acetonitrile), obtaining the polar extract and nonpolar extract, so that the nonpolar chemicals and the polars are well separated. Prior to analysis, usually conduct pre-treatment by solid-phase micro-extraction, solid-phase extraction, and affinity chromatography method. Plant metabolites, however, are vastly different. Many structures of these substances will change if being slightly disturbed. And the instruments suitable for the analysis and identification of different metabolites are different. At present, there is no one extracting method suitable for all the metabolites. Usually, the researchers can only based on the characteristics of the metabolite to be analyzed and the identification means to select appropriate method, while the extracting time, temperature, solvent composition and quality, and the experimenter's skills will also affect the level of sample preparation.

3.5 Perspectives

Like other analysis platforms, LC-MS will continue develop toward the improving of the separation efficiency, peak capacity, and sensitivity for LC and providing more accurate molecular weight and fragment cracking information of the analyte for MS, so as to meet the growing quantitative and qualitative need of metabolomics studies. The application of LC-MS in metabolomics studies is far behind that in proteomics, with a large development space. With the continuous improvement of separation capacity of LC and resolution and accuracy of MS as well as the fast updating of various metabolomics analysis software, there is no doubt that LC-MS will play an increasingly important role in plant metabolomics studies.

References

Antonio C, Larson T, Gilday A, Graham I, Bergstrom E, Thomas-Oates J. Hydrophilic interaction chromatography/electrospray mass spectrometry analysis of carbohydrate-related metabolites from *Arabidopsis* thaliana leaf tissue. Rapid Commun Mass Spectrom. 2008;22:1399–1407.

Bino RJ, Hall RD, Fiehn O, Kopka J, Saito K, Draper J, Nikolau BJ, Mendes P, Roessner-Tunali U, Beale MH, Trethewey RN, Lange BM, Wurtele ES, Sumner LW. Potential of metabolomics as a functional genomics tool. Trends Plant Sci. 2004;9:418–25.

Cubbon S, Antonio C, Wilson J, Thomas-Oates J. Metabolomic applications of HILIC-LC-MS. Mass spectrometry reviews; 2009.

Dixon RA, Strack D. Phytochemistry meets genome analysis, and beyond. Phytochemistry. 2003;62:815–16.

Duarte IF, Legido-Quigley C, Parker DA, Swann JR, Spraul M, Braumann U, Gil AM, Holmes E, Nicholson JK, Murphy GM, Vilca-Melendez H, Heatond N, Lindon JC. Identification of metabolites in human hepatic bile using 800 MHz H-1 NMR spectroscopy, HPLC-NMR/MS and UPLC-MS. Mol Biosyst. 2009;5:180–90.

Evans CR, Jorgenson JW. Multidimensional LC-LC and LC-CE for high-resolution separations of biological molecules. Anal Bioanal Chem. 2004;378:1952–61.

Fukusaki E, Kobayashi A. Plant meta-bolomics: potential for practical operation. J Biosci Bioeng. 2005;100:347–54.

Hagel JM, Facchini PJ. Plant metabolomics: analytical platforms and integration with functional genomics. Phytochem Rev. 2008;7:479–97.

Lindon JC, Nicholson JK, Wilson ID. Directly coupled HPLC-NMR and HPLC-NMR-MS in pharmaceutical research and development. J Chromatogr B: Anal Technol Biomed Life Sci. 2000;748:233–58.

Lindon JC, Nicholson JK, Sidelmann UG, Wilson ID. Directly coupled HPLC-NMR and its application to drug metabolism. Drug Metab Rev.1997;29:705–46.

Novakova L, Matysova L, Solich P. Advantages of application of UPLC in pharmaceutical analysis. Talanta. 2006;68:908–18.

Roessner U, Wagner C, Kopka J, Trethewey RN, Willmitzer L. Simultaneous analysis of metabolites in potato tuber by gas chromatography-mass spectrometry. Plant J. 2000;23:131–42.

Rogatsky E, Stein D. Evaluation of matrix effect and chromatography efficiency: new parameters for validation of method development. J Am Soc Mass Spectrom. 2005;16:1757–9.

Sheng LS, Su HH, Guo DB. Liquid chromatography and mass spectrometry coupling technique. Beijing: Chemical Industry Press; 2006.

Tanaka N, Nagayama H, Kobayashi H, Ikegami T, Hosoya K, Ishizuka N, Minakuchi H, Nakanishi K, Cabrera K, Lubda D. Monolithic silica columns for HPLC, micro-HPLC, and CEC. HRC J High Resolut Chromatogr. 2000;23:111–6.

Weckwerth W, Fiehn O. Can we discover novel pathways using metabolomic analysis? Curr Opin Biotechnol. 2002;13:156–60.

Weckwerth W, Loureiro ME, Wenzel K, Fiehn O. Differential metabolic networks unravel the effects of silent plant phenotypes. Proc Natl Acad Sci U S A. 2004;101:7809–14.

Yonekura-Sakakibara K, Saito K. Functional genomics for plant natural product biosynthesis. Nat Prod Rep. 2009;26:1466–87.

Chapter 4
Nuclear Magnetic Resonance Techniques

Fu-Hua Hao, Wen-Xin Xu and Yulan Wang

Nuclear magnetic resonance (NMR) refers to the physical process of the spin splitting of nucleus with nonzero magnetic moment under the applied magnetic field and the resonant absorption of radiofrequency (RF) radiation with certain frequency. Since its birth in 1945, NMR has experienced several important stages of development. Especially driven by modern computer technology, information technology and other high-tech technologies, NMR technology has been rapidly developing to an indispensable analytical tool in the field of physics, chemistry, life sciences, and medicines, etc. and is playing an increasingly important role.

4.1 An Overview of NMR Development

In 1924, Pauli predicted that certain nuclei have spin angular momentum and magnetic moment. This is the first recognition of the interaction between nuclei and magnetic field and the applied radiofrequency (RF) field, generating the theoretical basis of early NMR.

In 1945, the Block group in Standford University and the Purcell group in Harvard University independently observed NMR phenomenon and therefore won the 1952 Nobel Prize in Physics.

In 1953, the first 30 MHz continuous wave NMR (CW-NMR) spectrometer was produced. A few years later, 60 and 100 MHz NMR spectrometers appeared one after another.

F.-H. Hao (✉) · W.-X. Xu · Y. Wang
Wuhan Institute of Physics and Mathematics, Chinese Academy of Sciences,
Wuhan 430071, China
e-mail: haofuhua78@163.com

W.-X. Xu
e-mail: xwxshirley@gmail.com

Y. Wang
e-mail: yulan.wang@wipm.ac.cn

X. Qi et al. (eds.), *Plant Metabolomics*, DOI 10.1007/978-94-017-9291-2_4

In 1956, Knight discovered that the chemical environment of the nucleus affects NMR signal, and this effect is related to the molecular structure of material.

In 1965, Cooley and Tukey proposed the fast Fourier transform calculation method, and the pulse Fourier transformation NMR (PFT-NMR) technology began to rise.

In 1971, Jeener proposed the two-dimensional NMR methods, which plays a great role in promoting the analysis of material structure.

In the mid-1980s, Wuthrich developed the homonuclear two-dimensional NMR methods.

In the late 1980s, Bax et al. proposed the heteronuclear two-dimensional NMR methods.

In the early 1990s, high-field NMR spectrometers (750 MHz in 1993 and 800 MHz in 1995) appeared one after another, greatly improving the sensitivity of the instrument. At the same time, three-dimensional and four-dimensional heteronuclear NMR methods have been developing rapidly.

Currently, the high-field NMR, ultra-low-temperature probe and the coupling technology of chromatography and NMR have been widely applied in physics, chemistry, biology, medicine, and other disciplines and resulting in a large number of cross-disciplines.

4.2 NMR Fundamentals

When place a nucleus with nonzero spin quantum number such as ^{1}H, ^{13}C, ^{31}P, and ^{14}F in an external magnetic field B_0, the nuclear spin precession produces a weak magnetic field. Take ^{1}H, for example, the magnetic field has two opposite orientations according to the Boltzmann distribution law (Fig. 4.1). When the nuclear spin parallels with the direction of external magnetic field (Fig. 4.1a), the hydrogen

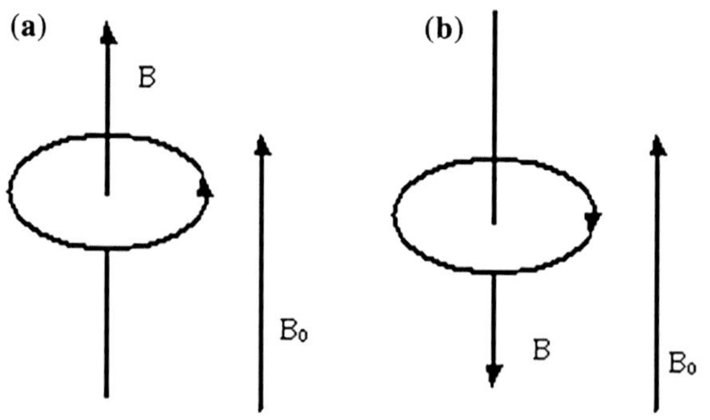

Fig. 4.1 The orientation of nuclei in the static magnetic field. **a** Spin parallel. **b** Spin antiparallel (B_0 is the applied magnetic field strength, B is the magnetic field strength generated by the hydrogen nuclear spin)

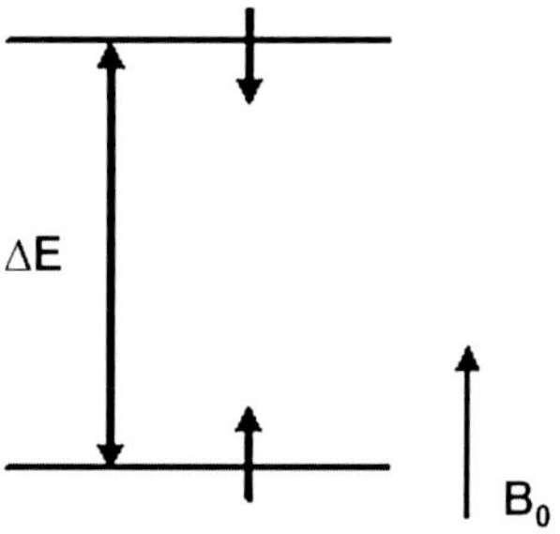

Fig. 4.2 The energy-level diagram of the hydrogen nuclei in the magnetic field

nucleus is in a stable low energy level; when the nuclear spin is opposite to the direction of external magnetic field (Fig. 4.1b), the hydrogen nucleus is in a relatively unstable high energy level. The energy difference ΔE between the above two energy levels (Fig. 4.2) is an important factor in deciding the nuclear distribution in various energy levels, which can be described by the Boltzmann's law (Eq. 4.1), wherein N_j is the number of nuclei in the high energy level, N_0 is the number of nuclei in the low energy level, K is Boltzmann's constant, and T is the absolute temperature.

$$\frac{N_j}{N_0} = \exp\left(\frac{-\Delta E}{\mathrm{K}T}\right) \tag{4.1}$$

Due to the generally small ΔE value (Eq. 4.2), the nuclei can easily complete transition when absorbing energy in the range of radio waves. However, only when the correlation between the absorption frequency v and the external magnetic field strength B_0 satisfies Eq. (4.3), the transition is able to occur. And this is the NMR phenomenon. The Eq. (4.4) can be obtained from the result of Eq. (4.3) substituting Eq. (4.2). It can be seen from Eq. (4.4) that the relationship among ΔE, B_0, and the radio frequency v is only correlated to the nucleus itself. y is a proportional constant, namely the nuclear gyromagnetic ratio.

$$\Delta E = hv \tag{4.2}$$

$$v = \frac{\gamma B_0}{2\pi} \tag{4.3}$$

$$v = \frac{\Delta E}{h} = \frac{\gamma B_0}{2\pi} \tag{4.4}$$

It can be easily seen from Eq. (4.4) that the NMR signal can be obtained in two ways: the first one is to keep radio frequency unchanged while changing the applied magnetic field strength (field sweep for short); the other one is to keep applied magnetic field strength unchanged while changing radio frequency (frequency sweep for short), then to receive a resonance signal via a coil, thereby obtaining the energy absorption curve. The frequency sweep is superior to the field sweep:

First of all, the field sweep needs a few minutes, while the frequency sweep only needs a 1–10 μs pulse and a few seconds for the signal acquisition; In addition, from the signal strength aspect, the signal strength obtained by field sweep is low, while multiple scans can be superimposed by frequency sweep to obtain a stronger signal. When a short but intense resonance RF pulse is applied to the sample, the curve of signal changing with time in the process of nuclear recovery to the balanced state after the pulse, i.e., the free induction decay (FID), is recorded. Then, Fourier transformation is performed to obtain the NMR frequency spectra. Thus, the modern NMR spectrometers generally belong to the latter: Fourier transform NMR spectrometers.

4.3 NMR Spectrometers

A variety of commercial NMR spectrometers appear in the market from the 1950s to date. According to the RF sources, they can be divided into continuous wave NMR spectrometers (CW-NMR) and pulse Fourier transform NMR spectrometers (PFT-NMR). The RF pulse of the PFT-NMR which appeared in the 1970s is equivalent to the multi-channel transmitter which can simultaneously stimulate all nuclei in the desired frequency range and record the FID, and then the NMR spectra is obtained by Fourier transformation, thus greatly improving the signal-to-noise ratio and speeding up the analysis.

Below is an introduction of the main components of NMR spectrometers (Fig. 4.3):

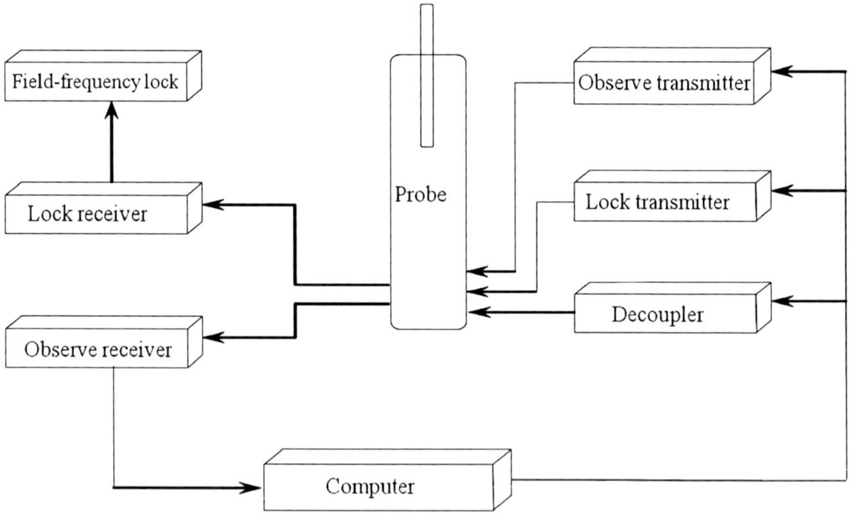

Fig. 4.3 The main components of NMR spectrometer

1. *Magnet.* Its role is to provide a stable high-intensity magnetic field, namely B_0. At present, the high-field-strength superconducting magnet is commonly used. Enough liquid helium and liquid nitrogen need to be added to the superconducting magnet to maintain a low temperature for the normal work of the spectrometer.
   The magnet contains a multiple sets of shim coils. The magnetic field inhomogeneity of the main magnet can be compensated by adjusting the current to make the shim coils constituting spatially mutual orthogonal gradient magnetic fields. The resolution can be improved and the high-quality NMR spectra can be obtained by repeatedly adjusting the current.
2. *Probe.* It is the core component of the NMR spectrometer. It is fixed to the center of the magnet and is cylindrical. The sample tube is placed on the center of the probe. The sending of the RF wave pulse to the sample and the transmitting of NMR signals generated by the sample to the signal receiving system are all completed by the probe. There are a variety of probes, such as positive probe, inverse probe, micro-probe, high-resolution magic angle probe (HR-MAS) and solid probe, which can be used for the detection of different samples (liquid, solid, semisolid, and biological tissue, etc.).
3. *RF transmitting system.* The RF sources with desired frequency required by the three channels of preliminary observe nucleus (such as ^{1}H, ^{13}C and ^{31}P), irradiated nucleus (such as the irradiated ^{1}H to eliminate the coupling effects to the observe nucleus) and lock nucleus (such as ^{2}D and ^{7}Li) is accurately synthesized by the frequency synthesizer in the RF transmitting system from a bundle of electromagnetic pulse with fixed frequency. The RF pulse transmitted by the RF source irradiates the sample by the transmitting coil on the probe for the excitation of all the nuclei in the desired frequency range.
4. *Signal receiving system.* When the RF pulse is emitted and transmitted to the sample, the FID signal generated by the sample is received by the signal receiving system. The signal receiving system actually shares the same set of coils with the RF transmitting system. The signal is amplified by the preamplifier, filtered and processed, and then converted to digital signal by the analog–digital conversion. And finally, the computer acquizite and records original FID signals.
5. *Acquisition and data processing system.* This unit is responsible for the control and coordination of the entire system, and the accumulation and Fourier transformation of the received FID signal.
6. *Accessories.* Accessories include variable temperature unit, autosampler, NMR sample tube, workstation, and software.

4.4 Chemical Shift

4.4.1 The Generation and Representation of Chemical Shift

According to the Eq. (4.3), the proton resonance frequency is determined by magnetic field strength and the nuclear gyromagnetic ratio γ. For the same type of nucleus, the gyromagnetic ratio is the same. But if the RF frequency is fixed, whether all the protons resonate in the exactly same frequency? It is not the case. This is because the proton resonance frequency is correlated with the chemical environment (the chemical environment mentioned here refers to the extranuclear electron cloud and the influence of neighboring protons). Electrons around the protons will directionally flow under the action of the external magnetic field to produce a secondary magnetic field B_1 with direction different with that of the external magnetic field. This type of secondary magnetic field generated by the extranuclear electron has shielding effect to protons, so the resonance frequency of the proton is not exactly equal to that of the applied magnetic field. This phenomenon is known as electronic shielding effect. If the direction of the secondary magnetic field B_1 is opposite to the direction of the external magnetic field B_0, the magnetic field strength B actually received by protons will be reduced, which can be expressed as Eq. (4.5):

$$B = B_0 - B_1 = B_0(1 - \sigma) \tag{4.5}$$

wherein σ is the shielding constant, which is related to the chemical environment of the proton. For example, if the functional group connected with a proton is an electron-repelling group, the nuclear extranuclear electron cloud density and the degree of shielding effect will increase, so that the protons must conduct resonance absorption in the external magnetic field with higher strength, and so the nuclear resonance signal is generated in the high magnetic field. Conversely, if the functional group connected with a proton is an electron-withdrawing group, the extranuclear electron cloud density and the degree of shielding effect will decrease, so that the protons must conduct resonance absorption in the external magnetic field with lower strength, and so the nuclear resonance signal is generated in the low magnetic field.

The difference of resonance frequencies of the same nucleus in different chemical environments produced by the different degree of shielding effect of extranuclear is small. It is difficult to accurately determine the absolute value. So in the actual operation, it is necessary to use a reference as a standard, thus introducing the concept of the chemical shift, i.e., the position difference between the absorption peak of a substance and the absorption peak of the standard is referred to as the chemical shift of the substance, which is often represented by δ:

$$\delta = \frac{\nu_s - \nu_r}{\nu_0} \times 10^6 \text{ (ppm)} \tag{4.6}$$

In the Eq. (4.6), δ and ν_s are the chemical shift and the resonance frequency of the protons in the sample, respectively; ν_r is the resonance frequency of the standard (usually $\nu_r = 0$); ν_0 is the frequency of the instruments. The δ value depends only on the shielding constant, independent of the magnetic field strength. For example, the δ of protons measured by 100 MHz NMR is 1, and the δ of protons measured by 800 MHz NMR is also 1.

4.4.2 The Selection of Chemical Shift Standard Substance

The chemical shift measured in the NMR spectra is the relative shift. In experiment, usually Tetramethylsilane (TMS), 2,2,3,3-d(4)-3-(trimethylsilyl) propionic acid sodium salt (TSP-d4) and sodium 3-(trimethylsilyl)-1-propanesulfonate (DDS) are used as standard substances. δ = 0. The structures of the three substances are as follows:

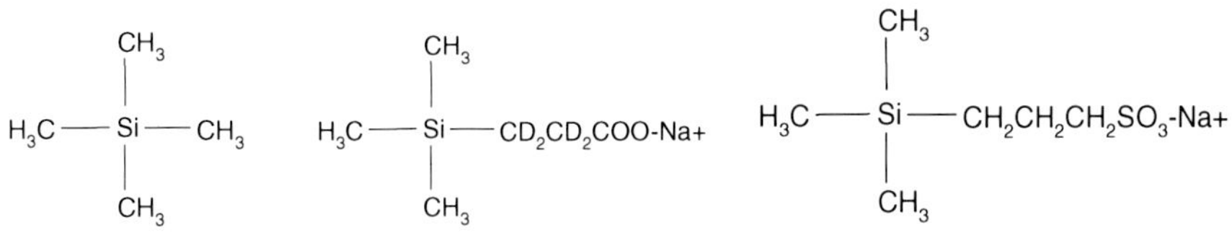

TMS TSP-d4 DSS

TMS as a standard substance is characterized by:

① Chemically inert and does not chemically react with other substances;
② Dissolve with many organic solvents each other;
③ Shielding constant σ is large, signal is generated in the high field, and does not interfere with the sample signal;
④ Due to the equivalent chemical environments of the 12 protons in the four methyl groups, only a sharp single peak appears in the NMR spectra, which is conducive for shimming;
⑤ Low boiling point (27 °C), easily removed, and is conducive for sample recycling;

Because TMS does not dissolve in water, commonly used TSP-d4 or DSS as a standard substance for aqueous samples, and the chemical shift value is also set to 0.

Adding standard substance and the sample to the solvent at the same time is referred to as the internal calibration standard. Sealing the standard substance in capillary tube then putting it into sample tube is known as the external calibration standard. Certain errors may be generated by both the internal calibration standard and the external calibration standard. Usually, use an internal calibration standard in NMR experiments.

4.4.3 The Affecting Factors of the Chemical Shift

The nuclear shielding effect imposed by the electron in the molecule is the reason for the formation of chemical shift. Therefore, factors capable of changing the extranuclear electron cloud density of hydrogen can affect the chemical shift of ^1H. If the extranuclear electron cloud density of hydrogen decreases, the position of peak moves toward low field (the left of the spectra), the chemical shift δ value increases, which is called the deshielding effect; Conversely, if the extranuclear electron cloud density of hydrogen increases, the shielding effect makes the position of peak moves toward higher field (the right of the spectra), the chemical shift δ value decreases.

Factors affecting the ^1H chemical shift are summarized as follows:

1. *Inductive effect.* The ^1H signal on the carbon connected with electronegative substituent, hybridized atoms and alkyl groups will move toward the low field, and the degree of moving increases with the increase of electronegativity, this effect is called inductive effect. Because of the inductive effect, the stronger the electronegativity of the substituent (electron-withdrawing group), the smaller the density of the electron cloud surrounding it, shielding effect is also decreased, and the H resonance signal on the carbon atom connected with the electronegative substituent moves to the lower field. If the substituent is an electron-repellant group, the electron density around it increases, and the chemical shift moves toward the higher field.
   For example, In CH_3X, there is significant dependency between the chemical shift δ value of protons and the electronegativity E_x of substituent X:

X	F	Cl	Br	I
E_x	3.92	3.32	3.15	2.9
δ	4.26	3.05	2.68	2.1

   The electron-withdrawing ability decreases with the decrease of the electronegativity of halogen X. The electron density around C increases, because H atoms are directly adjacent to C atoms, the electron density surrounding protons increases, and the shielding effect increases, the chemical shift value δ decreases, and vice versa.
   The inductive effect of the substituent can be reduced with the extension of carbon chain. The chemical shift of H on the α-C is greater than that on β- or γ-carbon atoms. For example,

Substituent	CH_3Br	CH_3CH_2Br	$CH_3CH_2CH_2Br$	$CH_3CH_2CH_2CH_2Br$
δ	2.68	1.65	1.04	0.9

When the number of substituent with large electronegativity increases, the chemical shift moves toward the lower field.

Substituent	CH_3Cl	CH_2Cl_2	$CHCl_3$
δ	2.68	5.33	7.26

2. *Conjugation effect.* In the molecular system having a multiple bond or a conjugate multiple bond, the transfer of π-electron leads to the change of electron density and shielding of a group, and this effect is called conjugate effect. There are mainly two conjugation effects: p–π conjugation effect and π–π conjugate effect. It is worth noting that the electron transfer directions of the two conjugation effects are opposite.
For example,

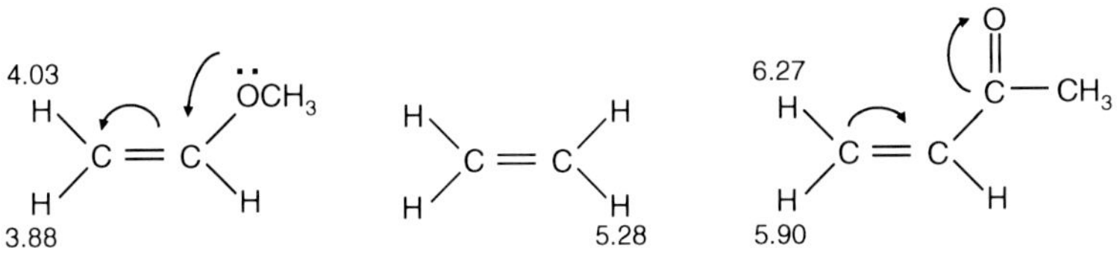

p-π conjugate effect (push electron to the ortho position) π-π-conjugate effect (pull electron from the ortho position)

Under normal circumstances, the chemical shift of the proton of vinyl is δ5.28. The O atom of the left molecule has a lone pair p electron, which constitutes a p–π conjugation with the vinyl double bond leading to the increase of electron cloud density of C and H in the ortho position; the magnetic shielding is also increased (positive shielding); thus, the chemical shift values reduce. The rightmost molecule belongs to the π–π conjugate, electron transfer results in the decrease of electron cloud density and magnetic shielding (de-shielding) of β C and H; thus, the δ value is increased.
Similarly, another example is, under normal circumstances, the chemical shift of benzene ring proton is δ7.21. The left molecule has a p–π conjugate (push electron to the ortho position), the electron cloud density of ortho H increases (n-shielding); and thus, the chemical shift value decreases. The right molecule has a π–π conjugate, the electron cloud density of ortho H decreases (de-shielding), and thus, the chemical shift value increases.

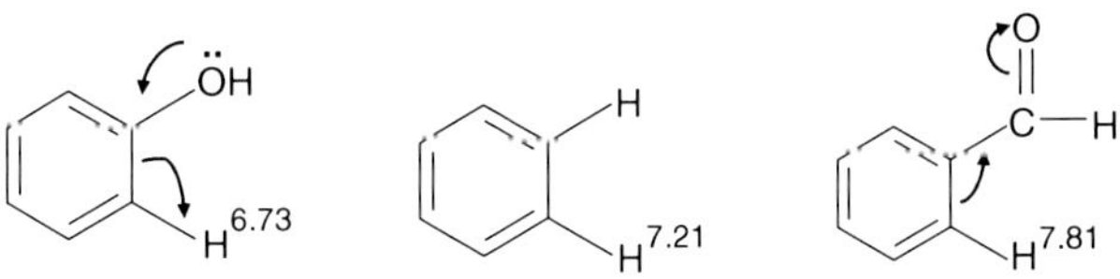

p-π conjugate effect (push electron to the ortho position) π-π- conjugate effect (pull electron from the ortho position)

3. *The magnetic anisotropy effect of the adjacent group.* The electronegativity of ethynyl group is greater than that of vinyl group. But the chemical shift of ethynyl group is δ2.8, which is smaller than that of vinyl and benzene ring

Fig. 4.4 The ring current of benzene ring is formed under the external magnetic field

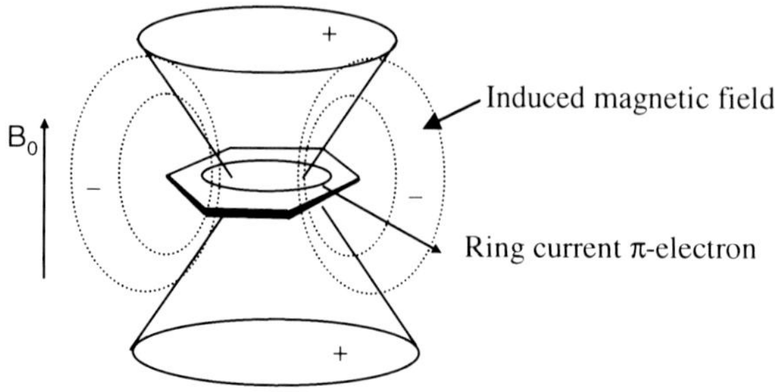

hydrogen nuclei. This is because the spatially interaction between H and some group in the molecule affecting the δ value of H, and this effect is called magnetic anisotropy effect, which can be divided into the following situations:

① Benzene ring. If only consider the hybridization, the chemical shift value of H on the benzene ring should be approximately δ5.7, but the actual chemical shift value of H on the benzene ring significantly moves toward the lower field ($\delta = 7.21$), this is due to the delocalization or fluidity of phenyl ring π-electron. Under the external magnetic field B_0, when the direction of B_0 is perpendicular to the benzene plane, π electron will flow along the benzene ring carbon chain to form a ring current. The electron flow generates a magnetic field, the direction of the magnetic lines is opposite to the direction of the magnetic lines above and below the benzene ring, i.e., the n-shielding area is formed at above and below the benzene ring plane, and the de-shielding area is formed at the side of the benzene ring plane. In other words, the ring current magnetic field makes the external magnetic field increase, the hydrogen nucleus is de-shielded, and the resonance peak position moves to the lower field. See the following diagram (Fig. 4.4):
The sample determined by NMR is in liquid state, and the sample molecule is in a state of constantly rolling in solution. Therefore, in consideration of the role of benzene ring π-electron ring current, various orientations of the benzene ring plane should be averaged. Only when the benzene ring plane is perpendicular to the external magnetic field, the ring current is generated, while when the benzene ring plane and the applied magnetic field are in the same direction, the applied magnetic field does not generate the induced magnetic field, hydrogen is not subject to the shielding effect. The result of averaging various orientations of the benzene ring plane is that the hydrogen nucleus is subject to the de-shielding effect.
In fact, not only the benzene ring, all the ring conjugate systems having $4n + 2$ delocalized π electrons have strong ring current effect. If H is strongly de-shielded at above or below the ring, the H peak will appear in the high-field region, and even its δ value is negative. If H is strongly de-shielded at the side surface of the benzene ring, the H peak will appear in the low field region and has a large δ value. For example,

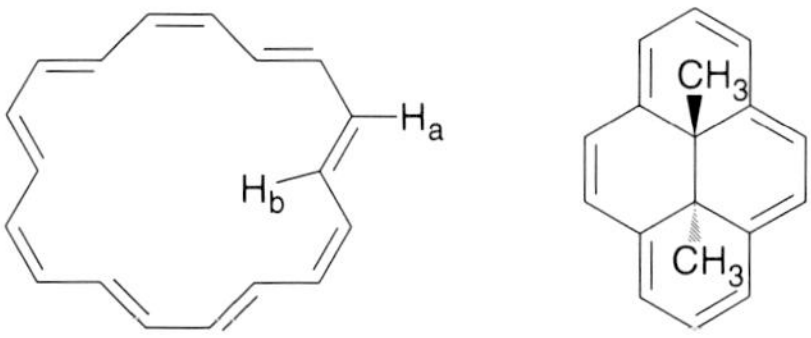

H_a =9.25 ppm CH_3 = −4.25 ppm
H_b =−2.9 ppm   环上氢 = 8.14 ppm

H_a of the left compound is strongly de-shielded at the side of the ring, the chemical shift is in the low field (δ9.25); while H_b of the left compound is strongly de-shielded at above and below the ring, the chemical shift is in the high field (δ2.9). The methyl group of the right compound is strongly de-shielded at above and below the ring, the chemical shift is in the high field (δ4.25); the protons are de-shielded at the side of the compound, so the chemical shift is in the low field (δ8.14).

② The carbonyl group (C=O) and carbon–carbon double bond (C=C). The π-electron clouds of carbonyl group (C=O) and the carbon–carbon double bond (C=C) are at above and below the double bonds like that of benzene ring. The direction of magnetic lines generated by the π-electrons of C=O and C=C is opposite to the applied magnetic field, respectively, and the side of the double bond is de-shielding area, which making the H of aldehyde in the de-shielding area moves to the lower field.

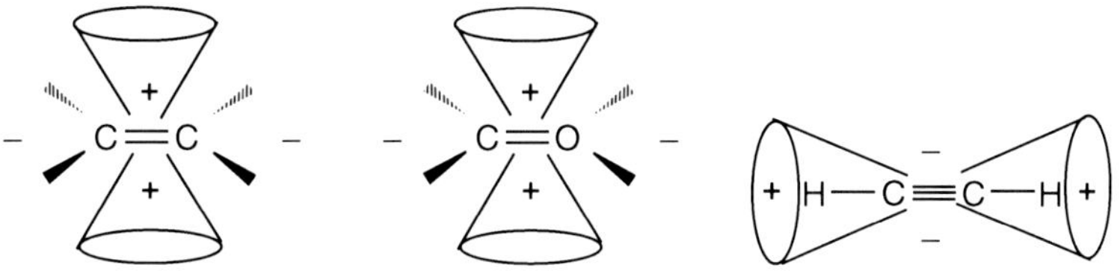

③ Carbon–carbon triple bond (C≡C). The shielding effects of alkynes are different with that of alkenes. Under the action of the applied magnetic field B_0, π electron rotates about the C≡C bond, the n-shielding area appears at the two ends of the triple bond. Therefore, the resonance peaks appear in the high-field region.

4. *Van der Waals effect.* When the space between the target H nucleus and nearby atom is less than the sum of the van der Waals radii, extranuclear electrons are mutually exclusive, so that the density of electron cloud around the hydrogen nucleus decreases, the shielding effect weakened, and the NMR signals moves toward the lower field. This is the van der Waals effect. For example,

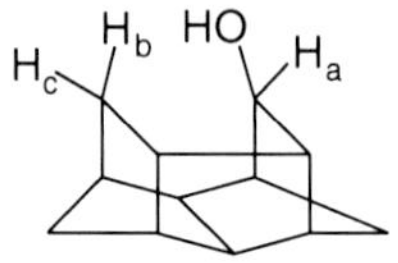

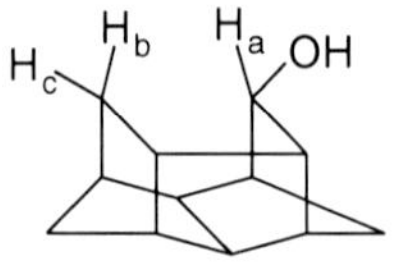

$H_a = 3.92$ ppm
$H_b = 3.55$ ppm
$H_c = 0.88$ ppm

$H_a = 4.68$ ppm
$H_b = 2.40$ ppm
$H_c = 0.88$ ppm

The van der Waals effect exists between H_a and H_b in the right compound. H_b will push the extranuclear electron of H_a to C, leading to the decrease of electron cloud density surrounding H_a, and the weakening of shielding effect and the chemical shift moves to the lower field relative to the left H_a. And the van der Waals effect exists in H_b in both the above two compounds, the only difference is that the hydroxyl group in the left compound is larger (relative to H_a), the van der Waals effect is more significant, and the degree of its chemical shift moving to the lower field is higher. Of course, the action of field effect cannot be excluded.

5. *Hydrogen bond.* The pattern of hydrogen bond is

$$\overset{-}{X} - \overset{+}{H} \cdots \overset{-}{Y}$$

wherein X and Y are usually elements with large electronegativity such as O, N, and F. The formation of hydrogen bond makes the hydrogen subjecting to the shielding effect (electron-withdrawing group), and the NMR signal moves toward the lower field.

Hydrogen bond can be formed intermolecularly or intramolecularly. The difficulty level of the formation of intermolecular hydrogen bond is correlated with the concentration of sample, the nature of solvent, and the temperature. If the concentration is reduced and the temperature is increased, the possibility of the formation of intermolecular hydrogen bond is reduced, the NMR signal moves toward the higher field.

For example, the intermolecular hydrogen bond is present in pure ethanol:

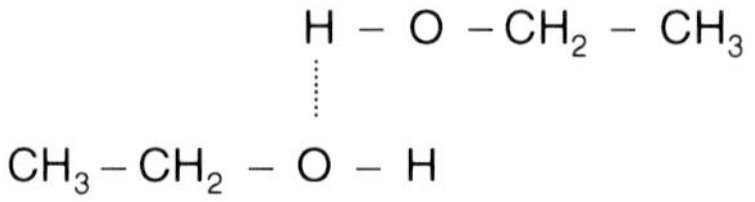

When the solvent is CCl_4, for different concentration of ethanol, C_2H_5 peak position is substantially constant, while the chemical shift of OH changes greatly, the chemical shift of hydroxyl proton in pure ethanol is $\delta 5.28$. When the ethanol is configured into a 5–10 % CCl_4 solution, the chemical shift range of hydroxyl proton is $\delta 3.0$–5.0; while when the concentration of ethanol in CCl_4 decreases to 0.5 %, the chemical shift of hydroxyl proton can move to about $\delta 1.0$.

At room temperature, the hydrogen bond presents in water molecule:

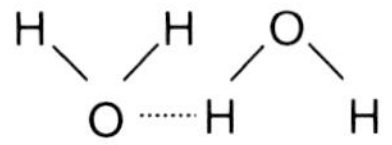

When temperature decreases, the hydrogen bond is enhanced, the spectral line of H_2O moves to the lower field. The spectral line observed in experiment is the average value of the chemical shift values of OH in various molecules (no hydrogen binding, bimolecular binding, and multi-molecular binding).
Because the exchange interaction occurs between OH in various molecules, if the exchange speed is greater than the operating frequency of the NMR spectrometer, only one spectral line can be observed.

6. *Solvent effect.* High-resolution NMR experiments often cannot be done without the solvent. The intermolecular interaction between the solvent and solute may cause the chemical shifts of solute different in different solvents. The effect of chemical shift change caused by different solvent is referred to as solvent effect. Experiments have proved that solvents $CDCl_3$ and CCl_4 have essentially no effect on the δ values of compounds, but aromatic solvents such as C_6H_6, C_6D_6, and C_5H_5N may cause large change of chemical shift of solute, especially for those with active hydrogen such as OH, SH, NH_2, and NH, the solvent effect will be more strong. This is because the approach of aromatic solvent molecules having magnetic anisotropy effect to the sample molecules will induce different shielding effect and de-shielding effect.
Different solvents have different levels of effect on the chemical shift of solute. The situation is complex and no regularity. Therefore, solvent effect should be minimized in practical work: Whenever possible, use the same solvent; make use of solutions with the same or similar concentration; under the premise of allowable test sensitivity, make use of the dilute solution as far as possible to reduce solute interactions.
7. *The influence of pH.* As for amino acids, the reason that the chemical shift of protons changes with pH is because the following dissociation equilibrium exists in the aqueous solution of amino acids:

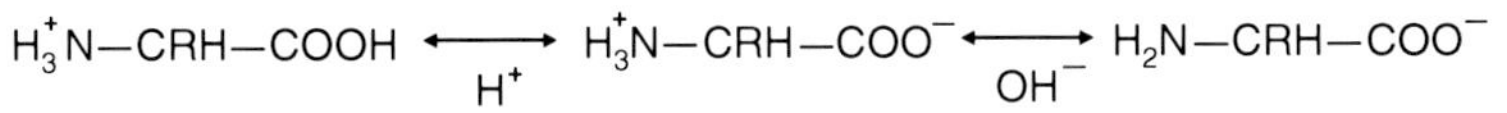

Because of the rapid exchange of protons in $-NH_3^+$ and –COOH with H_2O, the resonant absorption peak is generally cannot be observed.
When the pH increases, the carboxyl group will dissociate to produce negative charge density, thereby increasing the shielding effect on all protons, and reducing the chemical shift; when the protons of NH_3^+ dissociate, the shielding effect increases more greatly, and the chemical shift is smaller.

Figure 4.5 shows the pH titration curves of the chemical shifts of 2-CH in His and protons in Gly:

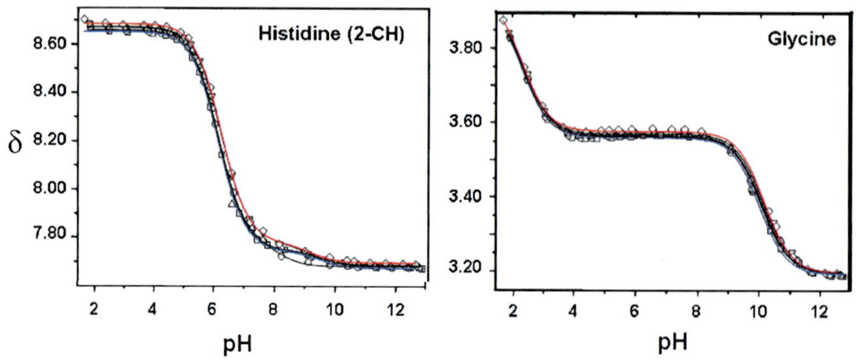

Fig. 4.5 The curve of chemical shift changing with pH

4.5 Spin Coupling

4.5.1 Spin Coupling and Spin Splitting

In the high-resolution NMR spectra, in addition to the chemical shift change caused by the density difference of electron clouds around the nucleus, the interaction between different nuclei in the same molecule can be observed. This inter-nuclear interaction is small. Though it does not affect the chemical shift of nucleus, it has an important impact on the spectral peak shape. Typically, there are two forms of interactions between the nuclear spins: one is the coupling interaction directly transmitted between the two nuclear magnetic moments, namely dipole–dipole interaction; the other one is the indirect interaction transmitted by bonding electrons, namely spin–spin coupling. The phenomenon of the increase of NMR spectral peaks caused by spin–spin coupling is known as spin–spin coupling and splitting, which has nothing to do with the external magnetic field B_0 or the spatial orientation of the nucleus. Only when the magnetic nonequivalent nuclei appear in the same molecule, the spin–spin coupling and splitting phenomenon occurs. Example, it can be observed in the NMR spectra of ethanol (CH_3–CH_2–OH) that the CH_3 and CH_2 show the triplet peaks and quartet peaks, respectively. The intensity ratio of the triplet peaks of CH_3 is 1:2:1, and the intensity ratio of the quartet peaks of CH_2 is 1:3:3:1 (Fig. 4.6).

Usually, the nuclear spin–spin coupling and splitting complies with the following rule:

① The $n + 1$ law: The number of splitting spectral peaks of ^1H nuclei arise from the coupling effect is $n + 1$, wherein n represents the number of ^1H nuclei producing coupling ($I = 1/2$). Strictly speaking, the number of splitting spectral peaks generated from the spin coupling of n nuclei with I spin quanta should be $2nI + 1$, namely the $2nI + 1$ law. The $n + 1$ law is only a special case of the $2nI + 1$ law ($I = 1/2$).

② The distance between each two adjacent peaks are all equal.
③ The intensity ratio between the spectral peaks is the coefficients of the expansion formula of $(a + b)^n$.

Take the previously mentioned ethanol for example, $n = 3$, so the intensity ratio of the quartet peaks is 1:3:3:1.

4.5.2 The Coupling Constant and Calculation

The distance between the splitting peaks generated by spin–spin coupling is known as the coupling constant, which is represented by J in Hz. When the spin–spin coupling occurs between two types of protons (represented by A and B, respectively), each of the resonance peaks in the NMR spectra will be split to multiplet peaks by the coupling protons. The splitting distance is equal; therefore, for protons A and B, the spin coupling constant J_{AB} is a constant. The size of the spin coupling constant reflects the strength of the spin–spin coupling effect and closely correlated with the chemical environment of the nucleus such as inter-nuclear distance, the number of chemical bond, and the dihedral. J is usually 5–8 Hz. Because the J coupling interaction is indirectly transmitted by electron clouds around the nucleus, which causing the generation of energy coupling between nuclear magnetic moments, it rapidly weaken with the increase of the number of chemical bond. Therefore, the coupling constant is a characteristic of the molecular structure of compounds, which is very important for the analysis of structure and assignment of the compound.

The size of the coupling constant can be measured from the NMR spectra, which is equal to the distance between the two spin splitting peaks. The coupling constant value is calculated by multiplying the chemical shift difference between the two adjacent splitting peaks to the working frequency (in MHz) of the NMR spectrometry. For example, in the 500 MHz NMR spectra, J_{AB} = the chemical shift difference of the two adjacent splitting peaks $\times 500 = (3.828 - 3.812) \times 500 = 8$ Hz.

4.6 The Chemical Shift Range of Common Groups

The chemical shift values of H are helpful in understanding the chemical environment of the hydrogen nuclei, predicting the kind of the proton as well as the surrounding chemical environment, thereby inferring the structure of the organic compound. The best understanding of the chemical shifts of common groups are necessary and helpful to the spectra analysis work.

1. *Saturated hydrocarbons*. The chemical shift range of saturated hydrocarbons H is generally δ0.7–1.7. Wherein the chemical shifts of methyl protons (CH_3) located in the high field: δ0.7–1.3, followed by that of methylene (CH_2):

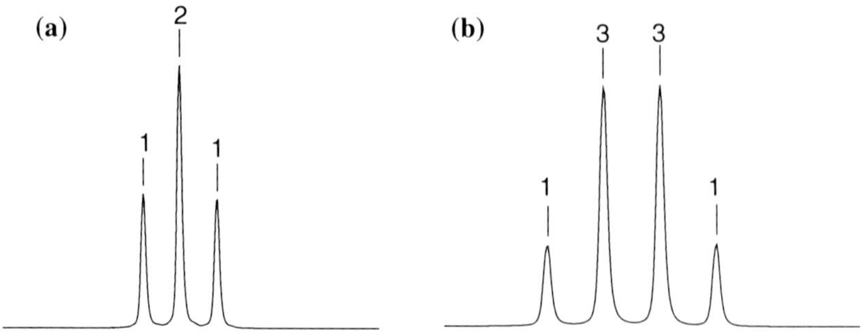

Fig. 4.6 The NMR spectra of ethanol. **a** The splitting spectral peaks of ethanol CH_3. **b** The splitting spectral peaks of ethanol CH_2

δ1.2–1.4 and that of methane is in a relatively low field: δ1.4–1.7. The CH and CH_2 in the long chain saturated hydrocarbons often overlap with each other, making it difficult to be distinguished. Only the CH_3 protons locate in the relatively high field (δ0.7–1.3), which can be separated from other peaks, and can be easily identified in the spectra. When it is connected to a group with large electronegativity, the chemical shift will move toward the low field. Figure 4.7 shows the H spectra of *n*-heptane.

2. *Alkenes*. There are two types of characteristic peaks in H of Alkene compounds. One is the H directly connected with the unsaturated carbons (C=C–H), their chemical shifts are often in the range of δ4.5–6.0; the other is the allylic H (C=C–CH_2–), their chemical shifts are at about δ1.6–2.6. When it is connected to a group with large electronegativity, the chemical shift will move toward the low field. Figure 4.8 shows the H spectra of 1-pentene:
3. *Aromatic compounds*. There are also two types of characteristic peaks in H of aromatic compounds like that in olefin compounds. The one is the H in benzene rings, their chemical shifts are in the range of δ6.5–8.0; the other is the H in the benzyl group, their chemical shifts are at between δ2.3–2.7. When it is connected to a group with large electronegativity, the chemical shift will move toward the low field. When the vinyl bond (or olefinic bond) is connected to the

Fig. 4.7 The spectra of *n*-heptane

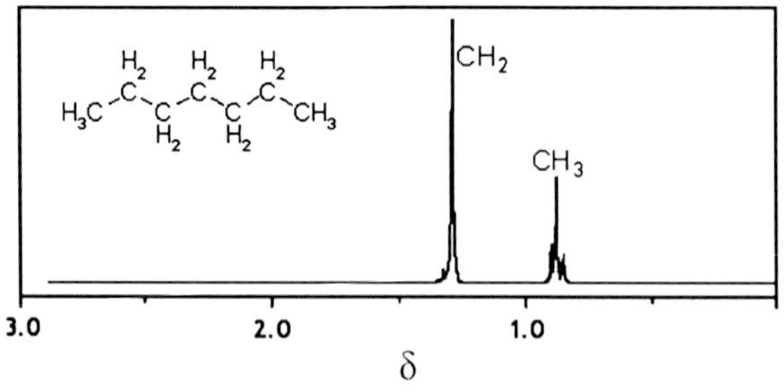

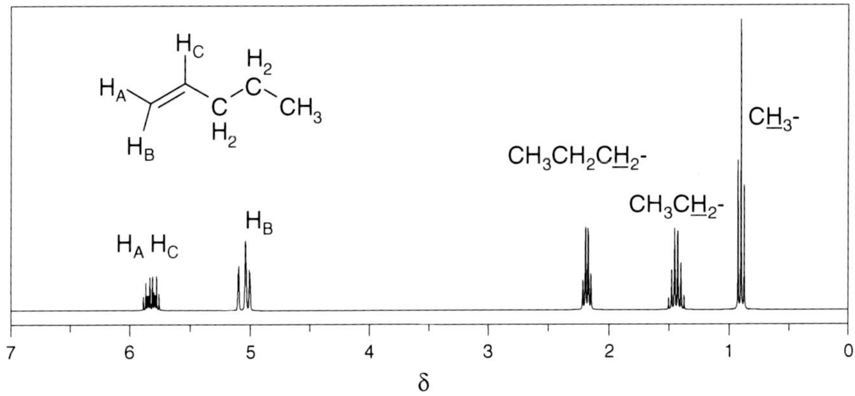

Fig. 4.8 The H spectra of 1-pentene

group with strong shielding effect, such as C=CH–OR, the chemical shifts of olefinic (or alkenyl) H will also fall between δ6.5–8.0. The following is the H spectra of ethylbenzene (Fig. 4.9).

4. *Heteroaromatic rings.* The heteroaromatic ring contains hetero atoms. The chemical shift of H is related to the location and type of the hetero atoms and is also influenced by the solvent. Figure 4.10 shows the chemical shift values of H in common heteroaromatic rings.
5. *Alkynes.* The chemical shifts of alkynes are generally in the range of δ1.7–3.1. When it is connected to an electron-withdrawing substituent, the chemical shift will move toward the low field; when it is connected to an electron-donating substituent, the chemical shift will move toward the higher field. The following is the H spectra of 1-hexyne (Fig. 4.11).
6. *Alcohols.* The hydroxyl hydrogen of alcohols is variable, and its chemical shift is related to the concentration, solvent, temperature, and the presence of trace amounts of water, acids, and alkali. The chemical shift value is generally at δ0.5–5.0. The hydroxyl hydrogen is an active proton, which can rapidly exchange with other protons, so the hydroxyl hydrogen does not coupling split with adjacent hydrocarbon hydrogen but producing a sharp single peak.

Fig. 4.9 The H spectra of ethylbenzene

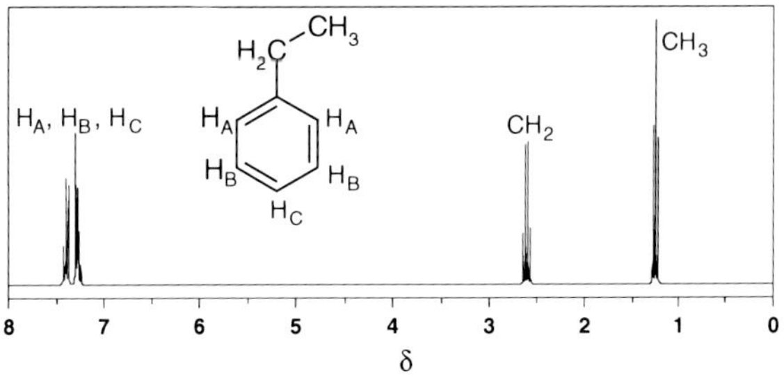

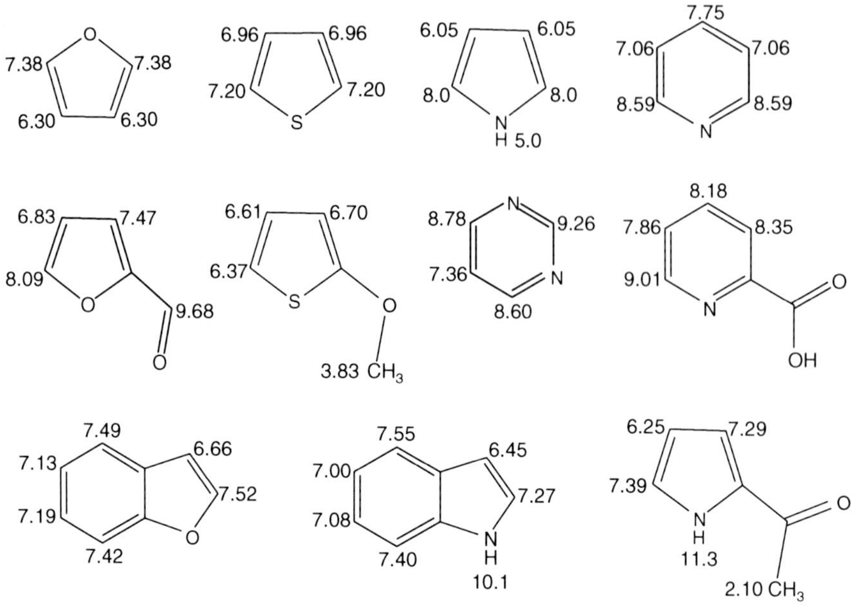

Fig. 4.10 The chemical shift values of H in common heteroaromatic rings

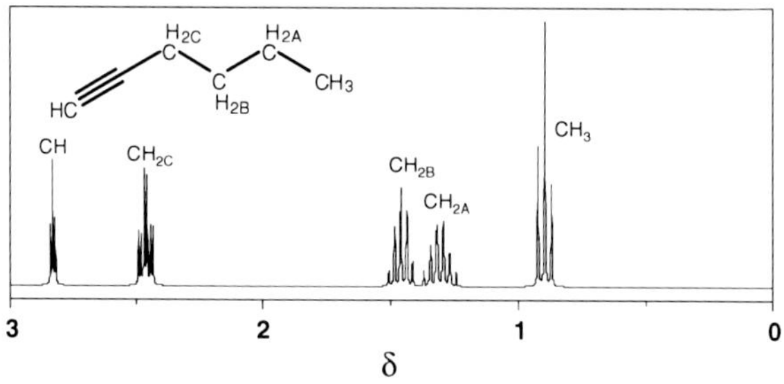

Fig. 4.11 The H spectra of 1-hexyne. The chemical shift of CH connected to the nitrile group (–CH–C≡N) is in δ2.1–3.0

In metabolomics study, hydroxyl hydrogen can rapidly exchange with protons of H_2O, and the saturation transfer effect causes the hydroxyl hydrogen signal is not observable in the pre-saturation pressure water sequence.

7. *Alkyl halides.* The halogen atom is an electron-withdrawing group, and therefore, the chemical shift of H adjacent to the halogen atom will move to the low field by shielding effect. The size of movement is related to the electronegativity of the halogen atom.

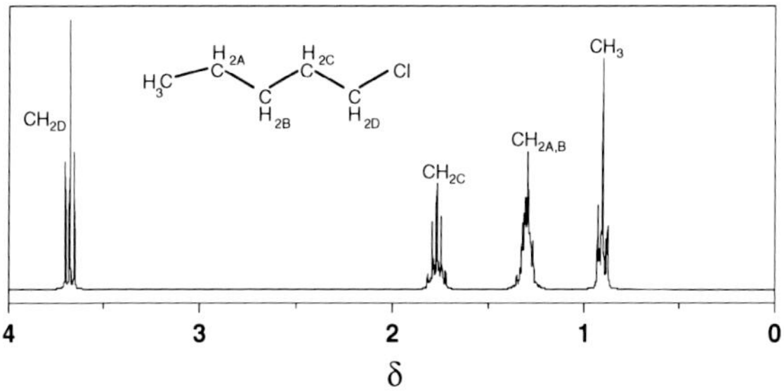

Fig. 4.12 The H spectra of 1-chloropentane

The chemical shift of iodide (CH–I) is at δ2.0–4.0, of bromide (CH–Br) is at δ2.7–4.1, of chloride (CH–Cl) is at δ3.1–4.1, of fluoride (CH–F) is at δ4.2–4.8. The following is the H spectra of 1-chloropentane (Fig. 4.12).

8. *Ethers.* Similar to alcohols, for ethers, the chemical shifts of the hydroxyl hydrogen connected with oxygen atom appear at δ3.2–3.8 by electron-withdrawing effect. The following is the H spectra of ether (Fig. 4.13)
9. *Amines.* Similar to the proton in hydroxyl group, the proton of NH in amines is active hydrogen, and the location of peak can be changed with concentration, solvent, temperature, and pH value. In addition to the changeable location, the H in NH often shows wide and weak peak shape. The chemical shifts of NH in such a compound are generally at δ0.5–4.0. The chemical shifts of hydrocarbon hydrogen connected with N atom (CH–N) appear at δ2.2–2.9. The following is the H spectra of 1-propylamine (Fig. 4.14).
10. *Aldehydes.* The magnetic anisotropy of carbonyl group (C=O) and the induction effect of oxygen atom make the chemical shifts of aldehyde hydrogen greatly moving to the low field, generally at δ9.0–10.0, which is the characteristic peak of the aldehyde hydrogen. Generally, no other types of hydrogen will appear in this area. The chemical shift of αH connected with carbonyl is at δ2.1–2.4. The following is the H spectra of 2-methylbutyraldehyde (Fig. 4.15).
11. *Ketones and esters.* The chemical shift of αH in ketones and esters (CH_{α}–C=O) is at δ2.1–2.4; of alkoxy (–COOCH–) is at δ3.4–4.8. Figure 4.16 is the H spectra of methyl propyl ester.
12. *Carboxylic acid.* The chemical shift of αH in carboxylic acid (–CH–COOH) is at δ2.1–2.4. The chemical shift of H in carboxylic acid is generally at

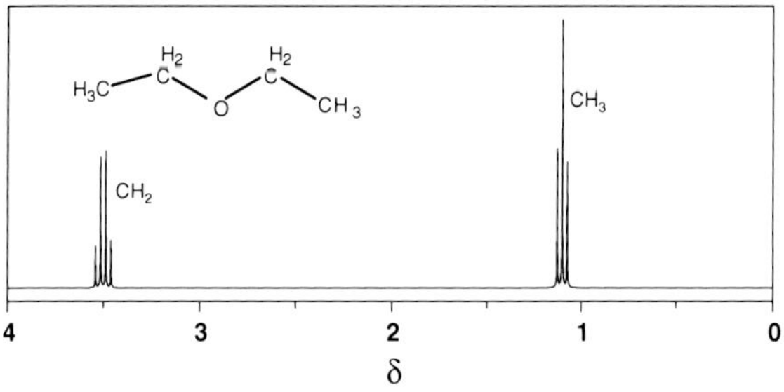

Fig. 4.13 The H spectra of ether

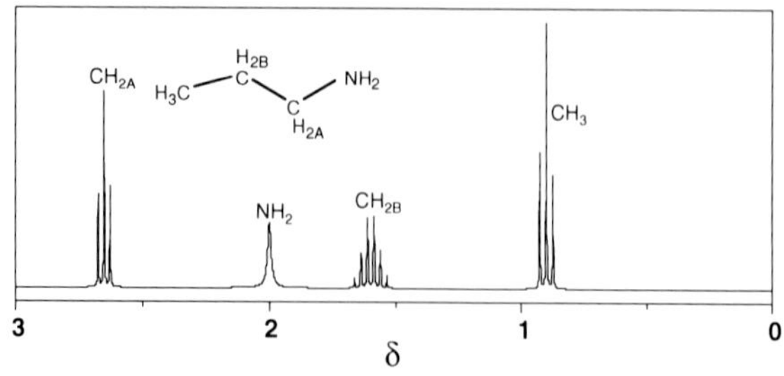

Fig. 4.14 The H spectra of 1-propylamine

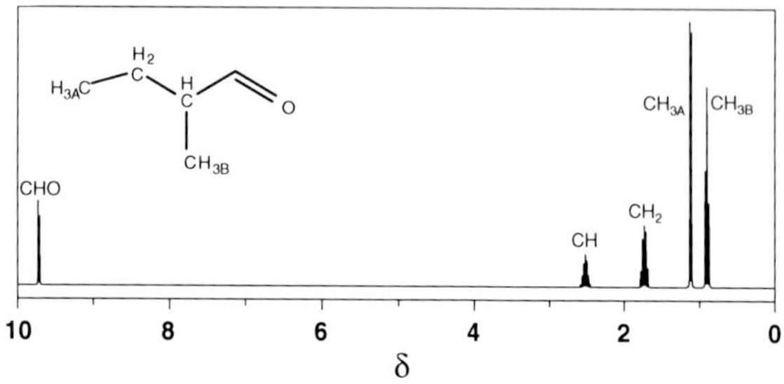

Fig. 4.15 The H spectra of 2-methylbutyraldehyde

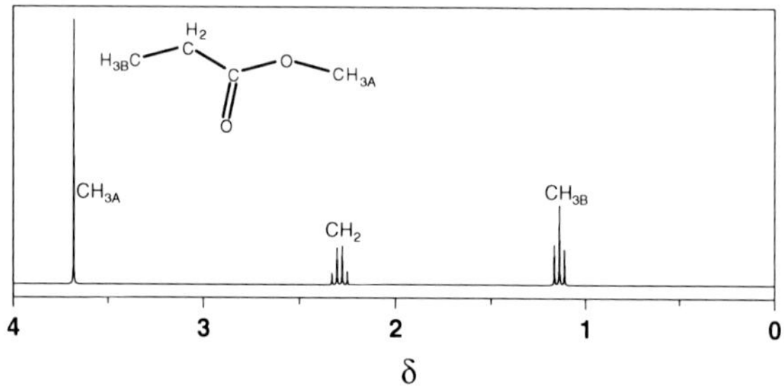

Fig. 4.16 The H spectra of methyl propyl ester

δ11.0–12.0, but the H is an active H, which can rapidly exchange with H/D in the H_2O or D_2O, leading to the disappearance of H in the carboxylic acid and no signal can be observed.

13. *Amides.* In amide-containing compound, the chemical shift of CONH proton is at δ5.0–9.0, of αH (CH–CONH) is at δ2.1–2.5, and of hydrocarbon hydrogen connected with N (CON–CH) is at δ2.2–2.9. The following is the H spectra of urea (Fig. 4.17).
14. *The chemical shift of active hydrogen.* As the active hydrogen exchange with each other in solution, and the chemical shift of them can be affected by hydrogen bond, temperature, concentration, and other factors, so the chemical shift value is unset. Table 4.1 shows some summarized chemical shift values of common active hydrogen.

    Fig. 4.18 shows a summary of the chemical shifts of above-mentioned groups.

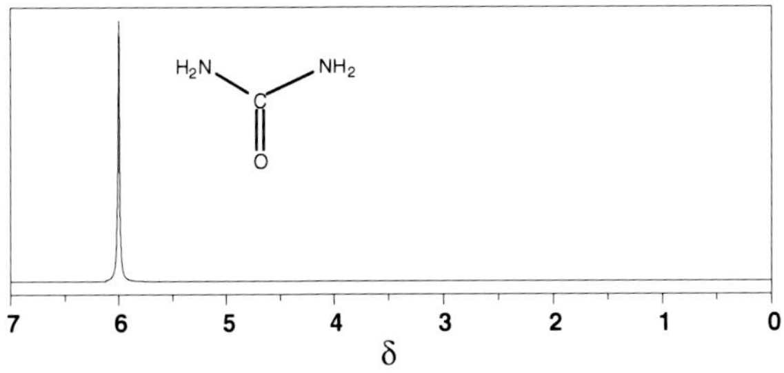

Fig. 4.17 The H spectra of urea

Table 4.1 Summarized chemical shift values of common active hydrogen

Compound type	δ	Compound type	δ
Alcohol	0.5–5.5	RSO_3H	11–12
Phenol	10.5–16	ArSH	3–4
Phenol (intramolecular hydrogen bond)	4–8	RNH_2	0.4–3.5
Enol (intramolecular hydrogen bond)	15–19	ArNHR	2.9–4.8
Carboxyl acid	10–13	RCONHR	6–8.2
Oxime	7.4–10.2	RCONHAr	7.8–9.4
Thiol	0.9–2.5	SiH	3.8

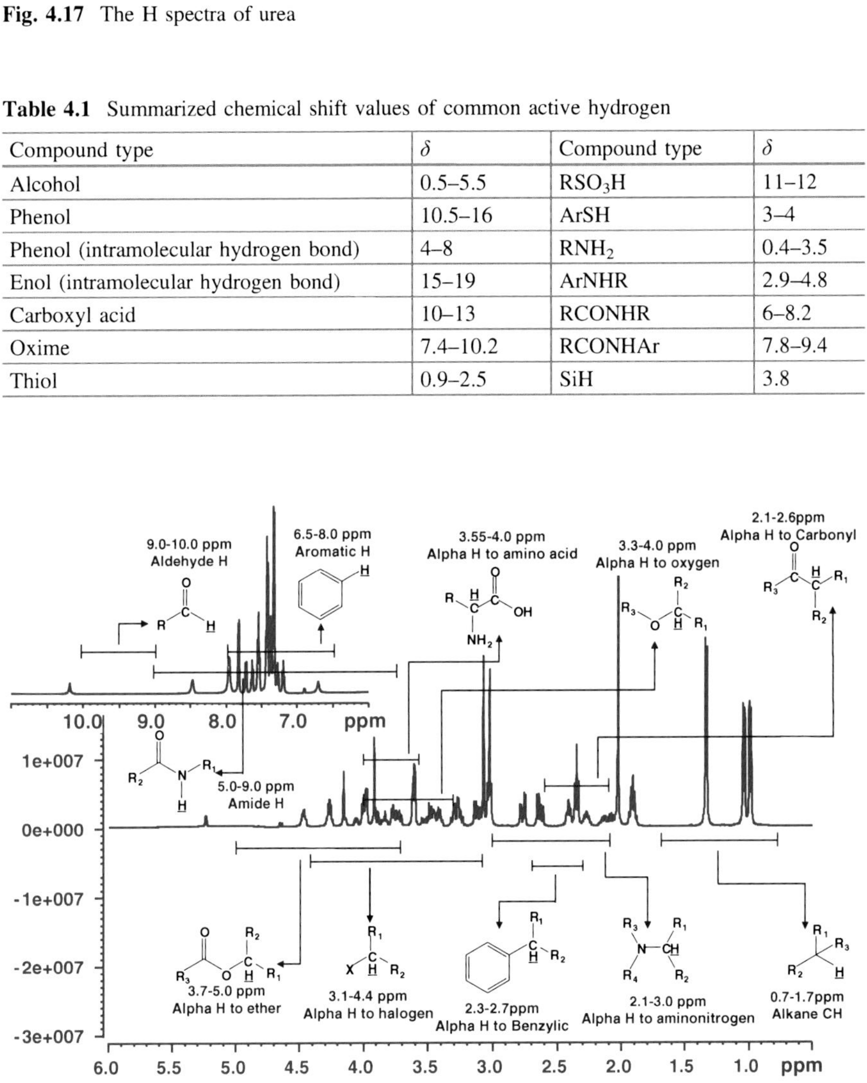

Fig. 4.18 Chemical shifts of common groups.

4.7 Commonly Used Two-Dimensional Spectroscopy in Metabolomics Study

The densely pack of peaks within a small area is often observed in the one-dimensional NMR spectra, so it is difficult to measure the chemical shifts and coupling constants and analyze structure of the compound relying on the one-dimensional NMR spectra alone. The emergence of two-dimensional spectroscopy opens a new way of thinking for compound structural analysis. The thinking of two-dimensional spectroscopy was earliest proposed in 1971 by Belgian scientist, Jeener. Then, through Ernst and Freeman group's efforts, a number of two-dimensional methods were developed and applied in physical, chemical, and biomedical research, providing a huge help for structural analysis and becoming an important branch of NMR.

Two-dimensional (2D) NMR spectra can be divided into three categories:

① Correlation Spectroscopy
② *J*-resolved spectroscopy
③ Multiple Quantum Coherence spectroscopy

The most commonly used 2D NMR spectroscopy in metabolomics research is the two-dimensional correlation spectroscopy, such as H-H-related 2D spectroscopy (H-H TOCSY), H,H-(homonuclear chemical shift) correlation spectroscopy (H,H-COSY), C–H correlation spectroscopy (HSQC and HMBC); 2D *J*-decomposition ^{1}H NMR spectroscopy. Let us turn to understand these commonly used two-dimensional NMR spectroscopy can provide what useful information which helps to analyze compound structures.

1. *Total correlation spectroscopy (TOCSY).* The spin lock is added to the pulse sequence of TOCSY. In the spin lock, all the hydrogen nuclei show the same chemical shift, i.e., the chemical shift is temporarily removed, making the original weak coupling between protons ($\Delta \nu > J$) become strong coupling ($\Delta \nu \gg J$), which gives cross-peaks between all the protons from the same spin system. Figure 4.19 is a schematic diagram of TOCSY, wherein both F1 dimension and F2 dimension are the chemical shift of hydrogen nuclei. There are two types of spectral peaks in the spectra, the one is the diagonal peaks fall on the diagonal line, and the other is the cross-peaks outside of the diagonal line. The emergence of cross-peaks indicates that protons are correlated with each other.
2. *H, H-COSY.* Departure from a hydrogen nucleus on the TOCSY spectra, the spectral peaks of all the hydrogen nuclei in the same coupling system can be found, but the connection relationship between different protons in the coupling system still cannot be obtained. H,H-COSY spectroscopy can give the cross-peak of protons connected on the same carbon or on adjacent carbons. The connection relationship between different protons in the same coupling system can be found by COSY spectra. Figure 4.20 is a schematic diagram of COSY.

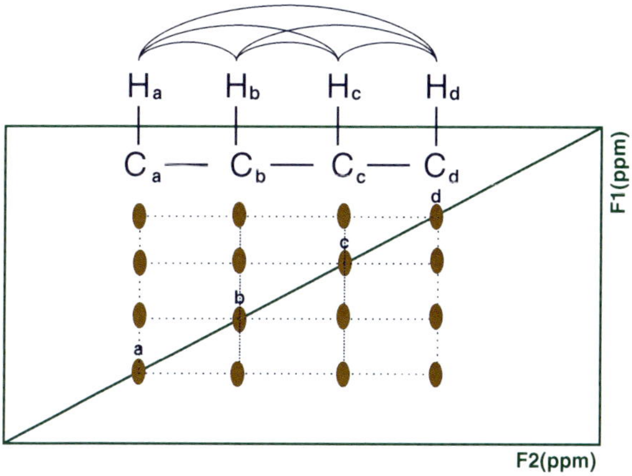

Fig. 4.19 The schematic diagram of visible cross-peaks on the TOCSY spectra

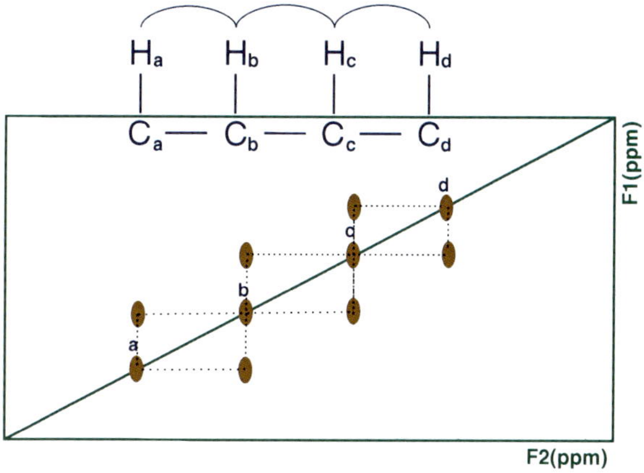

Fig. 4.20 The cross-peaks of adjacent protons showed on the COSY spectra

3. *Two-dimensional J-resolved spectroscopy ^{1}H NMR spectroscopy*. On the one-dimensional spectra, peak densely arranged at a small chemical shift range. The location, number, and coupling constants of the peaks cannot be obtained due to the overlap of the peaks. Two-dimensional *J*-decomposition ^{1}H NMR spectroscopy separates the chemical shift and coupling constant from F1 dimension and F2 dimension, wherein F2 dimension shows chemical shift, and F1 dimension shows the coupling splitting. The most commonly used in metabolomics research is the homonuclear two-dimensional *J*-decomposition H spectroscopy. Figure 4.21 shows the *J*-decomposition spectra of blood sample and corresponding one-dimensional H spectra.
4. *Heteronculear Single Quantum Coherence (HSQC)*. The HSQC shows the connection relationship of the detected H–C heteronuclei. The cross-peaks on the spectra show the ^{1}H nucleus and the directly connected ^{13}C nucleus. The F2 dimension of HSQC is the chemical shift of ^{1}H, and the F1 dimension is the chemical shift of ^{13}C.

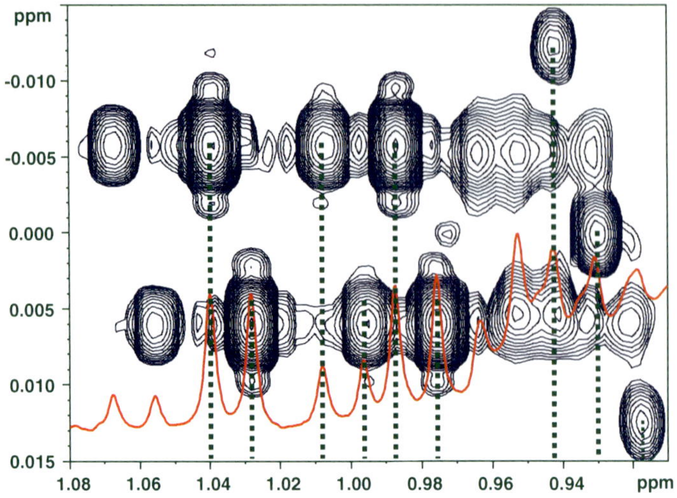

Fig. 4.21 The *J*-decomposition NMR spectra shows the split of protons

5. *Heteronuclear Multiple Bond Coherence (HMBC).* Similar to HSQC, the F2 dimension of HMBC is the chemical shift of 1H, and the F1 dimension is the chemical shift of ^{13}C. The HMBC correlates the ^{1}H nucleus with the remote coupled ^{13}C by heteronuclear multiple-quantum correlation experiment and can detect the remote coupling between protons and carbon within 2–3 bonds. The connection information of the two spin systems can be found from the HMBC. The cross-peak of C directly connected with H is not visible on the HMBC, but a satellite peak may appear in the same location. If cross-peak appears, it indicates that the same C exists in the symmetry position.
Figure 4.22 is the schematic diagrams of HSQC and HMBC.

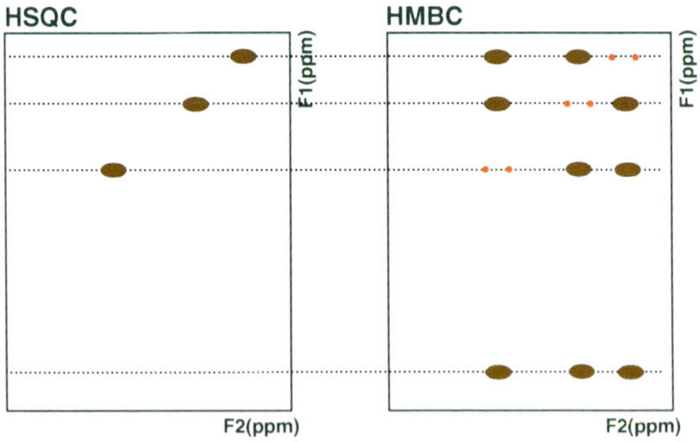

Fig. 4.22 The schematic diagrams of HSQC and HMBC

4.8 NMR Spectral Analysis

1. *Spectral analysis steps.* Metabolomics samples are generally body fluids or extract of tissues from plants and animals, which including a large number of metabolites involving dozens or even hundreds of kinds. Therefore, serious peak overlap exists in a relatively small range of chemical shifts. The attribution of peaks cannot be completed by one-dimensional NMR spectroscopy alone. The two-dimensional NMR spectrum not only contains information contained in the one-dimensional NMR spectra, but also contains more structural information. For the complex biological samples involved in metabolomics research, the above-mentioned two-dimensional NMR spectra are necessary to complete structural analysis.

Firstly, browse the one-dimensional NMR spectra. Although the one-dimensional NMR spectra can only provide very limited information on the compound structural analysis, its importance still cannot be ignored. An intuitive understanding of samples can be got from the one-dimensional NMR spectra. Then, experience such as the chemical shifts of the common groups mentioned in this chapter can be used to determine probable functional groups or structures of samples. And then, targeted attribution for each compound can be done.

Start from TOCSY to have an overall grasp of the spin system, understand the kind of protons and chemical shifts of the spin system, then by correlation spectroscopy (COSY) sequentially determine the connect order of them. And the help of *J*-resolved spectra is necessary to understand the splitting and coupling constant of these peaks, which are indispensable to spectral analysis.

HSQC is also needed to understand the chemical shift of carbon directly connected with hydrogen to better understand its chemical environment or chemical groups. In addition, HMBC can provide the chemical shift information of carbon with 2–3 bonds' distance to the proton and understand the connection relationship between the spin system and adjacent spin system.

The chemical structural formula can be determined by the integration of all the above information.

2. *Examples.* In practical operation, all the above knowledge needs to be comprehensively used to complete the structural analysis of compounds. The following are the NMR spectra of several common mixed solution of metabolites in metabolomics research, and the names and structures are completely unknown. The above-mentioned several commonly used two-dimensional NMR spectra in metabolomics will be used to do structural analysis of several metabolites.

Firstly, the spectral peak of the compound in the box at about $\delta 1.0$ in Fig. 4.23 can be intuitively understood. "The chemical shift range of common groups" in this chapter mentioned that the chemical shifts of CH_3 at about $\delta 0.7$–1.3. Therefore, this peak might be CH_3.

The *J*-resolved spectra (Fig. 4.24) show that there are two doublet peaks, the chemical shifts of which are $\delta 0.99$ and $\delta 1.04$, respectively.

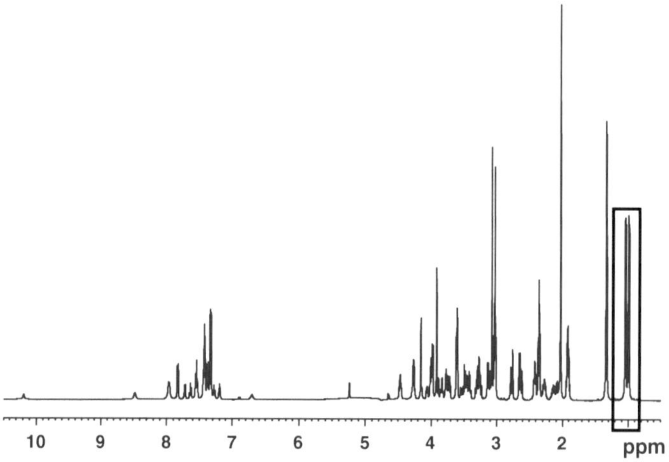

Fig. 4.23 A typical one-dimensional NMR hydrogen spectra

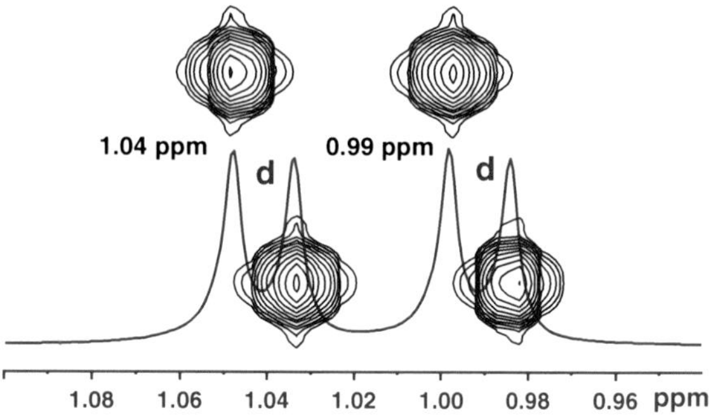

Fig. 4.24 The *J*-resolved spectra show the splitting mode of the peak in the box in Fig. 4.23

The coupling systems of the two peaks can be found from Fig. 4.25, which are in the dashed box. The amplified spectra of which are shown in Fig. 4.26.

It can be found that the hydrogen with chemical shift at $\delta 0.99$ and $\delta 1.04$ and the hydrogen with chemical shift at $\delta 2.27$ and $\delta 3.61$ have the same coupling relationship, indicating that the two hydrogen with chemical shift at $\delta 0.99$ and $\delta 1.04$ may be in symmetric structure, namely two CH_3, i.e., the following structure may exist.

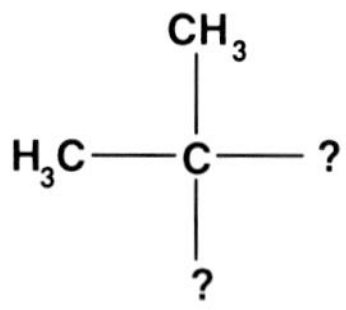

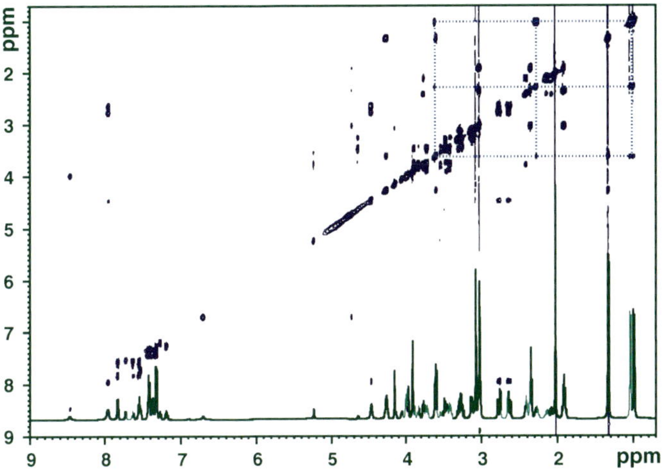

Fig. 4.25 The TOCSY spectra of Fig. 4.23 show the coupling system of peaks at $\delta 0.99$ and $\delta 1.04$

Fig. 4.26 The amplified effect of spectra in the *dashed box* in Fig. 4.25

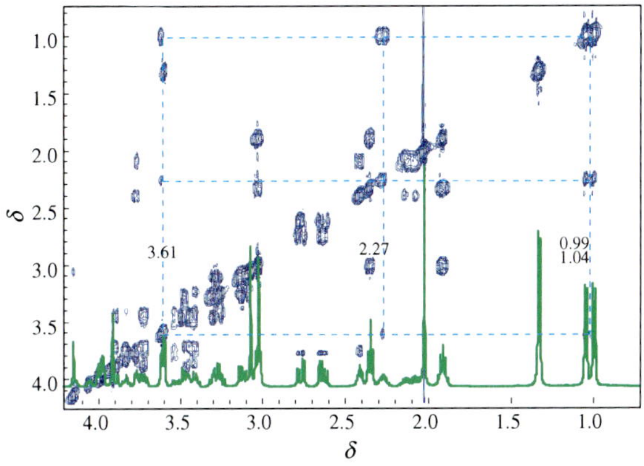

Then why such a small difference is present in the chemical shifts of the two H_3? How will the carbon chain extend? Let us further analyze by *J*-resolved spectra (Fig. 4.24).

The peaks of the two CH_3 are all doublet peaks. The $n + 1$ law indicates that the carbon connected with the two CH_3 connects only one hydrogen atom. And it can be seen from the following COSY and TOCSY spectra (Fig. 4.27) that the chemical shift of such proton is $\delta 2.27$.

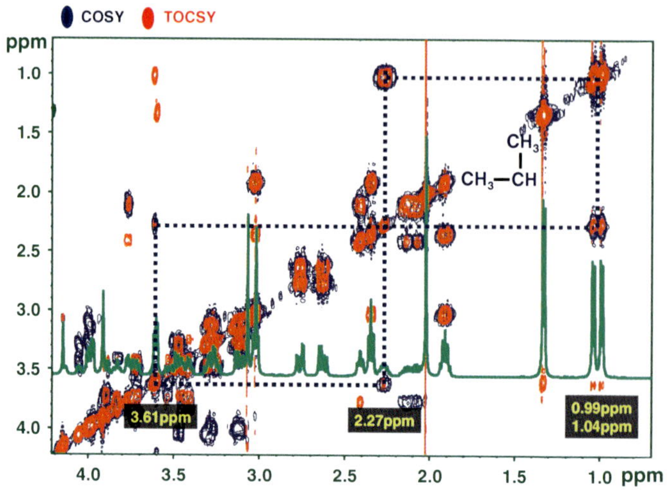

Fig. 4.27 The stacked diagram of TOCSY and COSY in Fig. 4.23

Therefore, the part of the compound structure can be determined as

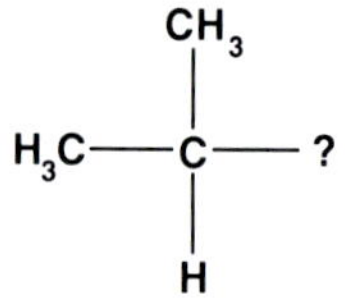

So, what kind of structure is also connected to it? As can be seen from the TOCSY and COSY, with the further extension of the carbon chain, the spin system should have another H with chemical shift at δ3.61. Then what about the information of C? Let us further analyze Fig. 4.28.

It can be seen from the HSQC spectra of Fig. 4.28 (in red) that the chemical shifts of carbon directly connected with protons with chemical shifts at δ0.99 and δ1.04 are δ19.5 and δ20.8, respectively. And the HMBC spectra (in blue) also have corresponding spectral peaks at the same chemical shifts, indicating that symmetric structures are present at this position, that is, there are two CH_3. This further confirms the judgment previously drawn from the TOCSY spectra.

It also can be seen that the chemical shift of carbons corresponding to the chemical shifts of the four kinds of hydrogen is the following:

	CH_3	CH_3	CH	CH–?
δ^1H (ppm)	0.99	1.04	2.27	3.61
δ^{13}C (ppm)	19.5	20.4	31.9	63.3

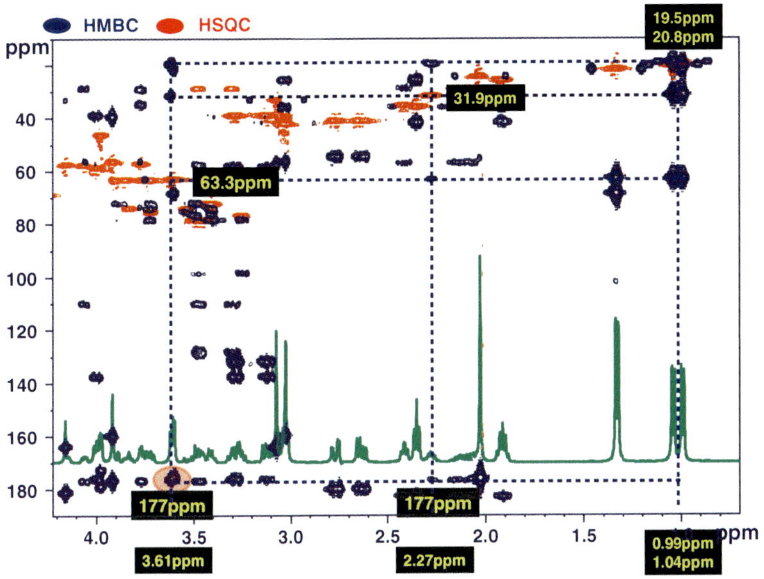

Fig. 4.28 The stacked diagram of HMBC and HSQC in Fig. 4.23

Further, you can see from the HMBC that, in addition connects to the carbons with chemical shifts at δ19.5, δ20.4, and δ31.9, the hydrogen with chemical shift at δ3.61 also connects another carbon with chemical shift at δ177, as shown in Fig. 4.28, which is marked in red circle.

The chemical shift of the carboxyl carbon is usually at δ160–180, so the carbon at this region is the characteristic peaks of the carboxyl carbon. Therefore, the carbon is connected with a carboxyl group. Then what is the other group? Is it possibly H? If it is H, then the structural formula of the substance should be

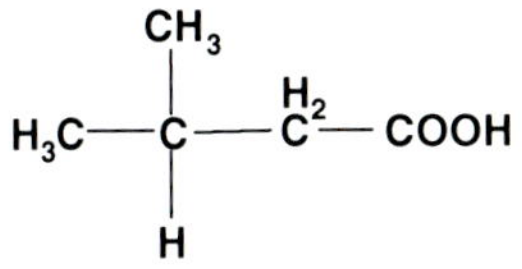

If this is the right formula, then the two methyl groups should be chemically equivalent, and the chemical shift should be the same. But in fact, the chemical shift of the two methyl groups is not the same, indicating that the four groups connected with the carbon with chemical shift at δ63.3 should be in asymmetric structure, but should not present two hydrogen atoms. The chemical shift of this H is at δ3.61. If it connects only one carboxyl group, the chemical shift of it should at between δ2.1–2.6, and the chemical of the C is impossible at δ63.3. While the chemical shift of amino acid αH is at δ3.5–4.0, and the chemical shift of the αC also matches. So the structure of this unknown compound is as follows:

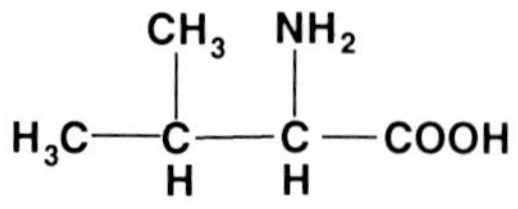

Valine

Now, let us analyze the structure of another compound. The aromatic peaks in the one-dimensional spectra of Fig. 4.29 will be used for analyzing.

Are the aromatic peaks in the one-dimensional spectra of Fig. 4.29 derived from the same compound? This question needs to be answered by TOCSY and COSY as shown in Fig. 4.30.

Such a spin system can be found in the TOCSY spectra of Fig. 4.30 (in red): the chemical shifts of H are at δ7.55, δ7.63, and δ7.83, respectively. The connection relationship of them can be obtained from the COSY spectra of Fig. 4.30 (in blue), which is H (δ7.83)–H (δ7.63)–H (δ7.55), i.e., H (δ7.63) is in the middle; H (δ7.83) and H (δ7.55) are on both sides. H at this chemical shift range should be protons directly connected with the benzene ring (see Sect. 4.6), the three different chemical shifts indicate the presence of three different hydrogen connected with the benzene ring. The following are the three possible structures:

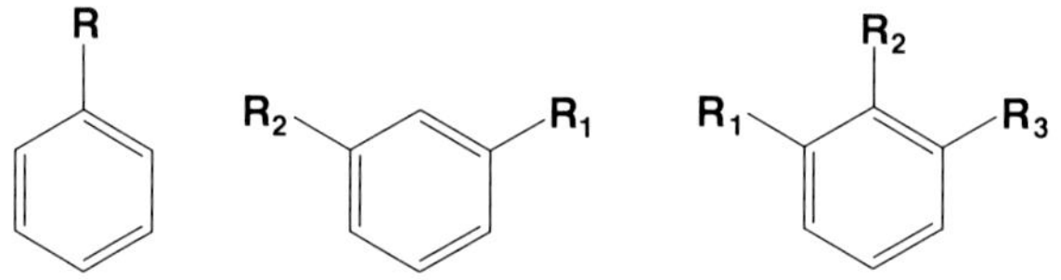

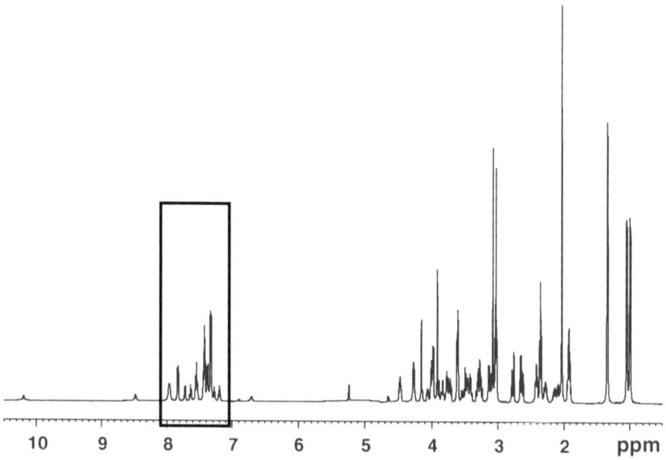

Fig. 4.29 The typical one-dimensional NMR H spectra

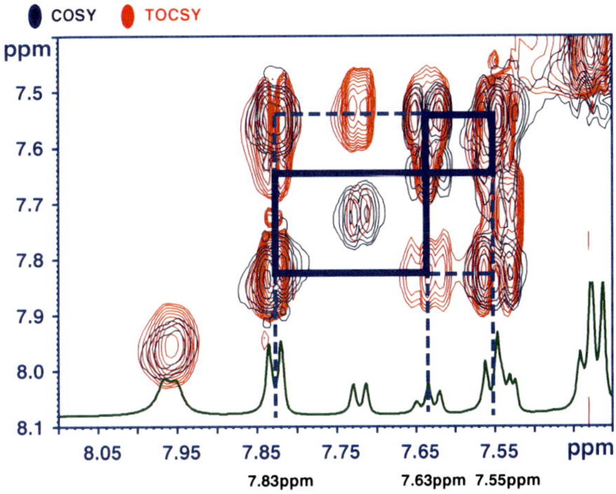

Fig. 4.30 The COSY and TOCSY spectra of the aromatic region in Fig. 4.29

Which one is correct? As shown in Fig. 4.31.

It can be seen from the *J*-decomposition spectra of Fig. 4.31 that the peak of these three hydrogen are all multiplet peaks, which may be due to the benzene ring π-bond; therefore, other spectra are necessary to help structural analysis as shown in Fig. 4.32.

It can be seen from the HSQC spectra of Fig. 4.32 (in red) that the chemical shifts of C directly connected with the three H are the following:

δ^{1}H	7.83	7.63	7.55
δ^{13}C	130.1	135.3	131.5

Also noteworthy is that of these three hydrogen in the HMBC spectra (in blue), only H (δ7.83) connects with the C with chemical shift at δ173.1. It has been mentioned earlier that the connection order of the three protons in the benzene ring is H (δ7.83)–H (δ7.63)–H (δ7.55), and HMBC can provide the C–H coupling

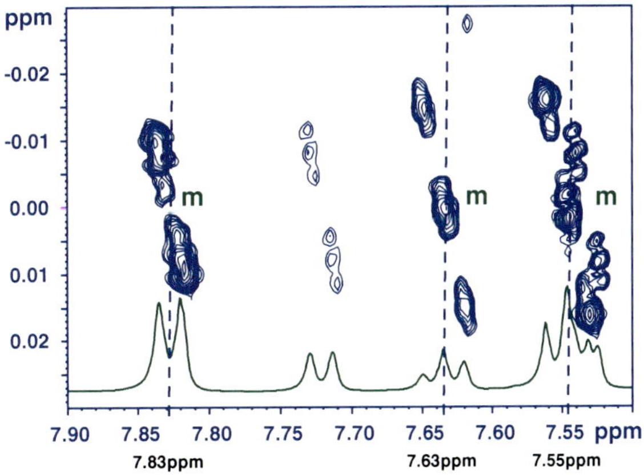

Fig. 4.31 The *J*-resolved spectra show the splitting mode of the aromatic region in Fig. 4.29. *m* represents the multiplet peak

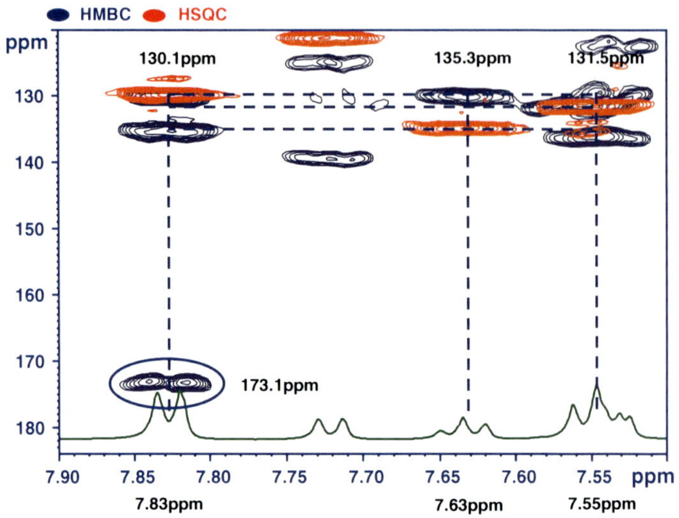

Fig. 4.32 The stacked diagram of HMBC and HSQC of the aromatic region in Fig. 4.29

relationship within 2–3 bonds. Therefore, it is possible that the ortho C of H (δ7.83) connects with a substituent, and the chemical shift of C in this substituent directly connected with benzene ring is δ173.1:

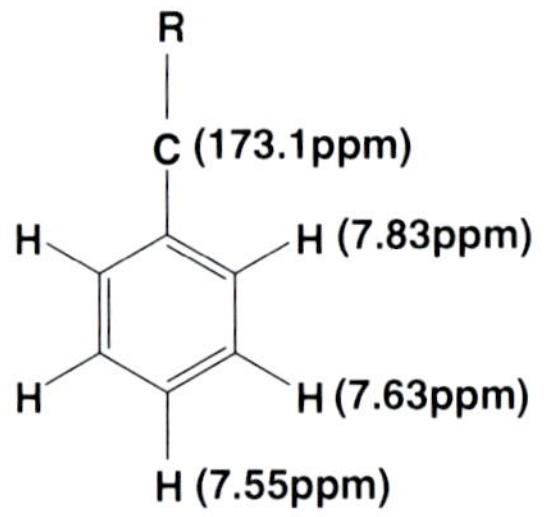

Then which group is further connected with this C (δ173.1)? Let us further observe the expanded spectra of HMBC and HSQC, as shown in Fig. 4.33.

It can be seen from the expanded spectra of HMBC and HSQC (Fig. 4.33) that the connection relationship exists in between the H with chemical shift at δ3.98 and the C (δ173.1) in the aliphatic region of HMBC. It is validated by observing the expanded diagram, as shown in Fig. 4.34.

To find out more information about the substituent, longitudinally expand the HMBC and HSQC at δ^{1}H3.98 (Fig. 4.35).

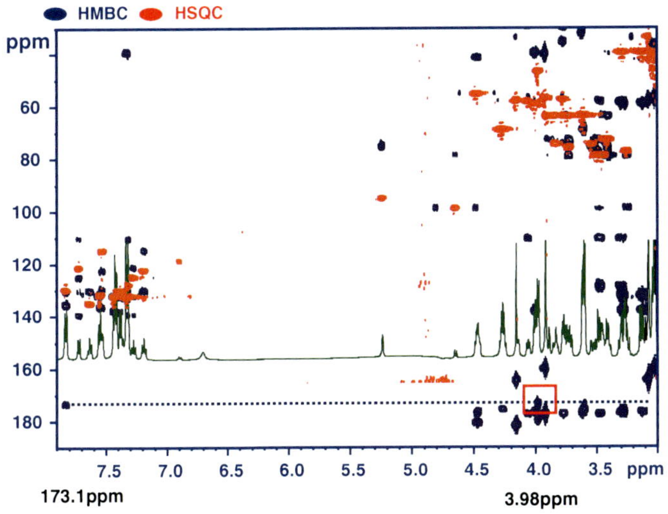

Fig. 4.33 The stacked diagram of the HMBC and HSQC spectra in Fig. 4.29

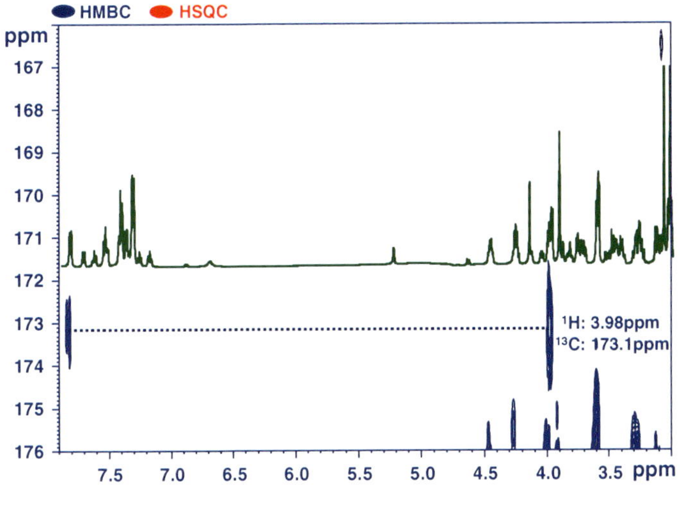

Fig. 4.34 The expansion of part spectra

It is found from the HSQC spectra of Fig. 4.35 (in red) that the chemical shift of C directly connect with the H (δ3.98) is 46.3, which also connect to C (δ173.1) and C (δ179.3).

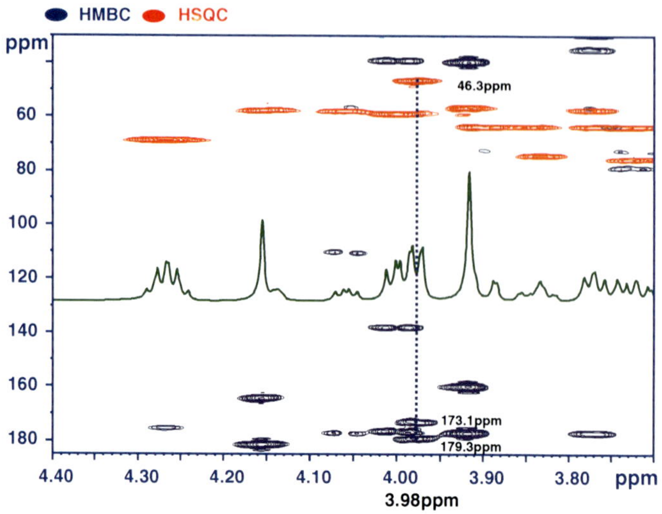

Fig. 4.35 The longitudinally expanded diagram of HMBC and HSQC at δ^1H3.98

The predicted chemical structure is

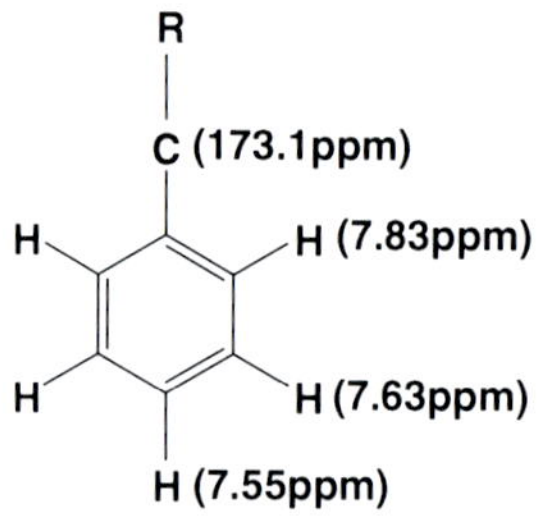

When the chemical shift of C is δ173.1, this C could be a carboxyl group, acid chloride group, an amide group or an ester group. But it can be seen from the HMBC that this C (δ173.1) is not located at the end, so it can be an amide group or an ester group. In addition, this C (δ173.1) connects to another C with chemical shift at δ46.3, and the chemical shift of H directly connected with this C (δ46.3) is 3.98. At the same time, the C (δ46.3) further connects another C with chemical shift at δ179.3. So the deduced structure formula is as the following:

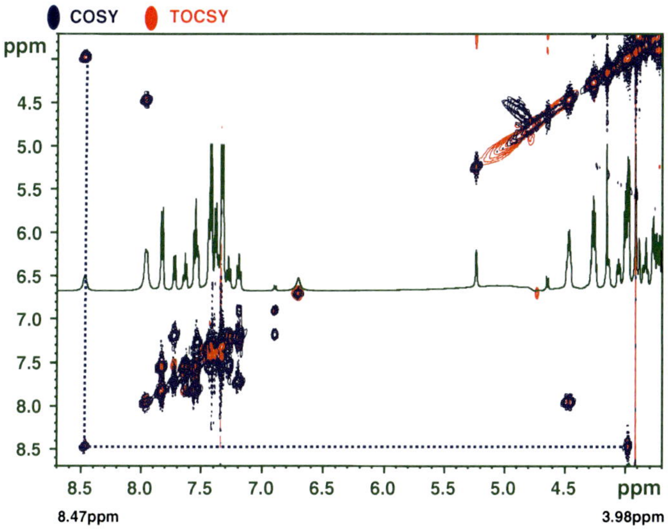

Fig. 4.36 The COSY and TOCSY spectra of Fig. 4.29

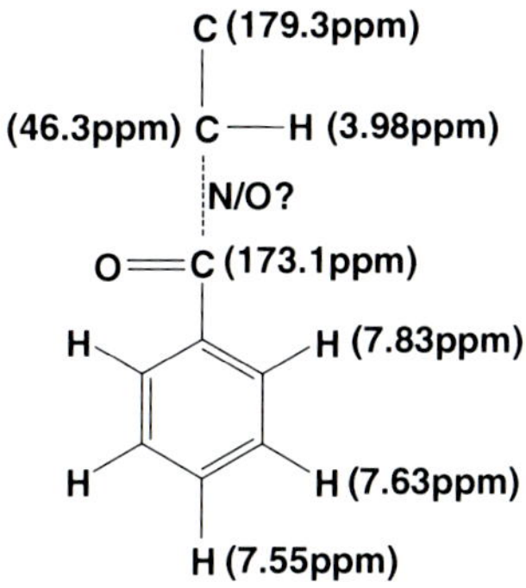

More information needs to be further found out. According to the Fig. 4.36, there is a coupling relationship between H(δ3.98) and H(δ8.47).

No chemical shift of C directly connected with the H (δ8.47) is found in the HMBC and HSQC spectra, indicating that this H (δ8.47) does not connect with a C. This indicates that there is an amide group rather than ester group in the structure. In addition, the C (δ179.3) does not connect with other atom, indicating that this C is in the end of molecule. So it should be carboxyl group. Finally, the structure of this molecule should as the following.

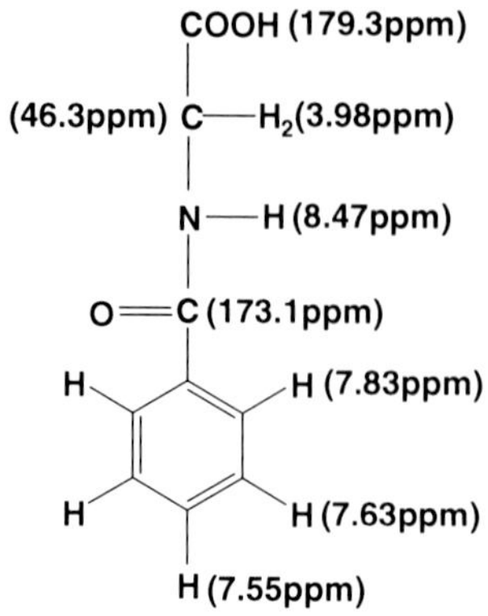

Hippuric acid

4.9 Common Problems and Solutions

NMR experiments involve many aspects, the parameters are many and complex, and so problems are inevitable in the course of the experiment. How to find the appropriate solution for the common problems in the experiment is one of the basic qualities that should possessed by the NMR spectrometer managers and operators.

The following is a summary for the common problems and solutions in the course of the NMR experiments:

1. Why the deuterated reagents are used in the test samples of NMR? The deuterium signal is needed for the field lock to reduce the drift of the field in the NMR detection.
2. How to choose the deuterated reagent in sample preparation? A principle usually followed in selecting the deuterated reagent is to select the deuterated reagent with polarity similar to the test sample. Commonly used deuterated reagents with their polarity in ascending order are as the following: benzene, chloroform, acetonitrile, acetone, dimethylsulfoxide (DMSO), pyridine, methanol, and water. Also, note that the chemical shift of the solvent peak is best to stay away from the sample peak. For example, D_2O is used as the solvent for the polysaccharide, which advantage is no interference of solvent signal in the ^{13}C-NMR detection. The deuterated dimethyl sulfoxide also can be used for solvent for the polysaccharide, because the solvent signal of deuterated dimethyl sulfoxide in the ^{1}H and ^{13}C spectra mostly does not overlap with the polysaccharide signal, which is convenient for the analysis of polysaccharide spectra. In addition, the solvent effect will bring the difference of the chemical shifts, which should be paid attention in the comparison and analysis with literature data.
3. How to determine the amount of the solvent in the sample preparation? Usually, the location and scale of the corresponding coil are marked on the graduated cylinder for detecting sample height, generally as long as the sample length is 3 mm more than the coil up and down the coil. Too little it will affect the

automatic shimming effect; too much it will not only waste solvent but also dilute the sample, reducing the effective amount of sample in the coil. Pay attention in this case to align the center of the sample liquid column with the coil center of the depth measuring cylinder. For example, when using a 5 mm NMR tube, the solvent amount of the sample is about 0.5 ml.

4. What should be paid attention when using the NMR tube? Firstly, best use high-quality NMR tube. If sample tube is too small or has cracks, it will be easily broken in the probe and pollute the probe. Therefore, the sample tube should roll on a plane to insure that it is straight before it is used; then, the presence of cracks is carefully detected under the light; to insure that the inserted rotor is not too tight or too loose. Further, select different size specifications of the NMR tube according to the sample volume. For example, available sizes of NMR tube are 5, 3, 2.5 and 1.7 mm, etc.
5. The deuterium lock signal cannot be detected after the sample is put in. There are many reasons may cause the loss of the deuterium lock signal. But the sample should be firstly checked. It should be verified that the deuterated reagent has been added into the sample; the sample height is appropriate; and the sample is placed in the detection coil range of the probe. If the deuterated reagent is added into the sample, the amount of deuterium should be roughly calculated. If the amount of the deuterated reagent added into the sample which is just putting into the magnet is much less than that of the previous samples, the lock power and gain should be considered. Or automatic lock could be used (modern spectrometers generally have automatic lock function).
   If the lock power and gain both have a reasonable value, but the deuterium lock signal still cannot be found, the magnetic field bias should be considered. When adjusting the magnetic field bias, the chemical shift of H spectra can be referenced. If the new sample solvent is deuterated chloroform, and the replaced sample solvent is D_2O, then the magnetic field bias should be reduced to check the deuterium lock signal. Because the solvent peaks of chloroform appear at the lower field. If the new solvent is deuterated DMSO, we should consider increasing the magnetic field bias, because the solvent peaks of DMSO appear at the higher field.
   If the deuterium lock signal still cannot be found by the above methods, it is necessary to consider the communication between the spectrometer control cabinet and the computer. You need to restart the instrument under the guidance of specialized technicians to check related causes.
6. The deuterium lock signal can be observed after the sample is put in, but cannot lock. The correctness of lock phase should be checked firstly. When the lock-phase deviation is large, effective lock and shimming cannot be completed. When checking the phase, unlocking observation is an intuitive method. Then, correct it by adjusting the lock phase.
7. The sample cannot be popped after the completion of the experiment. If the sample cannot be popped when inputting appropriate command after completing the experiment, firstly check if the compressed air is normal, if the value is

largely discrepant with required value. If the value is significantly smaller than the normal value, indicating that the pressure of the compressed air is not enough, then check the leaks. If there is no problem in the pressure of compressed air, you should consider if the sample is stuck in the probe. The fixed screws of the probe can be unscrewed, and the probe can be gently pulled out and then put back. Further blow air, the sample usually can be popped.

8. No signal is generated when acquiring spectra. The following reasons should be considered: ① The sample failed entering into the detection coil range of probe. In the experiment without lock (for example, many heteronuclear experiments do not need to lock field), acquired signal before the sample entering into the detection coil range of probe; ② if the sett observation nucleus is inconsistent with the desired observation nucleus, desired signal cannot be observed when acquiring signal; ③ when the concentration of the nuclei in the sample is not enough, for example, the sensitivity of ^{13}C is low, the signal cannot be observed within a short acquiring time. Or the solubility of the sample itself is relatively low, resulting in the signal unseen when the accumulating time is small; ④ the set spectral width and center frequency is not accurate. For common H spectra, the magnetic field bias changes a lot after changing the lock solvent; therefore, the frequency bias need to be correspondently adjusted; otherwise, signals cannot be generated within the defined spectral width; ⑤ it was un-tuned before acquisition. When replacing sample, it is necessary to tune according to the nature of the new sample. The tune range of ^{1}H spectra is very narrow; generally, signals can be observed without tuning. But for other nuclei other than H, sometimes signals cannot be seen without tuning. For example, sometimes signals of ^{13}C cannot be generated in the acquisition of two-dimensional HSQC experiment without tuning the ^{13}C; ⑥ parameters were set incorrectly in the decoupling experiment. When doing the decoupling experiment, the correctness of decoupling parameters should be checked, because the spectral quality will be very poor without decoupling, which may cause signals unseen in a short period of time; ⑦ the 90° pulse width is wrong. Each pulse is controlled by the RF field strength in the NMR. When the selected pulse width is the commonly used 90° pulse width, you should check whether the RF field strength is the usual value.
9. The spectral baseline is uneven. When acquiring the metabolic fingerprint of biological samples, the multi-pulse program is often used to achieve a good effect in suppressing the water peak. If the 90° pulse is set incorrectly, the spectral baseline will be uneven.

4.10 Sample Extraction Methods in Plant Metabolomics Study

There are a lot of sample extraction methods in plant metabolomics study. However, since there is a high degree of inconsistency in the physical and chemical properties of plant metabolites, so far, there is no one ideal method can be used to efficiently detect all categories of metabolites at the same time. Therefore, suitable plant metabolites extraction methods should be chosen according to the experiment purpose, the type of target metabolites and different parts of plant to be extracted. Standards in evaluating the quality of metabolite extraction methods are the following: ① maintaining original biochemical state of metabolites; ② extracting comprehensively; ③ the extraction process should not be selective and has no any physical and chemical modification; ④ good repeatability and operability. At present, the main steps of the commonly used plant metabolomics sample extraction method are: ① the crushing of plant material and the quenching of metabolic reactions. Plant materials are usually ground by liquid nitrogen or directly crushed by electronic homogenizer. The liquid nitrogen grinding is the "gold standard" in plant extraction, but the efficiency is low and very labor-intensive, and not suitable for the detection of a large number of samples. The efficiency of electronic homogenizer is high, but is not suitable for the extraction of a small number of samples. The purpose of quenching is to stop the metabolic reactions and prevent the degradation of metabolites. The traditional quenching method is processing by instantaneous low temperature (−40 °C methanol) or high temperature (boiling methanol/ethanol/water), or treated by strong acid (perchloric acid or trichloroacetic acid) or strong alkali (sodium hydroxide or potassium hydroxide) with extreme pH. Liquid nitrogen quenching is also a commonly used quenching method. ② Extraction of metabolites. Commonly used metabolite extraction solvents are methanol, ethanol, acetonitrile, perchloric acid, trifluoroacetic acid, potassium hydroxide, etc. According to the purpose of experiment, a single, binary, or ternary solvent can be used to extract primary or secondary metabolites. According to the type of metabolites, solvents with high extraction rate and good reproducibility should be chosen. Issues on solvent saturation and extraction times should be noticed in the cause of extraction. In addition, the ultrasonic-assisted extraction, supercritical fluid extraction, and solid phase extraction methods are also widely used in plant metabolomics study. ③ Concentration. Because metabolites are diluted in the cause of extraction, the concentration of metabolites is necessary. The freeze-dryer is generally used for aqueous-phase extract to be dried in low temperature and low pressure. The freeze-drying can remove water and prevent the thermal degradation of metabolites at the same time. The organic phase extract can be concentrated by reduced pressure evaporation or rotary evaporation. Of noteworthy is that the freeze-drying and solvent evaporation are not suitable for the study of volatile metabolites. ④ Preparing NMR samples. The NMR sample is prepared for analysis by weighting certain amount of concentrated powder then added it into 0.4–0.6 ml NMR buffer solution. The NMR buffer solution generally contains phosphate buffer with certain ionic strength to control pH, the deuterated reagent for locking (D_2O for the aqueous phase and CD_3Cl for the organic phase), and a small amount of internal standard for qualitative and quantitative analysis [trimethylsilylpropionic acid (TMSP) and tetramethylsilane (TMS)].

Below is a brief description of extraction steps in the NMR-based metabolomics study:

① The fresh plant sample is instantaneous frozen by liquid nitrogen and stored in −80 °C refrigerator for use. Ground the fresh sample in liquid nitrogen using a mortar, 1 ml A solvent (methanol: water = 1:1, v/v) is added to 100–250 mg ground powder. Shake for 30 s, and ultrasonic treatment in ice bath for 60 s. Repeat three times. 4 °C, 12,000 rpm centrifugation for 10 min. The supernatant is rotary evaporated to remove methanol. Freeze-drying in −40 °C. 0.6 ml 0.1 mol/L phosphate buffer containing 10 % D_2O (pH 7.4) is added to 2–10 mg extract for NMR analysis. 1 ml B solvent (methanol: chloroform = 1:2, v/v) is added to above solid residue. Shake for 30 s, and ultrasonic treatment in ice bath for 60 s. Repeat three times. 4 °C, 12,000 rpm centrifugation for 10 min. The supernatant organic phase is rotary evaporated to remove methanol and chloroform. Freeze-drying in −40 °C. 0.6 ml CD_3Cl containing 0.1 % TMS is added to 2–10 mg extract for NMR analysis.

② The fresh plant sample is instantaneously frozen by liquid nitrogen and stored in −80 °C refrigerator for use. Ground the fresh sample in liquid nitrogen using a mortar, freeze-drying in −40 °C freeze-dryer. 1 ml extraction solvent C (0.1 mol/L phosphate buffer containing 10 % D_2O and 2 % TSP, pH 7.4) is added to accurately weighed 25 mg freeze-dried sample. Shake 1.5 min at 20 Hz using tissue disrupter. Repeat three times. 4 °C, 12,000 rpm centrifugation for 10 min. 0.6 ml supernatant is used for NMR analysis.

4.11 New Technologies and Developmental Trends

NMR can provide information on the connection relationship between atoms, which can be used to efficiently understand metabolite structure and provide in situ molecular information of metabolites qualitatively and quantitatively. And the NMR technology has characteristics such as synchronization of detection, rich molecular information and noninvasive. Therefore, NMR analysis has been an important analysis tool in metabolomics. The core task of metabolomics is to rapidly and efficiently obtain in situ metabolic fingerprint information. Although there are many initial mature techniques and methods in metabolomics, structural determination methods for low-concentration unknown metabolites and quantitatively analysis methods for overlapping peaks are still lack.

The recently developed instrument coupling technology: high-performance liquid chromatography–diode-array detector–solid phase extraction–mass spectrometry–nuclear magnetic resonance (HPLC–DAD–SPE–MS–NMR) provides an efficient structural determination method for low-concentration unknown metabolites (as shown in Fig. 4.37).

The HPLC–DAD–SPE–MS–NMR instrument coupling system is mainly constituted by five parts: HPLC, DAD, SPE, MS, and NMR. Samples are firstly

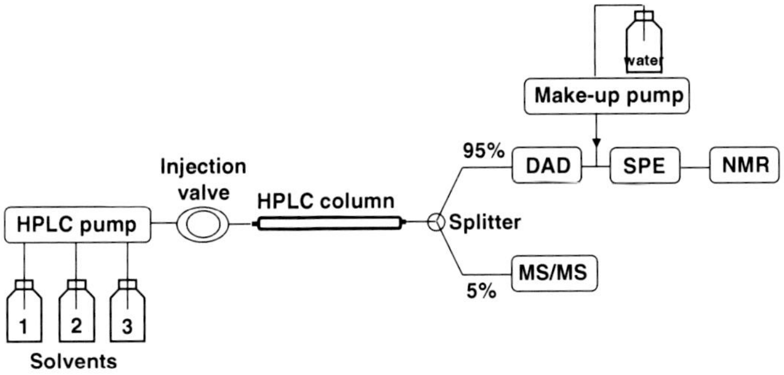

Fig. 4.37 Working diagram of HPLC–DAD–SPE–MS–NMR instrument coupling system

separated by HPLC chromatographic column. Samples eluted from the column are divided into two parts: 95 % of which flows into the DAD, while only 5 % of which flows into MS to obtain information on molecular weight. To take such a split-flow fashion is due to the relatively high detection sensitivity of MS. Samples flowed into DAD go through a procedure of adding water by water supply pump, then flow into SPE. Of particular noteworthy is that suitable SPE column should be chosen according to the chemical properties of metabolites to retain interested metabolites in the SPE column. The purpose of adding water is to reduce the content of organic solvent so that to validly retain the metabolites.

Samples retained in the SPE column is dried and eluted by corresponding deuterated reagent, then directly fed into NMR spectrometer for the acquisition of various NMR spectra. The first advantage of HPLC–DAD–SPE–MS–NMR instrument coupling system is that it integrates information on molecular polarity, functional group, molecular weight, and the connection relationship between atoms to achieve the purpose of resolving the structure of unknown metabolites; the second advantage is that samples can be repeated enriched by multiple SPE columns to increase the concentration of test metabolites and overcome the drawback of low detection sensitivity of NMR. The third advantage is that the test sample is dried and eluted by full deuterated reagent solvent and then fed into the NMR system, thus avoiding the need for multi-solvent peak suppression in NMR experiments. Finally, the instrument coupling system avoids the access of deuterated solvent into the MS system, enabling obtaining more accurate molecular weight information of metabolites. If the trace probe or ultra-low-temperature probe is configured in the NMR system, the detection sensitivity will be greatly improved.

This instrument coupling system has been applied in the identification of plant secondary metabolites, such as the identification of carnosic acid and *cis*-4-glucoside coumaric acid in rosemary. Due to the relatively large molecule and weak polarity of metabolites, generally the C_{18} chromatographic column is used to achieve better separation and enrichment effect. However, for the separation and enrichment of molecules with strong polarity, such as metabolites in animal body fluids or tissues and cells, still a lot of problems need to be further resolved.

References

Derome AE. Modern NMR techniques for chemistry research. New York: Pergamon Press; 1987.

De Koning W, van Dam K. A method for the determination of changes of glycolytic metabolites in yeast on a subsecond time scale using extraction at neutral pH. Anal Biochem. 1992;204:118–23.

Gao HB, Zhang ZF. NMR principles and experimental methods. Wuhan: Wuhan University Press; 2008.

Kaiser KA, Barding GA, Larive CK. A comparison of metabolite extraction strategies for ^{1}H-NMR-based metabolic profiling using mature leaf tissue from the model plant *Arabidopsis thaliana*. Magn Reson Chem. 2009;47:S147–56.

Lambert JB, Holland LN. Nuclear magnetic resonance spectroscopy: an introduction to principles, applications, and experimental methods. New Jersey: Pearson Education Inc; 2006.

Mao XA. Modern NMR practical technologies and the applications. Beijing: Scientific and Technical Documentation Press; 2000.

Mattoo AK, Sobolev AP, Neelam A, Goyal RK, Handa AK, Segre AL. Nuclear magnetic resonance spectroscopy-based metabolite profiling of transgenic tomato fruit engineered to accumulate spermidine and spermine reveals enhanced anabolic and nitrogen-carbon interactions. Plant Physiol. 2006;142:1759–70.

Nicholson JK, Lindon JC, Holmes E. 'Metabolomics': understanding the metabolic responses of living systems to pathophysiological stimuli via multivariate statistical analysis of biological NMR spectroscopic data. Xenobiotica. 1999;29:1181–9.

Qiu ZW, Pei FK. Nuclear resonance spectroscopy. Beijing: Science Press; 1987.

Stefan B, Siegmar B (translators: Tao JX, Li Y, Yang HJ). 200 and more NMR experiments, a practical course. 3rd ed. Beijing: Chemical Industry Press; 2007.

Tang HR, Wang YL. Metabolomics: a revolution in progress. Prog Biochem Biophys. 2006;33:401–17.

Tang HR, Wang YL. High-resolution NMR spectroscopy in human metabolism and metabonomics. In: Webb GA, editor. Handbook of modern magnetic resonance. Berlin: Kluwer Academic Publishers; 2006.

Xia YL, Wu JH, Shi YY. NMR principles and applications in biology. Lecture notes.

Xue S. NMR spectroscopy. Lecture notes; 2004.

Yang CR, Li XC, Wang DZ. New NMR technologies and the applications in the structural analysis of natural organic compounds. Kunming: Yunnan Press and Publication; 1990.

Zhang H. Modern organic spectral analysis. Beijing: Chemical Industry Press; 2005.

Chapter 5
Multivariate Analysis of Metabolomics Data

Jun-Fang Wu and Yulan Wang

5.1 Introduction

Due to the huge number of samples, the complexity of the data information as well as the high degree of correlation between variables in the multidimensional data matrix of metabolomics information derived from NMR and MS methods, data information cannot be extracted using traditional univariate analysis method. Thus, the mining and refining of potential relevant information between metabolites from these massive data plays an important role in the subsequent finding of biomarker groups and the interpretation of biological significance. At the same time, the selection of appropriate data analysis methods is also crucial for the correct extraction of metabolomics information.

Chemometrics is a chemistry subdiscipline that is first proposed by a Swedish chemist S. Wold in 1970s (Wold 1995). It utilizes methods and means that combine mathematics, statistics, computer science, and chemistry to design and select optimal chemical measurement methods, to resolve chemical measurement data, and to maximize access to the information contained in the measured data (Yu et al. 1991). The primary means of chemometrics in resolving the classification and biomarker search in complex system is pattern recognition. The primary idea of pattern recognition is the compression, dimensional reduction, and classification analysis of the collected multidimensional and massive amounts of information by computer, and the classification, feature selection, and internal characteristic finding of certain properties (usually implicit) of samples in the sample group by chemical measurement data matrix as a multivariate analysis technique (Ni 2004; Weckwerth and Morgenthal 2005). It generally includes unsupervised methods and supervised methods. The unsupervised pattern recognition method is based on the nature of the

J.-F. Wu (✉) · Y. Wang
Wuhan Institute of Physics and Mathematics, Chinese Academy of Sciences,
Wuhan 430071, China
e-mail: jennifer.wjff@gmail.com

X. Qi et al. (eds.), *Plant Metabolomics*, DOI 10.1007/978-94-017-9291-2_5

data itself to determine whether the sample belongs to a different category (Hu 1997), such as principal component analysis (PCA) and hierarchical cluster analysis (HCA). The supervised pattern recognition method is to randomly divide the samples of known category into two sets (the training set and the test set), modeling based on the training set of known category, and then characterizing the modeling performance by the accuracy of the test set. Commonly used supervised methods include partial least squares (PLS), linear discriminant analysis (LDA), and K-nearest neighbor analysis (KNN) (Qiu et al. 2005).

In general, the pattern recognition procedure usually includes three aspects: the preprocessing of the data sets, data feature extraction and selection (including unsupervised and supervised pattern recognition method), and the establishment and validation of the data model.

5.2 The Preprocessing of the Data Sets

In order to obtain data that can go through subsequent analysis, the original data set usually need to be preprocessed, which includes data normalization, peak alignment, and data scaling.

5.2.1 Data Normalization

After the correction of phase and baseline and the calibration of chemical shift of complex original NMR spectra, for the comparability between samples of different concentrations, it is often necessary to quantitatively analyze spectra data by the subsection integral of the original spectra, e.g., the data preprocessing by normalization. The early normalization method often calculates the proportion of the integration value of the metabolite within the subsection interval (bucket size) in the total spectral valid signals. The bucket size is mostly δ 0.04 (Waters et al. 2005). Thus, the original 32 K spectrum is divided into hundreds of data points. But because the δ 0.04 is too large and severely reduces the resolution of the spectra and affects assignment and subsequent analysis of metabolites with overlapping peaks (Fig. 5.1a), with the increasing development of computer technology, the bucket size is now to be taken as δ 0.002 or smaller (Fig. 5.1b). In the meantime, the increased subsection data points better maintain the information of original spectra.

Currently, there are mainly three normalization methods. The first one is described above, which provides the sum of total spectral valid signals is 1 and the proportion of subsection integral value in the total spectral valid signals as a normalized variable. Such a normalization method is mainly used to eliminate the concentration difference between different samples in a moderate range. For example, this method can be used in urine samples of the general animal model to eliminate the difference of urine concentration caused by water and other factors between different animals. However,

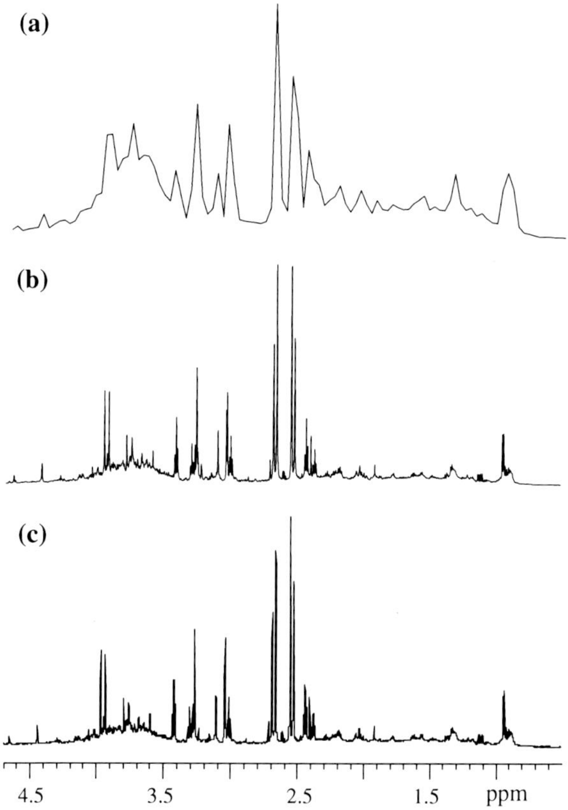

Fig. 5.1 The normalized spectra and original spectra with different bucket size of one-dimensional ^{1}H-NMR spectrum. **a** The normalized spectra with bucket size of δ 0.04; **b** the normalized spectra with bucket size of δ 0.002; **c** original spectra

this method is not applicable if the concentration change of certain metabolites in the sample is too large. This is because the significantly increased metabolite would make the concentration of other metabolites with no change relatively decreased, leading to the generation of false results in the subsequent data analysis. For example, if the first normalization method is used in the analysis of body fluid of the diabetic patient (the glucose concentration is very high), the proportion of other metabolites other than glucose in the total spectra will be relatively decreased. In this case, the most suitable method is a second method, i.e., take the ratio between the subsection integral value and the relatively unchanged peak area of metabolite as the normalization variable. This normalization method affects little the finding of the variation of metabolites other than glucose and the interpretation of biological significance. The best method for the analysis of plant metabolome is the normalization based on the dry weight before extraction; results obtained in this way are the absolute value of metabolites variation. Furthermore, the Dieterle group also proposed a method namely probabilistic quotient normalization, which is by looking for the most probable dilution factor to eliminate the spectra difference in subsequent analysis caused by sample elution in the biological samples (Dieterle et al. 2006).

5.2.2 Peak Alignment

During the experiment, the pH and concentration and other factors of the sample can easily cause the chemical shift of certain functional groups. Although the subsection integral of spectra can ease the tiny spectra shift within bucket size, it is not applicable for metabolites with large chemical shift, and may cause the distortion of the line loading plot in subsequent data analysis. As shown in Fig. 5.2, due to the three carboxyl groups of citric acid in the sample, and the similar level 3 ionization constant (3.13, 4.7, 6.4) (Xiao et al. 2009), the chemical shift is very sensitive to solution environment. Without peak alignment operation, significant distortions appear on a variety of metabolite loading plots, and the change trends of metabolites adjacent to citric acid (such as demethylamine) cannot be correctly presented because they are covered by that of nitric acid (Fig. 5.2b). For this reason, Stoyanova et al. (2004) proposed a method of automatic correction NMR peak alignment. In addition, NMR peak alignment can be done by beam search (Lee and Woodruff 2004), genetic algorithm (Forshed et al. 2003), and other mathematical operations and then by normalization to provide data information closer to original spectra (Fig. 5.2c).

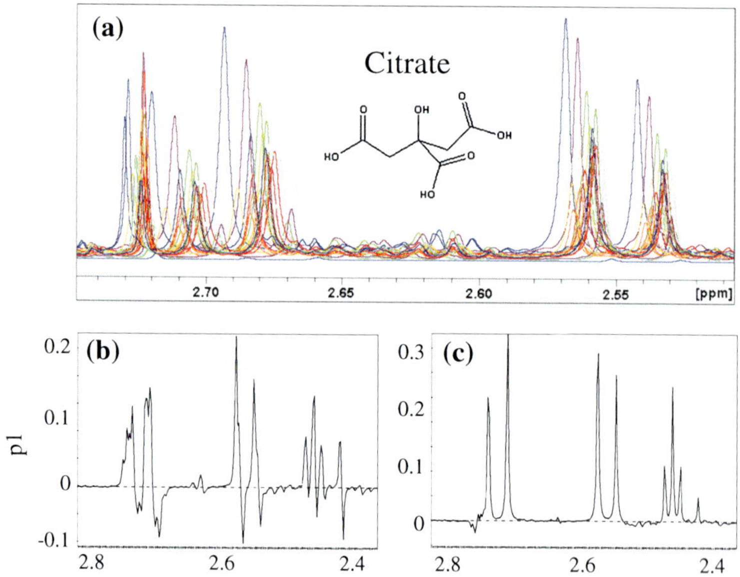

Fig. 5.2 The effect of peak alignment on data analysis. **a** The original spectra of citric acid region in the urine sample; **b** the loading plot before peak alignment; **c** the loading plot after peak alignment

5.2.3 Data Scaling

The variable dimensions of NMR data might be different, or the relative concentrations vary widely, sometimes even reach to several orders of magnitude. After the multivariate analysis of these unprocessed data, signals with high concentration will be highlighted, and signals with low concentration will be lost at the same time, thus significantly affecting the subsequent modeling and the finding of specific biomarkers. In order to overcome the impact of the dimension differences and concentration differences to the results, to improve the predictive ability of the model and optimize the extraction of data information, it is necessary to preprocess the data set. Such preprocess mainly includes data scaling, weighting, and back-transformation (Meloun et al. 1992). The projection principle-based cluster analysis pattern recognition method is very sensitive to the methods of data preprocessing. Different preprocessing results may result in different final analytic results (Vandenberg et al. 2006). And thus, the selection of preprocessing means is particularly important.

There are a variety of data scaling mode, which mainly include mean center scaling, unit variance scaling, and Pareto scaling (Vandenberg et al. 2006), with the purpose of minimizing concentration differences between samples or giving all variables the same weighting. The mean center scaling refers to the matrixing of the differences between each variable and the mean value of all the variables in the current column, and the establishment of a new coordinate system. The original point of the new coordinate system coincides with the center point of the matrix group points, minimizing the dynamic range of data while maintaining the original data relationships (Qiu et al. 2005) (Fig. 5.3a). The line loading plot of NMR obtained by mean center scaling is most similar to the original spectra. But the

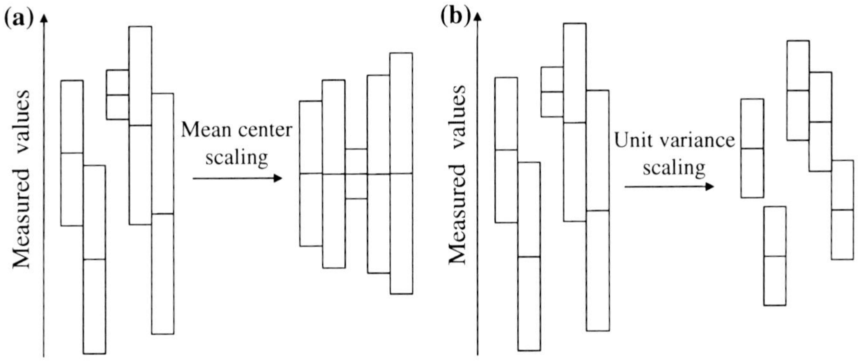

Fig. 5.3 The schematic diagram of NMR data scaling modes (Eriksson et al. 2006). **a** The schematic diagram of mean center scaling; **b** the schematic diagram of unit variance scaling (Copyright © Umetrics Inc. Reprinted with permission)

covariance calculation of metabolites obtained by this method affect little the separation function of model, and thus has a poor interpretation ability to model (Cloarec et al. 2005a, b). The formula of mean center scaling is as follows:

$$y_{np} = x_{np} - \overline{x_p} \tag{5.1}$$

where in y_{np} is the matrix descriptor after the mean center scaling; x_{np} is the original data of the original matrix; $\overline{x_p}$ is the mean value of the descriptors within the data set.

Unit variance scaling is a new matrix scaling method which is eliminating the standard deviations of variables in each column in the data matrix that has been going through mean center scaling. Its advantage is that it gives variables in each column the same weighting, which is conducive to the analysis of the change trend of metabolites with low concentration (Fig. 5.3b). Although the distortion of NMR line loading plot obtained by unit variance scaling is very serious (Fig. 5.4a), the change of metabolites obtained from this method reflects the actual changes more accurately in the substance (Cloarec et al. 2005a, b). The formula of unit variance scaling is as follows:

$$y_{np} = \frac{\left(x_{np} - \overline{x_p}\right)}{\mathrm{Std}_p}, \quad \mathrm{Std}_p = \sqrt{\frac{1}{n-1}\sum_{n=1}^{n}\left(x_{np} - \overline{x_p}\right)^2} \tag{5.2}$$

where in y_{np} is the matrix descriptor after unit variance scaling; x_{np} is the original data of the original matrix; $\overline{x_p}$ is the mean value of the descriptors within the data group; Std_p is the standard deviation of the descriptors in the data set.

Pareto scaling is an intermediate scaling method between mean center scaling and unit variance scaling; thus, it possesses the advantages and disadvantages of the above two methods. There is also distortion in the NMR line loading plot obtained by this scaling method, but it is not very serious. The range of relevance between the metabolite change obtained from the line loading plot and the model is intermediate

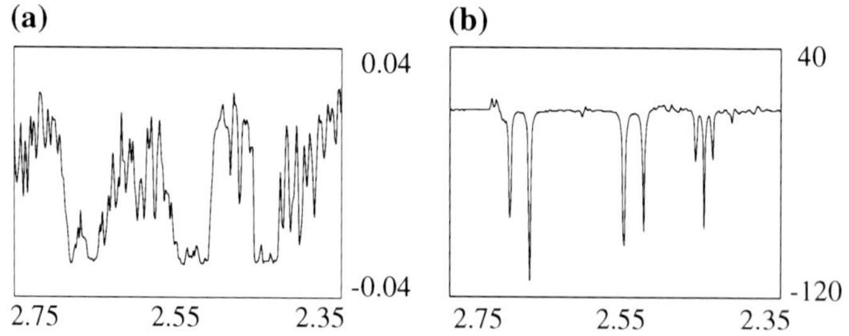

Fig. 5.4 The effect of back-transformation to the visualization of data. **a** The loading plot by unit variance scaling without data back-transformation; **b** the loading plot by unit variance scaling with data back-transformation

between the above two methods. The formula of Pareto scaling is as follows (wherein the meaning of each variable is the same as that of unit variance scaling):

$$y_{np} = \frac{\left(x_{np} - \overline{x_p}\right)}{\sqrt{\mathrm{Std}_p}} \quad \mathrm{Std}_p = \sqrt{\frac{1}{n-1}\sum_{n=1}^{n}\left(x_{np} - \overline{x_p}\right)^2} \tag{5.3}$$

5.2.4 *Back-Transformation of Data*

The back-transformation of data refers to certain transformation of the distorted loading plot to intuitive visualized spectra similar to NMR spectra, which maintains the peak shapes of NMR spectra, thereby improving the interpretation degree of the loading plot. The back-transformation is closely related to the scaling mode in the data preprocessing (Cloarec et al. 2005a, b). If the data are preprocessed by mean center scaling, the loading plot will not be distorted, and so the data can be directly used in subsequent analysis without transformation, while the loading value exported from unit variance scaling need to be multiplied by the standard deviation value (Std$p$) of variables in each column to obtain visualized spectra in subsequent data analysis (Fig. 5.4b). And the loading value exported from Pareto scaling need to be multiplied by the square root value of the standard deviation (Std$p$) of variables in each column to conduct data transformation.

5.3 The Extraction and Selection of Data Features

5.3.1 *Unsupervised Pattern Recognition Methods*

The unsupervised methods include dimensional reduction of data points and cluster analysis, wherein the dimensional reduction of data points refers to the representation of the whole data by several primary components in the multidimensional space, and cluster analysis refers to the finding of common features of samples with clustering trend (Goodacre et al. 2007). Early metabolomics mainly used unsupervised methods to process the data. Its main advantage is in the case of without any sample grouping information to realize the visualization of the intrinsic relationship of sample metabolic phenotypes by the dimensional reduction of high dimensional data, aiming at understanding potential characteristics lying in the multidimensional space data points and the mutual relationships among them. This approach can show the similarities and differences, separation trends, and the presence of outliers between samples in the Dataset as much as possible, without the overfitting problems of the model. Cluster analysis mainly includes PCA and HCA (Wagner et al. 2003), of which PCA is used more common (Ciosek et al. 2005).

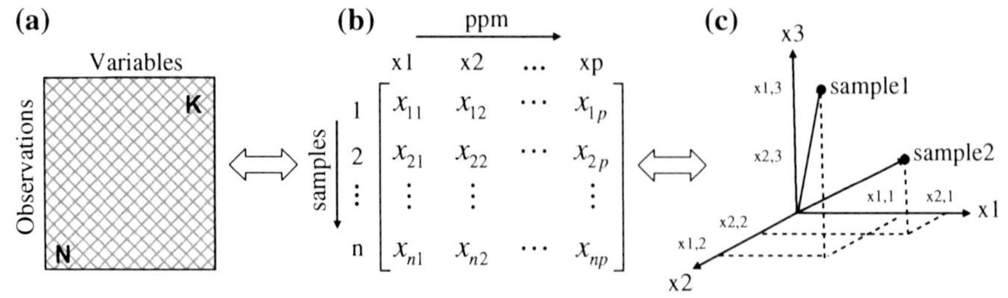

Fig. 5.5 The original matrix of metabolomics data and the representation of data in the three-dimensional space. **a** The schematic diagram of data matrix; **b** the entity matrix diagram of PCA data; **c** the simulation diagram of data in the three-dimensional space

Normalized data are often expressed as a data matrix of N (rows) × K (columns) (Fig. 5.5a), where in N represents observations, which can be biological samples, compound structures, or different time points of continuous reaction in the data model; K represents variables, which can be the original subsection integral variables of the spectra (NIR, NMR, and HPLC) and measurement values of various processing (pressure, temperature, and flow rate). In the NMR data matrix, N represents different samples in the model, and K represents each integral variable in the subsection integration region. For example, x_{11} represents the relative integral area of sample 1 in the x_1 integration region (Fig. 5.5b), and the integral intensity values of many integration region in the first row together constitute the characteristic distribution of sample 1 in the multidimensional space. Thus, each point in the multidimensional space represents a physical sample, but each data is actually a vector in the p-dimensional space (Fig. 5.5c).

Mathematically, PCA first directly decomposes the sample measurement data matrix, secondly projects the principle components (score vectors), and then performs discriminant analysis. There are several methods for the principal component decomposition of sample measurement data matrix, in which the nonlinear iterative PLS method and the singular value decomposition in linear algebra are commonly used. The data matrix is expressed as follows:

$$
\begin{aligned}
\mathrm{PC}_1 &= u_{11}x_1 + u_{21}x_2 + \cdots + u_{p1}x_p \\
\mathrm{PC}_2 &= u_{12}x_1 + u_{22}x_2 + \cdots + u_{p2}x_p \\
&\cdots\cdots \\
\mathrm{PC}_\mathrm{p} &= u_{1\mathrm{p}}x_1 + u_{2p}x_2 + \cdots + u_{pp}x_p
\end{aligned}
$$

where in the first principal component axis (PC_1) obtained from PCA is the maximum variance direction of the data matrix, which maximizes the description of variables in the data set; the second principal component represents the axis of the second variance in the data set; and so on. The principal components are independent to each other, and these principal component axes are orthogonal to each other, and together cross through the center of the space coordinate system, thereby guaranteeing the maximum retaining of effective information in the projection of high-dimensional space to low-dimensional space. Although theoretically a number of principle components can be obtained from the calculation of the input variables, the truly effective principle components are only a few of them from the very beginning of calculation (Wold et al. 1987). Therefore, the description of data in the lower-dimensional space achieves the aim of data dimensional reduction and can visually study characteristics of samples and the correlation between samples in the dimensional reduced space. More simply, PCA extracts sample vectors from the sample matrix and then extracts variable vectors from them.

The PCA data can be represented by scores plot and loading plot, including scatter loading plot and line loading plot (Fig. 5.6). The score plot of principal components mainly represents the distribution trend of data points (Fig. 5.6a), in which each point represents a sample, and the similarities and differences between samples are reflected by the dispersion trend or aggregation degree of samples in the score plot. The aggregation of points represents the high-degree similarity of observation variables, and the discretization of points represents the significant difference of observation variables. It should be noted that because PCA analysis is an unsupervised method, the classification information of the sample is not involved in the modeling, the difference in color in the plot is just for the sake of visual recognition. The scatter loading plot and line loading plot represent variables and metabolites leading to the difference in samples, in which the longer the space distance of variables to the origin point, and the more constant to the sample discretization axis of the score plot, the greater the influence of this variable to the model (Teague et al. 2007).

The main application of the PCA method is in the observation of the discretization trend between groups in the experimental model and the presence of outliers.

As shown in Fig. 5.7, the discretization degrees in different color symbols represent the distribution trend of samples in control group and treatment group at the PC_1 axis and PC_2 axis, respectively. It is visible from the PCA score plot that there is an outlier sample significantly away from the other samples in the control group, indicating the metabolome of this outlier sample is significantly different with other samples. By consulting the original spectra, it is found that this animal's urine sample was contaminated by fecal samples in the sample collection process;

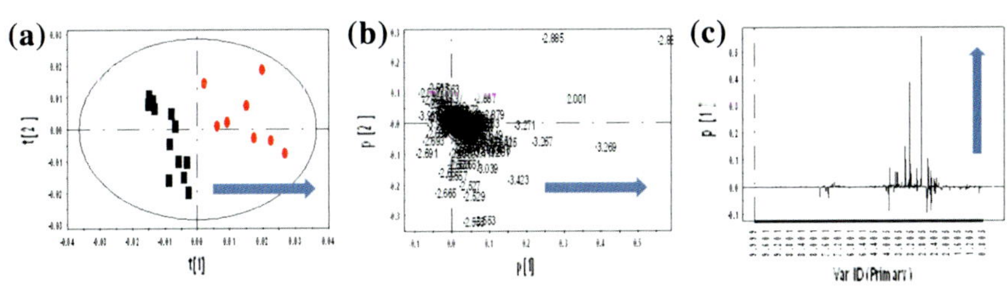

Fig. 5.6 The graph representation of PCA results. **a** The score plot; **b** the scatter loading plot; **c** the line loading plot

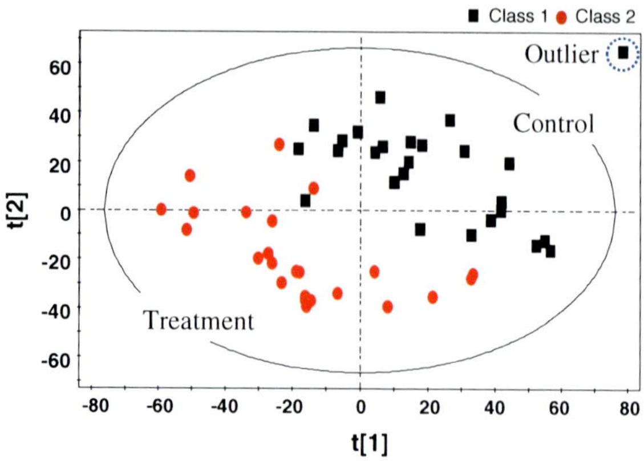

Fig. 5.7 PCA analysis of the discretization trends and outliers of the model

thereby, the abnormal deviation trend emerged in the PCA score plot and became the outlier point of the whole model. In order to eliminate differences in interpretation, after the clear reason of the emergence of the outlier is found, this outlier sample can be got rid of in the subsequent data analysis. At the same time, significant discretization trend between the two groups can be seen within the 95 % confidence interval by PCA method, indicating the significant influence of treatment factors to the metabolic patterns of samples in the two groups.

In addition, (Hubert and Engelen 2004) proposed the robust PCA method, which combines the projection principle and the low-dimensional space estimation to improve the classification results of the sample and the ability to better recognize the outlier samples. Scholz et al. (2004) proposed the independent component analysis method, which is based primarily on the distribution of Kurtosis to determine the optimal number of principal components, extract independent component, and perform cluster analysis, effectively resolving the inconsistency in the difference between the samples and the study focus.

5.3.2 Supervised Pattern Recognition Methods

The supervision pattern recognition method refers to the finding of difference of variables in known classification groups and the finding of the relationship between variables under the premise of understanding the classification information of samples. This method is mainly to find out the largest difference of sample classifications, which include PLS or projection to latent structure (PLS), Bayesian probabilistic approaches, LDA, and neural networks (NN), (Holmes et al. 1994, 2001). The modeling method varies with the study system. The PLS method is most commonly used in metabolomics data processing, which is based on the principle of the least squares method; that is, for a large number of experimental data, the deviation of the fit function is not required to be exactly zero, but in order to make the approximate curve reflect as much as possible the change trend of data points, the minimum sum of square of deviations is required. This approach can reflect the maximum difference between classification groups in the experimental data. Further, a correlation between the NMR data and other variables associated with the classification model can be observed by the PLS method.

We know that the organism is a complex whole. There are a high degree of individual differences at the metabolite level, which can be easily influenced by the genetic level, environmental change, and other factors. Thus, the elimination of random information (also called structured noise) uncorrelated with the experimental observation variables and the maximization of the extraction of changes correlated with experiment itself is an important issue in the metabolomics data analysis.

Scholars engaged in chemometrics then proposed an orthogonal to partial least squares (O-PLS) on the basis of PLS (Wold et al. 1998; Trygg and Wold 2002). This method can remove changes in *X* matrix uncorrelated with *Y* matrix and maximize the relationship between *X* matrix and *Y* variables (Wold et al. 1998; Trygg and Wold 2002). It is noted that there is no significant difference between PLS and O-PLS in the predictive ability of the model. The PLS method compensates the influence of the structured noise mainly through the extraction of principle components (Fig. 5.8a), while O-PLS compensates the influence of structured noise through the rotating coordinate system, thereby maximizing the difference between the control group and the treatment group (Fig. 5.8b). But the use of O-PLS must base on the model validation of PLS model by model validation, which is only used in the finding of metabolites leading to the difference of models. PLS models without model validation can also get two groups with complete separation by the O-PLS processing, but the metabolites found may have no biological significance.

The PLS (or O-PLS) data are ultimately manifested by coefficient plot (Fig. 5.9). Because the metabolite variables leading to the difference of models obtained from the PCA loading plot may also be due to the high concentrations of certain

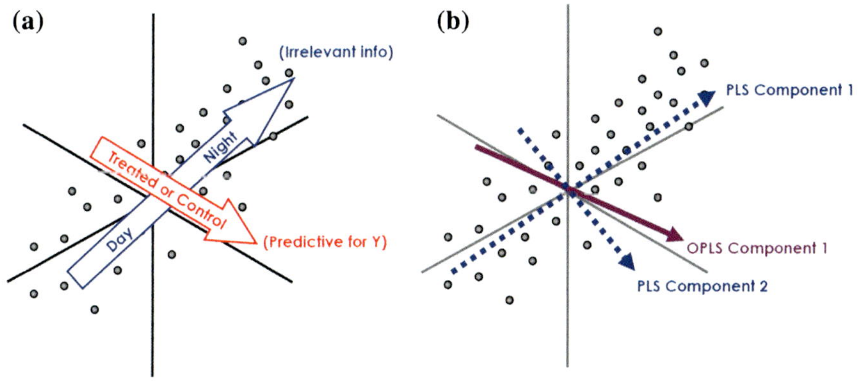

Fig. 5.8 The structured noise processing of PLS and O-PLS and differences between them

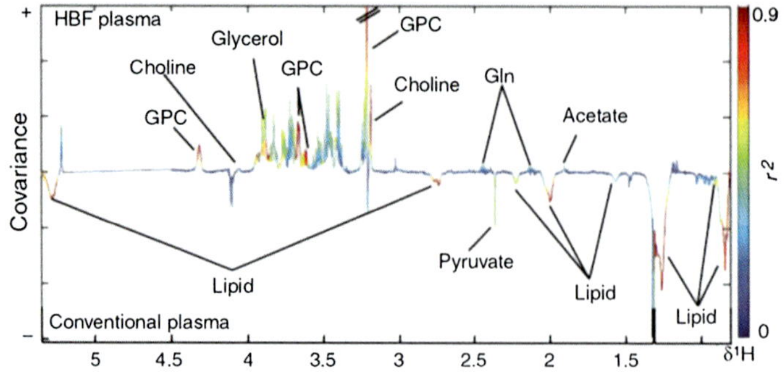

Fig. 5.9 The graph representation of coefficient plot (Reprint from Martin et al. (2007))

metabolites themselves, the representation method of coefficient plot is also proposed on this basis (Cloarec et al. 2005a, b), in which the chemical shift (variable) of NMR spectra is the horizontal axis, one vertical axis is the same as the line loading plot, representing the loading value of the variable, and another vertical axis (cold color) represents the coefficient values of all the variables correlated with the model, wherein metabolites with high coefficients are rendered as hot color and metabolites with low coefficients are rendered as cold color. Then based on the coefficient table, provides that metabolites (variables) exceed the threshold value is of significance to the distinction of models. This representation method takes the coefficients of variables as reference, avoiding changes caused by concentration differences of different metabolites in NMR spectra, and variables obtained from the coefficient plot are the biomarkers (or biomarker groups) of metabolic responses of organism to endogenous or exogenous stimuli.

5.4 The Identification and Validation of Data Model

The reliability validation by multivariate analysis modeling is an important prerequisite for the subsequent data analysis. We know that the focus of metabolomics research is the modeling, finding specific biomarkers, and finding metabolic pathways correlated with specific biological system. But the validity of the model and the ability to subsequent analysis and deduction on the basis of this model remain to be verified. Thus, the validation of quality control and statistical algorithm of data gradually received attention in recent years of study. There is a growing recognition that the cross-validation of model in each stage of data processing and analysis is necessary for both unsupervised and supervised pattern recognition analysis.

There are three model validation methods in the multivariate analysis. The first one is an internal validation method; that is, the Q^2 value fitted by the model represents the predictive ability of the model. Usually, $Q^2 > 0.9$ indicates a strong predictive ability of the model, $Q^2 > 0.5$ indicates a general predictive ability of the model, and Q^2 smaller than 0.5 indicates a poor predictive ability of the model. In general, the size of the Q^2 value is correlated with the sample number in the experimental study. The smaller the sample number, the greater the Q^2 values required by the meaningful model; the larger the sample number, such as that in the population cohort study, its $Q^2 > 0.3$ can be considered an acceptable predictive ability. But the internal validation method is associated with a variety of factors (such as the number of hidden variables in the PLS and the training time of neural network) in the modeling and often improves the predictive ability of the final model leading to the over fitting trend of the model, and thus, it cannot be an alternative to external model validation. The most commonly used external validation methods include the leave-one-out method and the permutation test.

The basic elements of the leave-one-out method include three parts, i.e., the training set, the validation set, and the test set. A number of possible models can be obtained from the training set, the quality of the training set can be evaluated by the validation set, and the predictive ability of the optimized model is tested by the independent test set data. In the conventional leave-one-out method, the training set and validation set are merged and are then randomly divided into two data sets. Each time of dividing, a part of samples are randomly rejected and then the predictive ability of the remaining samples is observed, and thus a cycle begins till finally obtaining the prediction value of the model (Brown et al. 2005) to evaluate the model (Fig. 5.10).

The permutation test as another method of external model validation is mainly used in the validation of the fitting degree of PLS/DA model (User guide to SIMCA-P$^+$ 2008), which observes the difference between the model of multiple random permutated y variables and the model of original y variables by randomly changing the order of y variable. Then, a regression line is made between the randomly generated R^2 values and Q^2 values and the original cumulative R^2 values and Q^2 values (Fig. 5.11); the intercept between the regression line and one vertical axis is one indicator for testing the model quality, wherein the randomized R^2 and Q^2 values represent the interpretation degree of data in the model of randomized y variables and the predictive ability of the model (Clayton et al. 2006; Slupsky et al. 2007), respectively. The model validation of the permutation test can be verified by

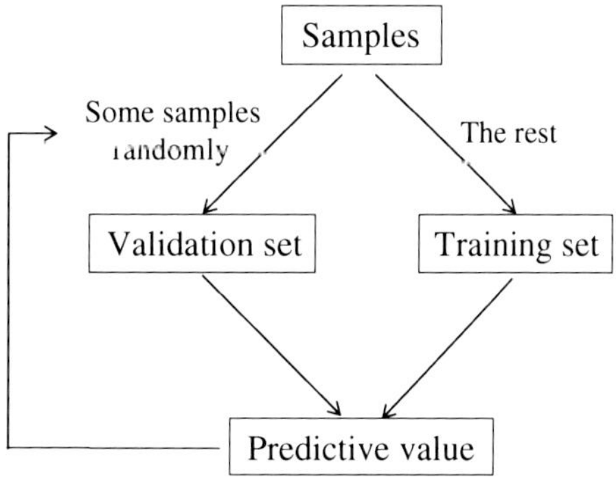

Fig. 5.10 The schematic diagram of the leave-one-out method

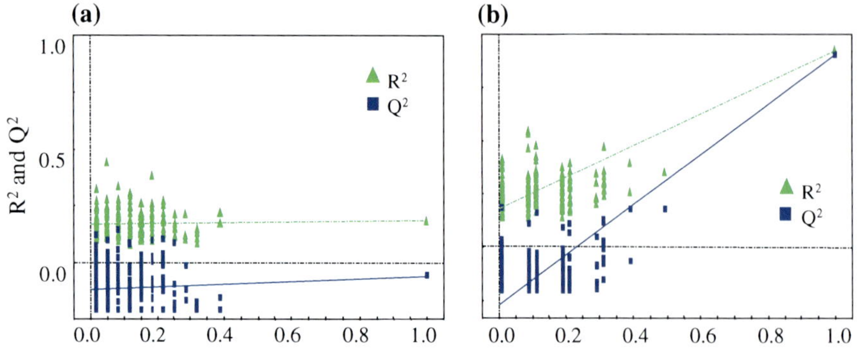

Fig. 5.11 Graph representation of the permutation test. **a** The meaningless data model validated by permutation test; **b** the meaningful data model validated by permutation test

the following two aspects: The first one is the slope of the regression line—the greater the slope of the regression line and the smaller the intercept between the regression line and the vertical axis, the more the data for interpreting the model and better the predictive ability of the model—and the second one is the difference of randomized R^2 and Q^2 values, respectively—the smaller the difference between the two, respectively, the smaller the difference between the interpreted data of the model and the predicted data of the model and better the quality of the model (Clayton et al. 2006; Slupsky et al. 2007). In short, if the predictive ability of original model is greater than the predictive ability of any model of random permutation y variables, then the model is of good quality and can be a premise for the subsequent finding of biological marker group of the model (Fig. 5.11b); otherwise, the model is of poor quality and is not suitable for subsequent analysis Fig. 5.11a).

5.5 The Statistical Total Correlation Spectroscopy

One of the NMR-based analytical methods is the statistical total correlation spectroscopy (STOCSY), which is a new data analysis tool used in metabolomics and developed in recent years (Cloarec et al. 2005a, b). It makes use of "related analytical methods in mathematical statistics to look for links between different variables, realize graphics display by computer, provide greater visibility of the data, explore information such as molecular structure and physiological relations" and help to find new biological metabolic profiling information (Zhu 2008). A pseudo two-dimensional spectrum is obtained by the statistical analysis of a series of one-dimensional spectra, but the cross-peak is not derived from the mutual coupled atoms like that of two-dimensional NMR spectra, but from the one-dimensional NMR signals with similar peak area (concentration) change; thus, it can be considered from the NMR signals of different protons in the same molecule. This method is helpful in the assignment and identification of spectra peaks,

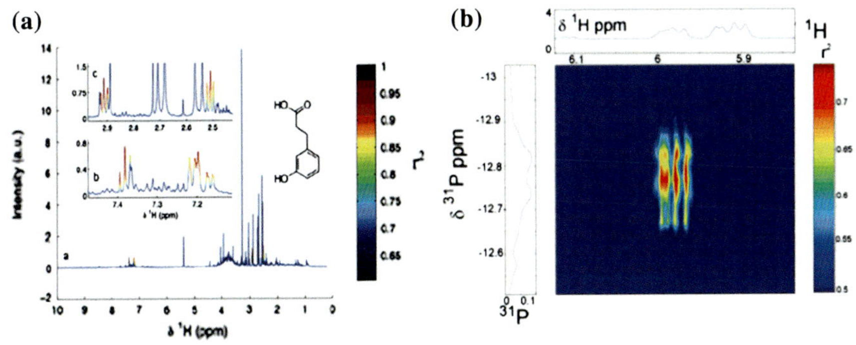

Fig. 5.12 The applications of statistical total correlation spectroscopy. **a** The homonuclear statistical correlation spectroscopy (Reprinted from Cloarec et al. (2005a, b). Copyright © 2005 American Chemical Society. Reprinted with permission); **b** the heteronuclear statistical total correlation spectroscopy (Reprinted from Coen et al. (2007). Copyright © 2007 American Chemical Society. Reprinted with permission)

especially for that of single peaks. At the same time, different molecules may be involved in the common biochemical metabolic pathway including the interdependence of concentration or the common reaction-regulating mechanism. Thus, the exploration of the relationships between different metabolites in the same metabolic pathway can provide important information.

With the further development of the method, the various forms of statistical correlation spectroscopy analysis methods are gradually being excavated. In addition to the homonuclear STOCSY, the heteronuclear statistical total correlation spectroscopy (HET-STOCSY) (Coen et al. 2007; Wang et al. 2008) and the STOCSY between different data collection methods of the same sample [such as LC-NMR-STOCSY (Cloarec et al. 2007) or UPLC-MS-STOCSY (Crockford et al. 2006)] are also becoming the regular effective multivariate data mining tools in metabolomics and providing new development space for metabolomics methods (Fig. 5.12). Likewise, they also make use of the correlation between signals from the same molecule or between signals of different molecules from the same metabolic pathway to integrate data obtained by different detection means or the detection of different nuclear and establish a correlation network of molecules, thereby further obtaining information correlated with structure or metabolic pathway.

5.6 Summary

In conclusion, the selection of suitable multivariate data analysis method is of significance to metabolomics data mining and the correct extracting of information. The data analysis procedures can be summarized in Fig. 5.13. After the original one-dimensional ^{1}H NMR spectra are obtained, it goes through manual correlation of phase and baseline and scaling, and then goes through subsection integration.

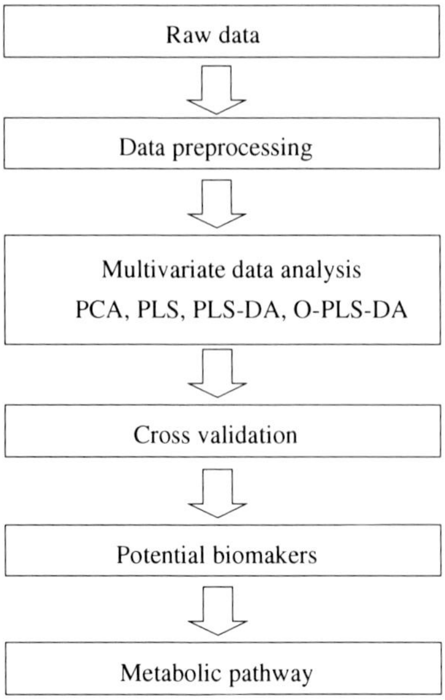

Fig. 5.13 The flowchart of metabolomics data analysis

The data processed by appropriate normalization is imported into software for the multivariate data analysis of pattern recognition, in which PCA is mainly used to observe the distribution trend of the sample as a whole and the presence of outliers; PLS is used to find the correlation between NMR data (*x* variables) and other variables (*y* variables); PLS-DA uses the leave-one-out method to cross-validate the quality of the model, and with the obtained R^2x and Q^2 values (representing the predictable variables of the model and the predictive degree of the model, respectively) to validate the effectiveness of the model. After this, the effectiveness of the model is further validated by permutation test to randomly and repeatedly change the order of the classification *y* variable and obtain corresponding different randomized Q^2 values. Then, O-PLS-DA is used to maximize the differences between different groups within the model and further summarize the metabolites with statistical significance by the analysis of coefficient. Finally, the focus is given on the analysis of metabolic pathways that are involved by metabolites with significant differences by literature reviewing and the biological significance of them is explained in detail.

References

Brown M, Dunn WB, Ellis DI, Goodacre R, Handl J, Knowles JD, O'Hagan S, Spasić I, Kell DB. A metabolome pipeline: from concept to data to knowledge. Metabolomics. 2005;1:39–51.

Ciosek P, Brzoka Z, Wrolewski W, Martinelli E, Di Natale C, D'Amico A. Direct and two-stage data analysis procedures based on PCA, PLS-DA and ANN for ISE-based electronic tongue-effect of supervised feature extraction. Talanta. 2005;67:590–6.

Clayton TA, Lindon JC, Cloarec O, Antti H, Charuel C, Hanton G, Provost JP, Le Net JL, Baker D, Walley RJ, Everett JR, Nicholson JK. Pharmaco-metabonomic phenotyping and personalized drug treatment. Nature. 2006;440:1073–7.

Cloarec O, Campbell A, Tseng LH, Braumann U, Spraul M, Scarfe G, Weaver R, Nicholson JK. Virtual chromatographic resolution enhancement in cryoflow LC-NMR experiments via statistical total correlation spectroscopy. Anal Chem. 2007;79:3304–11.

Cloarec O, Dumas ME, Craig A, Barton RH, Trygg J, Hudson J, Blancher C, Gauguier D, Lindon JC, Holmes E, Nicholson J. Statistical total correlation spectroscopy: an exploratory approach for latent biomarker identification from metabolic ^{1}H NMR data sets. Anal Chem. 2005a;77:1282–9.

Cloarec O, Dumas ME, Trygg J, Craig A, Barton RH, Lindon JC, Nicholson JK, Holmes E. Evaluation of the orthogonal projection on latent structure model limitations caused by chemical shift variability and improved visualization of biomarker changes in ^{1}H NMR spectroscopic metabonomic studies. Anal Chem. 2005b;77:517–26.

Coen M, Hong YS, Cloarec O, Rhode CM, Reily MD, Robertson DG, Holmes E, Lindon JC, Nicholson JK. Heteronuclear ^{1}H-^{31}P statistical total correlation NMR spectroscopy of intact liver for metabolic biomarker assignment: application to galactosamine-induced hepatotoxicity. Anal Chem. 2007;79:8956–66.

Crockford DJ, Holmes E, Lindon JC, Plumb RS, Zirah S, Bruce SJ, Rainville P, Stumpf CL, Nicholson JK. Statistical heterospectroscopy, an approach to the integrated analysis of NMR and UPLC-MS data sets: application in metabonomic toxicology studies. Anal Chem. 2006;78:363–71.

Dieterle F, Ross A, Schlotterbeck G, Senn H. Probabilistic quotient normalization as robust method to account for dilution of complex biological mixtures. Application in ^{1}H NMR metabonomics. Anal Chem. 2006;78:4281–90.

Eriksson L, Johansson E, Kettaneh-Wold N, Trygg J. Multi- and megavariate data analysis part II: advanced applications and method extensions. Umetrics Inc. 2006.

Forshed J, Schuppe-Koistinen I, Jacobsson SP. Peak alignment of NMR signals by means of a genetic algorithm. Anal Chim Acta. 2003;487:189–99.

Goodacre R, Broadhurst D, Smilde A, Kristal B, Baker J, Beger R, Bessant C, Connor S, Capuani G, Craig A, Ebbels T, Kell D, Manetti C, Newton J, Paternostro G, Somorjai R, Sjöström M, Trygg J, Wulfert F. Proposed minimum reporting standards for data analysis in metabolomics. Metabolomics. 2007;3:231–41.

Holmes E, Foxall PJ, Nicholson JK, Neild GH, Brown SM, Beddell CR, Sweatman BC, Rahr E, Lindon JC, Spraul M, Neidig P. Automatic data reduction and pattern recognition methods for analysis of ^{1}H nuclear magnetic resonance spectra of human urine from normal and pathological states. Anal Biochem. 1994;220:284–96.

Holmes E, Nicholson JK, Tranter G. Metabonomic characterization of genetic variations in toxicological and metabolic responses using probabilistic neural networks. Chem Res Toxicol. 2001;14:182–91.

Hu YZ. An concise course of chemometric. China medical science press 1997.

Hubert M, Engelen S. Robust PCA and classification in biosciences. Bioinformatics. 2004;20:1728–36.

Lee G-C, Woodruff DL. Beam search for peak alignment of NMR signals. Anal Chim Acta. 2004;513:413–6.

Martin FP, Dumas ME, Wang YL, Legido-Quigley C, Yap IK, Tang HR, Zirah S, Murphy GM, Cloarec O, Lindon JC, Sprenger N, Fay LB, Kochhar S, van Bladeren P, Holmes E, Nicholson JK. A top-down systems biology view of microbiome-mammalian metabolic interactions in a mouse model. Mol Syst Biol. 2007;3:112–27.

Meloun M, Militky J, Forina M. Chemometrics for analytical chemistry Volume: PC-aided statistical data analysis. 1st ed. Ellis Horwood Limited 1992.

Ni YN. The applications of chemometrics in analytical chemistry. Science press 2004.

Qiu YJ, Sha SA, Ye CH, Liu ML. Pattern recognition methods in biomedical magnetic resonance. Chinese J Magn Resonan. 2005;22:99–111.

Scholz M, Gatzek S, Sterling A, Fiehn O, Selbig J. Metabolite fingerprinting: detecting biological features by independent component analysis. Bioinformatics. 2004;20:2447–54.

Slupsky CM, Rankin KN, Wagner J, Fu H, Chang D, Weljie AM, Saude EJ, Lix B, Adamko DJ, Shah S, Greiner R, Sykes BD, Marrie TJ. Investigations of the effects of gender, diurnal variation, and age in human urinary metabolomic profiles. Anal Chem. 2007;79:6995–7004.

Stoyanova R, Nicholls AW, Nicholson JK, Lindon JC, Brown TR. Automatic alignment of individual peaks in large high-resolution spectral data sets. J Magn Reson. 2004;170:329–35.

Teague CR, Dhabhar FS, Barton RH, Beckwith-Hall B, Powell J, Cobain M, Singer B, McEwen BS, Lindon JC, Nicholson JK, Holmes E. Metabonomic studies on the physiological effects of acute and chronic psychological stress in Sprague-Dawley rats. J Proteome Res. 2007;6:2080–93.

Trygg J, Wold S. Orthogonal projections to latent structures O-PLS. J Chemom. 2002;16:119–28.

User guide to SIMCA-P^+. 2008. Version 12. Umetrics Inc.

Vandenberg RA, Hoefsloot HCJ, Westerhuis JA, Smilde AK, Vander Werf M. Centering, scaling, and transformations: improving the biological information content of metabolomics data. BMC Genomics. 2006;7:142–56.

Wagner C, Sefkow M, Kopka J. Construction and application of a mass spectral and retention time index database generated from plant GC/EI-TOF-MS metabolite profiles. Phytochemistry. 2003;62:887–900.

Wang YL, Cloarec O, Tang HR, Lindon JC, Holmes E, Kochhar S, Nicholson JK. Magic angle spinning NMR and ^1H-^{31}P heteronuclear statistical total correlation spectroscopy of intact human gut biopsies. Anal Chem. 2008;80:1058–66.

Waters NJ, Waterfield CJ, Farrant RD, Holmes E, Nicholson JK. Metabonomic deconvolution of embedded toxicity: application to thioacetamide hepato- and nephrotoxicity. Chem Res Toxicol. 2005;18:639–54.

Weckwerth W, Morgenthal K. Metabolomics: from pattern recognition to biological interpretation. Drug Discov Today. 2005;10:1551–8.

Wold S. Chemometrics; what do we mean with it, and what do we want from it? Chemometr Intell Lab. 1995;30:109–15.

Wold S, Antti H, Lindgren F, Ohman J. Orthogonal signal correction of near-infrared spectra. Chemometr Intell Lab. 1998;44:175–85.

Wold S, Esbensen K, Geladi P. Principal component analysis. Chemometr Intell Lab. 1987;2:37–52.

Xiao CN, Hao FH, Qin XR, Wang YL, Tang HR. An optimized buffer system for NMR-based urinary metabonomics with effective pH control, chemical shift consistency and dilution minimization. Analyst. 2009;134:916–25.

Yu RQ. An introduction to chemometrics. Hunan educational publishing press 1991.

Zhu H. Several NMR based new data analysis methods and the applications in metabolomics studies. Doctoral dissertation of Graduate School, Chinese Academy of Sciences 2008.

Chapter 6
Metabolomic Data Processing Based on Mass Spectrometry Platforms

Tian-lu Chen and Rui Dai

In order to have full access to potential information from all kinds of instrument test data, to find metabolites effectively representing differences from different plants, and eventually interpret the biological significance contained in data, metabolomic data processing mainly adopts chemometrics methods (Shi et al. 2010) to analyze huge amounts of data. This process is generally divided into three steps: data extraction, peak extraction, and preprocessing, as well as statistical pattern recognition (Xu and Shao 2004). This chapter will focus on introductions of the idea and mathematical principle of each method to provide reference and basis for practical applications. What have to be stated is that all the mathematical methods are merely supplementary analysis tools. The final results can only be drawn from researchers with a full integration of their practice knowledge in specific field, the characteristics of analysis objects, and various analysis methods.

6.1 Data Extraction

Spectrogram generated from previously mentioned devices such as GC or LC, which is based on mass spectrometry (MS) platform may contain hundreds to thousands of chromatographic peaks, each has at least 3–5 scans per second, some precise instruments even scan 20–50 times. Therefore, one chromatogram may contain thousands of or tens of thousands of mass spectrograms (Dettmer et al. 2007; Richard 2003), and how to dig out useful information from such huge amounts of data is hence quite important. As the first step of data processing, data extraction converts the spectrograms generated from various instruments into analyzable data forms such as Network Common Data Form (Net CDF) or American Standard Code for Information Interchange (ASCII) format for subsequent processing and analysis. If an experiment is required to complete data processing of the bulk samples, the

T.-l. Chen (✉) · R. Dai
Center for Translational Medicine, Shanghai Jiao Tong University Affiliated Sixth People's Hospital, Shanghai 200233, China
e-mail: chentianlu@sjtu.edu.cn

X. Qi et al. (eds.), *Plant Metabolomics*, DOI 10.1007/978-94-017-9291-2_6

workload of the step of data extraction has already been quite large, hence simple manual analysis is basically impossible. With the development of computer information system integration, automation of analysis techniques has greatly improved the performance and efficiency of instruments. Taking the example of GC-MS from Perkin Elmer Ltd., in USA, the data extraction first takes the integration via the Turbomass software of the instrument then matches each peak in the spectrogram by retention time and the mass-to-charge ratio to generate complete analysis data (Qiu et al. 2010). Researchers also use the DataBridge function of Turbomass software to convert spectrogram data from other instruments into Net CDF or ASCII formats. Of course, such method of operating each spectrogram separately is time-consuming and is not optimal for batching a large number of samples. Most commonly used instruments, such as gas chromatography time-of-flight–mass spectrometry (GCTOF-MS) by Leco Company, ultraperformance liquid chromatography coupled to quadrupole-time-of-flight–mass spectrometry (UPLC-QTOF-MS) by Waters Company and so on, are all equipped with specialized data saving and exporting modules, so that the spectrograms of batch samples can be converted into standard Net CDF or ASC II format (Qiu et al. 2009, 2010; Xie et al. 2010). As the development of metabolomics, the self-embedded functions of batch data extraction and transformation of various instruments are enjoying rapid development and improvement.

6.2 Data Preprocessing and Peak Extraction

Metabolites in biological samples are quite sensitive to the environment; in the meantime, there exist errors during the process of sample collecting, storing, production, and preparation, as well as in procedures of equipment testing (Qiu et al. 2007). Therefore, in order to eliminate the impacts of external factors or human disturbances, appropriate methods should be applied to improve data quality and enhance stability. On the other hand, the huge amounts of data extracted from the spectrogram also bring to statistical analysis a great burden. Therefore, there is need to further extract useful information (such as peak intensity) from spectrograms for subsequent analysis (Ni et al. 2008). The step of preprocessing and peak extraction is the process of generating a peak data set/table via transformation and integration of spectrogram data. Such data set consists of a series of peaks identified by three-dimensional marks of retention time, mass-to-charge ratio, and signal strength. While the amount of data is significantly reduced, its complexity will not be greatly decreased due to the inextricably inherent relationship between the peaks and the samples. In addition, most metabolomics studies understand physiological processes information through comparisons of spectrograms of a large number of sets or various kinds of samples; meanwhile, they have to deal with multiple samples simultaneously. Therefore, data preprocessing and peak extraction of gas chromatography–mass spectrometry in metabolomics are much more complex than the traditional single sample processing (Richard 2003; Jonsson et al. 2004, 2005).

(a)

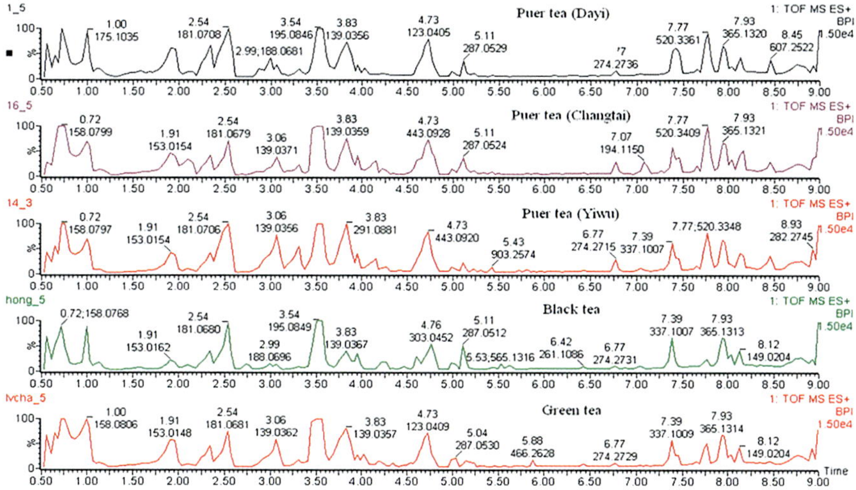

(b)

index	Ret. Time	Mass	1_1	1_2	1_3	1_4	1_5	2_1	2_2	2_3	2_4	2_5
1	1.7837	227.0807	67.2972	66.7161	35.9132	45.5696	42.5316	42.5316	42.0033	28.8429	50.6329	28.3544
2	0.874	165.0555	105.3165	173.2848	167.3102	154.2169	159.8553	141.2271	133.2032	115.3646	146.7635	124.8468
3	1.311	165.0557	79.2089	137.9526	134.3422	133.8036	147.874	140.8879	0	113.2214	127.0766	0
4	1.7786	165.0121	48.6076	49.288	53.455	47.1909	47.4788	58.7203	51.5294	47.5949	50.6329	45.5696
5	3.2915	261.0779	72.7762	60.0789	60.2773	55.1681	46.6206	66.2203	49.6876	46.4471	52.2433	87.3202
6	0.7185	261.0124	53.6709	42.5316	43.1447	43.5443	52.2398	52.2206	56.1724	49.0837	41.519	52.6582
7	0.8747	133.0318	55.2457	103.2103	84.7257	76.6733	83.1945	54.8433	53.4719	60.7595	108.2279	72.2194
8	2.7233	133.09	0	0	0	0	0	0	0	0	0	38.4618
9	9.3955	133.0865	10.9625	20.1943	0	22.3206	0	0	29.669	0	0	0
10	2.13	105.0342	0	52.355	0	0	0	32.4051	20.2212	0	0	32.9365
11	4.7369	105.0324	53.848	0	39.8943	43.9394	0	0	54.5603	60.962	56.5558	59.0219
12	2.8669	270.1457	71.3441	71.8987	60.2846	73.3823	86.0214	76.962	68.5364	49.1295	88.4323	78.7875
13	4.0612	267.0158	27.3418	18.2278	0	0	26.3291	21.2658	16.381	26.0224	0	24.02
14	0.6618	134.0986	36.4467	22.0769	20.9264	15.4378	21.1692	29.1784	23.2911	26.7364	15.4927	25.682
15	1.7782	205.0981	645.4165	935.9224	583.6372	695.3232	671.2256	523.6299	599.2124	554.145	684.2383	597.5005
16	5.0731	205.0876	64.4747	69.1493	81.5714	73.9913	76.8387	66.6495	74.9957	80.5263	72.2197	79.5167
17	3.4754	365.2766	30.7105	21.7191	32.7663	24.665	16.0476	19.4577	14.1772	0	0	0

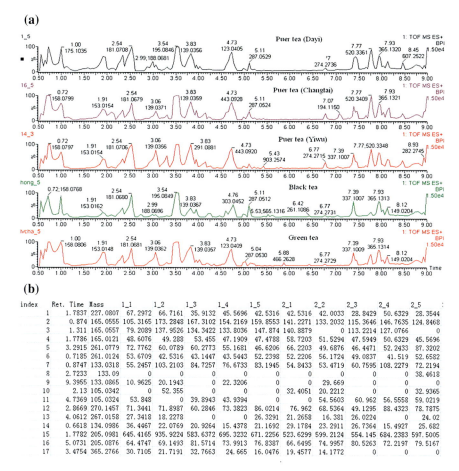

Fig. 6.1 Input (**a**) and output (**b**) of typical metabolomics data (from the UPLCQTOFMS instrument) preprocessing *X* axis: time; *Y* axis: strength; ES+: positive ion mode; BPI: base peak (base peak intensity; Puer tea (Dayi): Pu-Erh tea (benefit); Pu-Erh tea (Changtai): Pu-Erh tea (Changtai); Pu-Erh tea (Yiwu): Pu-Erh tea (Yiwu)

As shown in Fig. 6.1, the inputs of metabolomics data preprocessing are usually raw data from MS or chromatography (such as a large number of LC-MS data of the original format), the outputs are usually peak features in the form of tables. Each row is a peak (could be a substance, noise, or impurities), and each column is a sample.

Typical metabolomics preprocessing primarily consists of singularity elimination, denoising and baseline alignment calibration, peak alignment, peak identification, peak feature extraction, and results summarizing (Fig. 6.2) (Trygg et al. 2007). Data processing methods and theories are complex and diverse, and can be applied to almost every field. According to different types of analytical instruments,

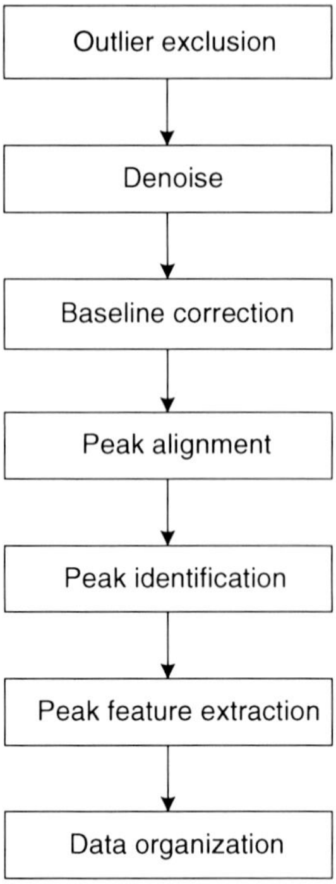

Fig. 6.2 Typical procedures of metabolomics data preprocessing

data characteristics, and requirements, various preprocessing procedures and methods should be applied to maximize the reliability of finally provided information of the peaks.

6.2.1 Characteristics of Mass Spectrometry and Chromatogram Data

Understanding the characteristics of data to be processed helps selecting appropriate approach. Metabolomics MS and chromatogram data usually consists of large numbers of peaks with a lot of noise and aliasing. There are several ways of digital description of spectral peaks, such as the center location of peaks (or the location of peak point), the peak height, peak width (or half-peak width), the peak area, and so on. Among which, the peak height and peak area are most commonly used. The relationship between these two characteristics of peaks largely depends on the shape of peaks. Most cases can be covered by three typical peak shapes, which are described by the combination of two fitting methods (Richard 2003).

6.2.1.1 Gaussian Peak Fitting

Gaussian shape is a most common form of peak shape. The mathematical expression for Gaussian shape is

$$x_i = A\ \exp\left[-(x_i - x_0)^2 \Big/ S^2\right] \tag{6.1}$$

where x_i denotes every sampling point; x_0 is peak time; A is the peak height at time x_0; S is a parameter related to the peak width.

In fact, Gaussian function is consistent with normal distribution. Here, x_0 is equivalent to the mean, and S can be regarded as the standard deviation. It can be calculated that half-peak width is $\Delta_{1/2} = 2S\sqrt{\ln 2}$, and the peak covers an area of $\sqrt{\pi}AS$.

6.2.1.2 Lorentzians Peak Fitting

Lorentzians peak corresponds to statistical functions of Cauchy distribution. It is a most prevalent one apart from Gaussian peak, especially in the NMR data. The simplified math expression is given as

$$x_i = A\Big/\left[1 + (x_i - x_0)^2/S^2\right] \tag{6.2}$$

where A is the height of central location of the peak; x_0 is the central location; S is a parameter related to the peak width, so it is similar with the Gaussian shape. The half-peak width is $\Delta_{1/2} = 2S$, and the covered area is πAS.

Gaussian and Lorentzian shapes are shown in Fig. 6.3. It can be seen that when the two are of the same half-peak width, Lorentzian shape has wider tails.

There are three typical spectral peaks. Apart from the standard Gaussian shape, there are two other shapes corresponding to the combination sequence of Gaussian and Lorentzian shapes, as shown in Fig. 6.4, first half of (a) is Gaussian-shaped, the latter part is the Lorentzian; on the contrary, the first half of (b) is of Lorentzian shape and the second is Gaussian-shaped.

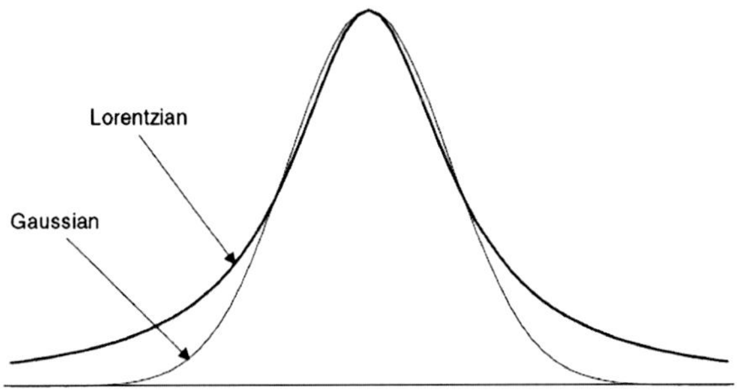

Fig. 6.3 Typical Gaussian and Lorentzian shapes

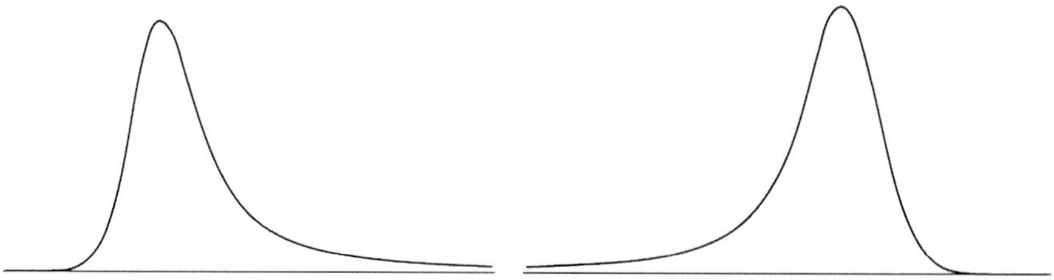

Fig. 6.4 Typical non-Gaussian peak shape

Fig. 6.5 An example of overlapping peaks

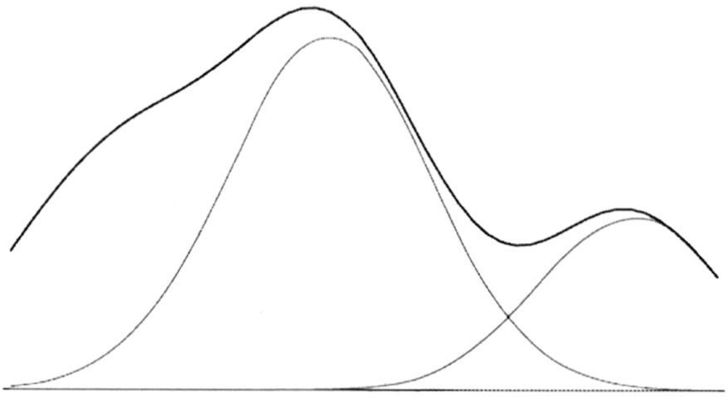

Due to the characteristics of biochemical substances and the limit of instrument resolution, many peaks are usually overlapped with each other. As illustrated in Fig. 6.5, the final spectral peak is actually formed by the superimposition of three separate peaks. Direct-extracted characteristics of peak height and peak area are not capable of representing the actual conditions of a sample if without peak separation, and this will be inimical to subsequent pattern recognition and biochemical explanation. At present, the identification and separation of overlapping peaks have become a bottleneck for data processing of high-throughput analytical instruments.

6.2.2 *Basic Steps and Methods of Preprocessing and Peak Extraction*

As shown in Fig. 6.2, the basic steps of preprocessing (including peak extraction) include singular point elimination, denoising and smoothing, baseline alignment calibration, peak alignment, peak identification, peak feature extraction and result summarizing, and so on.

6.2.2.1 Singular Point Elimination

If spectrograms of some specific samples are significantly different from others in the same group or one spectrogram almost have no peak, then after taking into consideration the sample information and experimental operations, they can be regarded as singular points to be removed and will not participate in subsequent processing. While eliminating singular point in favor of generating optimal analysis results of the desired data, this step should be taken with caution to avoid accidental deletion of important information or ignorance of an "unconventional" phenomenon which is of research value.

6.2.2.2 Denoising and Smoothing

In biochemical experiments, noise and errors are unavoidable due to imperfections of the instruments, sample preparation procedure, and means of measurement.

Noise in data includes measurement noise and random noise. Denoising and smoothing are an essential step in data processing. Denoising methods for MS and chromatography data include threshold limit, matched filtering, moving window average (we generally take five or seven points) filtering, median filtering, Savitzky–Golay polynomial fitting, discrete wavelet transform methods, and so on. Theories of matched filtering and moving window average filtering are relatively simple, and lots of software are embedded with such function, so they are more widely applied.

6.2.2.3 Baseline Alignment Calibration

Ideally, all spectral intensity values should be based on zero, in other words, when there is no material being eluted, the mentioned intensity values should be zero. However, due to the present inadequate measurement methods, instrument being contaminated or its features being unstable, the actual detected spectral intensity is often greater than zero, hence there is need to calibrate its baseline. Commonly used baseline calibration method is to subtract the minimum value from all the data values in the same spectrogram so as to ensure the baseline to be zero. This is of great help in subsequent work of searching for the starting and finishing points of peaks.

6.2.2.4 Peak Alignment

With the development of detection theory and electronics techniques, the accuracy, stability, and repeatability of analytical instruments have significantly improved. However, it is certain that increased number of samples leads to an increase of measure time; consequently, the accumulated time offset has significant impact on data processing. Peak alignment is such step to fix the time offset so that the retention times of peaks representative of the same substance are unanimous. This facilitates batch processing and statistical analysis, the extraction of meaningful variables with comparability.

Peak alignment methods can be divided into two categories. One is to artificially insert several standard components during the sample preparation procedure, identify the peaks of these standard components, and then adjust the retention times with piecewise linearity in the process of preprocessing. The other is completely dependent on mathematical methods to make alignment after data acquisition. Such method can be subdivided into linear and nonlinear alignments. The representative method of linear alignment is similar to that of inserting standard components, one can select peaks that appear in all samples and whose retention times and intensity are comparatively ideal as the mark substances, and then adjust various samples within each section; or one can select a sample as reference, and then adjust other samples to align with the reference. Nonlinear alignment methods include correlation optimized warping (COW) by Nielsen, covariance method by Par Jonsson, fast Fourier transform method by Jonsson et al. (2005), and so on. COW uses several optimal parts of windows found through correlation to calibrate, but this requires time offset between samples that was not greater than the time interval between two adjacent peaks. Covariance method makes peak alignment by finding the maximum covariance in the spectrogram; it is very effective when samples have great similarity. In current metabolomics studies, spectrograms of contrast samples are usually very similar, and most of the main peaks appear in all samples, thus the covariance method is widely used.

6.2.2.5 Peak Identification

Peak identification is defined as the determination of the starting and the finishing points of the peak. Generally, one could use the method of derivation (the turning point where the second derivative is zero) or the second-order Gauss filtering.

6.2.2.6 Peak Feature Extraction

Peak feature extraction mainly means to extract peak height or to calculate the area covered by the peak. For calculation of the covered area, one can use the rectangular integration method, the trapezoidal integration method, or the Romberg integration method, which is of higher accuracy (trapezoidal integration with variable step size).

6.2.2.7 Characteristic Variable Consolidation

This part is also introduced in previous chapter "metabolomics data processing based on NMR." Readers who have read this part can go directly to Sect. 6.2.3.

(1) *The integrity of data* missing data will have certain impacts on the data analysis results, so data should be completed in advance. It is more appropriate

to replace the missing data with the average value of all samples in the group rather than to merely use the zero value. It can be removed from the original data if a characteristic variable is highly relevant with other variables, or if there are some apparently redundant variables, constant variables, and so on.

(2) *Normalization* in some cases, normalizing a data vector is a simple and efficient data preprocessing step (Trygg et al. 2006, 2007), the three commonly used types of normalization are as follows:

① The norm of vector can be normalized to 1:

$$x_{ik}{}^{*} = \frac{x_{ik}}{\|x_k\|} \tag{6.3}$$

where x_{ik} is the source data vector, $x_{ik}{}^{*}$ is the normalized data vector, and the norm of vector is $\|x_k\| = \sqrt{x_{1k}^2 + x_{2k}^2 + \cdots + x_{nk}^2}$.

② Area can also be normalized to 1:

$$x_{ik}{}^{*} = \frac{x_{ik}}{\sum_{i=1}^{n} x_{ik}} \tag{6.4}$$

③ Maximum value can also be normalized to 1:

$$x_{ik}{}^{*} = \frac{x_{ik}}{\max(x_{ik})} \tag{6.5}$$

(3) *Centralization* in order to eliminate the constant offset of the data, one can coordinate the origin. General step is the mean centralization, which uses the following formula: for each variable x_{ik}, subtract it with its mean value x_k.

$$x_{ik}{}^{*} = x_{ik} - \bar{x}_k \tag{6.6}$$

$$\bar{x}_k = \frac{1}{n}\sum_{i=1}^{n} x_{ik} \tag{6.7}$$

(4) *Simple transformation* to convert the raw data into one easier to deal with, or adjust its distribution, one can process the raw data with operations of multiplication, division, addition, subtraction, or log algorithm.

(5) *Scaling* variables represent different natures of samples or objects. In most cases, these natures vary significantly, thus values between the columns may be of great difference; in other words, different variables have different absolute values and ranges (variances). Both cases will affect multivariate methods based on statistics. However, these differences can be eliminated by proportion adjustment. There are two main methods of proportion adjustment: One is called range scaling which makes adjustment according to the range of values; the other is the method of automatic adjustment, which is based on standard deviation.

① range scaling

$$x_{ik}^{*} = \frac{x_{ik} - x_k(\min)}{x_k(\max) - x_k(\min)} \quad 0 \le x_{ik}^{*} \le 1 \tag{6.8}$$

② automatic adjustment

$$x_{ik}^{*} = \frac{x_{ik} - \overline{x_k}}{s_k} \tag{6.9}$$

$$s_k = \sqrt{\frac{\sum_{i=1}^{n} \left(x_{ik} - \overline{x_k}\right)^2}{n - 1}} \tag{6.10}$$

where n denotes the number of objects, S_k is the standard deviation. Automatic adjustment makes adjustments to the data so that the mean equals 0, variance is 1, and the norm of vector becomes $\sqrt{n-1}$.

Typically, one may take uniform treatments to all data and choose the automatic adjustment method. Also, for data with comparatively large noise, one can make adjustments using the formula $x_{ik}^{*} = \frac{x_{ik} - \bar{x}_k}{\sqrt{S_k}}$. Appropriate ignorance of parts of low intensity "noise" data will help to improve the signal-to-noise ratio (SNR) of the data set.

After the above-mentioned adjustment methods, the complexity of biological samples, the human and instrument error involved during preparation process and testing procedure will have a certain degree of improvement or decrease. For the long time testing of large bulk samples, the impacts of status of detection instrument on the results cannot be ignored. In order to further increase the credibility of the results, one can apply some quality control (QC) strategies (Frans et al. 2009) when necessary. Commonly used QCs are appropriately added blank, pooled standards, or pooled samples during the injection process. According to the samples and compound properties, choose one or two known substances and uniformly add a certain amount to each sample. Because the amounts of these standard substances (internal standard) in all samples are the same, theoretically their peak intensities should be the same. As a result, the change of peak intensity can be used to monitor equipment status and the error introduced during sample preparation. Normalizing other peaks in the sample by the internal standard peak can, to a certain extent, eliminate these errors. In addition, some research institutions insert multiple pins of mixing samples during the injection process to monitor instrument status. Although such method increases the burden of testing and analyzing, there is room for extension and improvement and it is noteworthy.

6.2.3 Overlapping Peak Identification

In actual detection spectrograms, because of the close retention time of material or excessive peak width, peaks representing a wide range of substances often overlap with each other and cannot be artificially identified or separated. In order to generate accurate and reliable results, one has to separate each peak from the overlapping peak and then extract their characteristics, which is where the difficulty of peak extraction is.

Most early algorithms of peak identification are based on statistical distributions and multi-order derivative smoothing, the effect of which is not ideal. Then, there develops the deconvolution method. For example, such method is adopted in both the free software automated mass spectral deconvolution and identification system (AMDIS) developed by National Institute of Science and Technology (NIST) and Chromat by Leco Company. However, early deconvolution method is only suitable for dealing with a single sample; the speed of processing and the degree of automation are less than ideal. In recent years, some researchers attempted to use multivariate curve resolution (MCR), which basically achieves the identification and separation of overlapping peaks of bulk sample (Jonsson et al. 2005).

6.2.3.1 Deconvolution

Deconvolution is the inverse operation of convolution.

An original signal $x(n)$ passing through a system (with $h(n)$ as its eigenfunction) becomes the output signal $y(n)$, such process can be described by the convolution operation, that is,

$$y(n) = x(n) * h(n) \tag{6.11}$$

The inverse operation of convolution consists of the following two situations:

① Given output $y(n)$ and the eigenfunction $h(n)$ of the system, try to compute the original signal $x(n)$. This case refers to the overlapping signal extraction.

② Given output $y(n)$ and input $x(n)$, try to compute the eigenfunction $h(n)$ of the system. This is used to study the system performance.

According to the convolution theorem "convolution in the time domain equals multiplication in the frequency domain," the process of deconvolution is stated as follows.

① Convert the given output $y(n)$ and input $x(n)$ (or the eigenfunction $h(n)$ of the system) into the frequency domain by Fourier transform, so that one gets the corresponding frequency response $Y(W)$, $X(W)$, and $H(W)$. Their relationship is

$$Y(\omega) = X(\omega) \cdot H(\omega) \tag{6.12}$$

② Via $X = Y/H$ or $H = Y/X$, one gets the frequency response of the original signal or the eigenfunction of the system.

③ Take the inverse Fourier transform to generate the original signal or the eigenfunction of system in the time domain.

The matrix description of deconvolution is stated below.

Convolution of a discrete-time system can be expressed as

$$y(n) = \sum_{m=0}^{n} x(m)h(n-m) \ (n \text{ is the sampling time}) \tag{6.13}$$

Expand it as

$$\begin{bmatrix} y(0) \\ y(1) \\ y(2) \\ \cdots \\ y(n) \end{bmatrix} = \begin{bmatrix} h(0) & 0 & 0 & \cdots & 0 \\ h(1) & h(0) & 0 & \cdots & 0 \\ y(2) & h(1) & h(0) & \cdots & 0 \\ \vdots & \vdots & \vdots & \vdots & \vdots \\ h(n) & h(n-1) & h(n-2) & \cdots & h(0) \end{bmatrix} \begin{bmatrix} x(0) \\ x(1) \\ x(2) \\ \cdots \\ x(n) \end{bmatrix} \tag{6.14}$$

where $x(i)$ is the original signal at time i.

Deconvolution is the reverse process that calculates $x(n)$ and $h(n)$.

It can be seen from the expanded matrix that $x(0)$ only depends on $y(0)$ and $h(0)$. $x(n)$ depends on y and h at time n and all time before n. As a result,

$$\begin{aligned} x(0) &= y(0)/h(0) \\ x(1) &= [y(1) - x(0)h(1)]/h(0) \\ x(2) &= [y(2) - x(0)h(2) - x(1)h(1)]/h(0) \\ &\cdots \\ x(n) &= \left[y(n) - \sum_{m=0}^{n-1} x(m)h(n-m)\right] \bigg/ h(0) \end{aligned} \tag{6.15}$$

Similarly, $h(n)$ is given as

$$h(n) = \left[y(n) - \sum_{m=0}^{n-1} h(m)x(n-m)\right] \bigg/ x(0) \tag{6.16}$$

The graphical representation of the function of deconvolution is shown in Fig. 6.6. The total ion chromatographic peaks on the left (black solid line) actually consist of three separate but superimposed components (marked by green, red, and blue colors, respectively). The overlay effect in mass spectrogram has also been

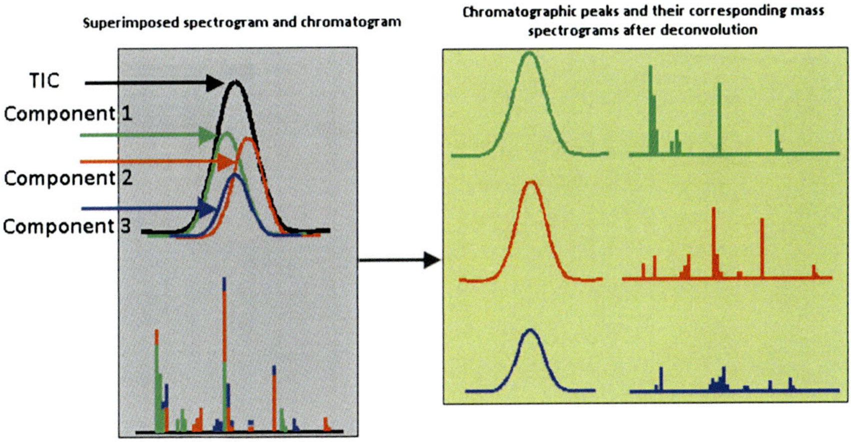

Fig. 6.6 Deconvolution diagram (on the *left* is the original, superimposed spectrogram and chromatogram, on the *right* is the separated chromatographic peaks and their corresponding mass spectrograms after deconvolution)

marked with the corresponding colors. After deconvolution, three independent chromatographic peaks and the corresponding mass spectrogram (matrix, interference, and target) will be separated out as well. The number of separated components is determined by the rank of the matrix.

6.2.3.2 Multiple Curve Resolution

Large sample volume and complex three-dimensional data in metabolomics require advanced data analysis methods. Traditional advanced data analysis methods based on iterative fitting, eigenvalues, or eigenvectors require the data to meet the trilinear eigenstructures, otherwise a reasonable solution cannot be obtained. In common experiments, acquisition of data that completely meet the trilinearity is very hard. In 1993, Tauler proposed multiple curve resolution (MCR), which is based on alternating least squares iterative optimization, and can be applied to non-trilinear data analysis (Garrido et al. 2008). This method uses the results derived from evolving factor analysis method as initial values of alternating least squares iterative optimization and enables simultaneous analysis of multiple non-trilinear data matrices, so it is widely applied in several fields, such as chemical component identification, overlapping peaks identification, higher-order data resolution analysis, and chemical imaging analysis. Advantages of this method have become increasingly prominent in metabolomics data preprocessing.

The mathematics idea of MCR is shown in Fig. 6.7. The first row shows the matrix description Data = $CS^T + E$, while in the second row, there is a graphical representation. For a matrix D consisting of three-dimensional data (within a certain period of retention time Windows) made from N samples (sample 1, sample 2, …, sample N),

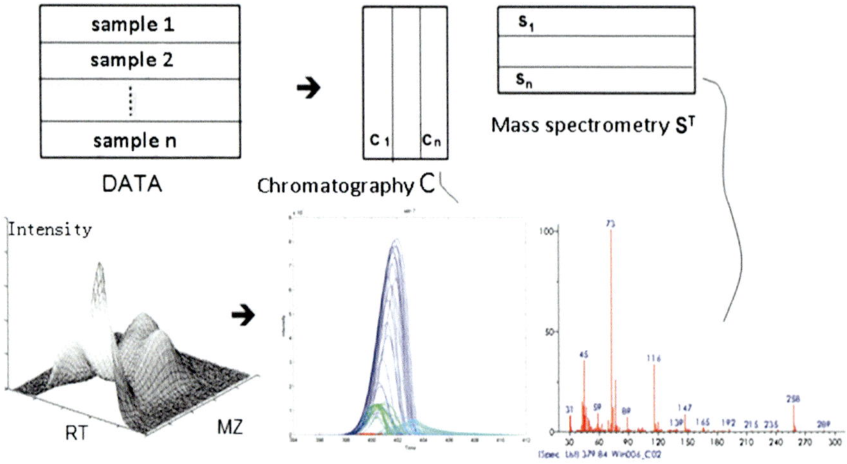

Fig. 6.7 Main idea of MCR (Data = $CS^T + E$)

one can obtain via MCR two matrices C and S. C is the matrix of separated chromatographic peaks; each column is a peak (each independent component marked by a different color); S^T is the mass spectrogram data matrix that corresponds to each chromatograph; each row is a mass spectrogram data. The number of columns in C equals to that of rows in S^T; it represents the number of extracted peaks.

Key steps of MCR mainly include singular value decomposition (SVD) that determines the number of independent components, initial value estimation, and optimization. The number of eigenvalues resulting from the SVD is the number of independent components to be separated. Generally, the initial value of S is the score vector of the first principal component derived from principal component analysis (PCA). ALS optimization is a typical conditional extreme value problem. The two constraints are as follows (D stands for data, C and S have the physical meaning of chromatography and MS, respectively).

$$\min_{\widehat{C}} \left\| \widehat{D}_{\mathrm{PCA}} - \widehat{C}\widehat{S}^T \right\|$$

$$\min_{S^T} \left\| \widehat{D}_{\mathrm{PCA}} - \widehat{C}\widehat{S}^T \right\|$$

MCR can be divided into three types: iterative, non-iterative, and hybrid. Among which, the non-iterative and most hybrid approaches are difficult to achieve automation, hence they are not suitable for data analysis of a large number of samples in metabolomics. Iterative approaches have alternating regression (AR) and target transformation factor analysis (TTFA) as their representatives. Some researchers have attempted to apply AR to the extraction of overlapping peaks in chromatograph and mass spectrogram, and the corresponding software HDA is also being further perfected and upgraded.

Compared with the deconvolution method, algorithm using MCR as its core is able to handle a large number of samples simultaneously. Besides, looking at the big picture, such algorithm can interpret sample data from two-dimensional aspects of retention time and the mass-to-charge ratio, respectively, and meet metabolomics application demand. In addition, according to the principle that fluctuations of substance concentration in the same set of samples are small, such method is able to identify and separate two substances (peaks) even if their retention times are all the same, because it has comprehensive information of multiple samples that are highly correlated. This is what all overlapping peak processing methods are unable to achieve in the past.

6.2.4 Software Tools

According to the developer and level of openness, software achieving preprocessing and peak extraction can be mainly divided into three types.

(1) developed by analytical instrument company, the instrument comes with packages, such as MassLynx (Fig. 6.8) by Walters Ltd, Chromatof (Fig. 6.9)

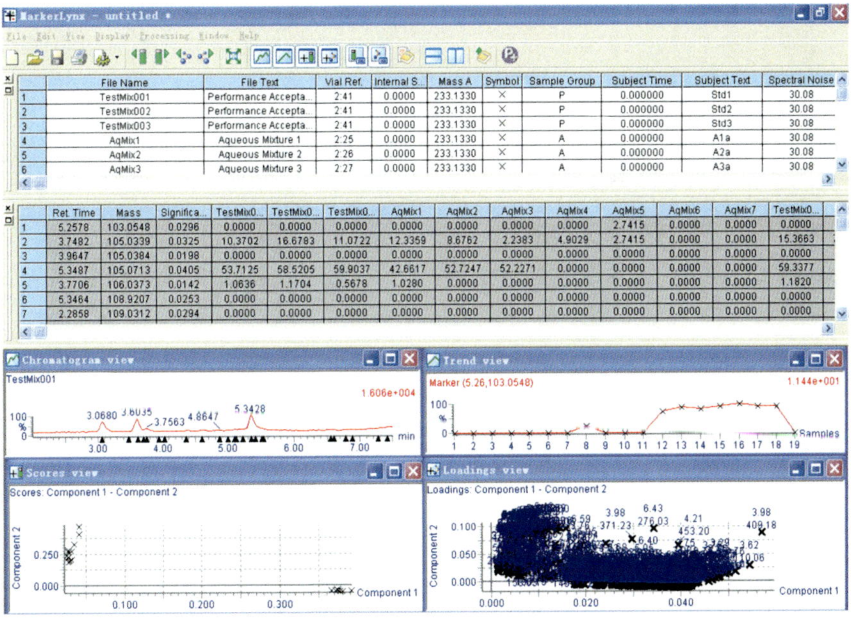

Fig. 6.8 MassLynx main interface

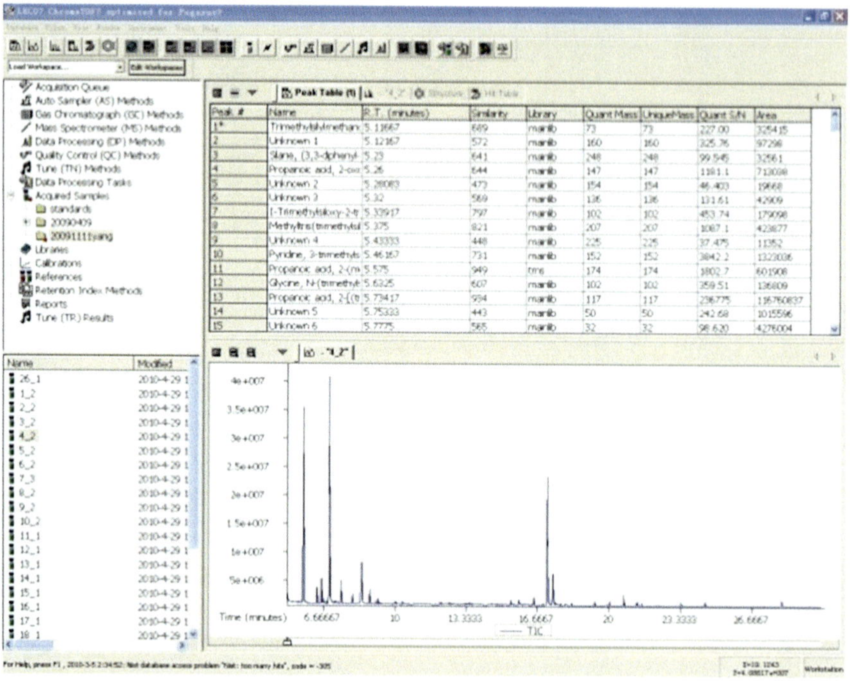

Fig. 6.9 Chromatof main interface

by Leco Company, as well as Turbomass by Perkin Elmer Ltd (manual peak extraction only), and so on. These softwares only support the use of the corresponding instruments; the internal algorithms are not open to public, and hence, they are subject to certain restrictions.

(2) Open, software packages available for free download, for example, XCMS based on the *R* platform (Fig. 6.10) (http://metlin.scripps.edu/download/), AMDIS developed by NIST, and TagFinder by the Max Planck Institute of Molecular Plant Physiology (http://www-en.mpimp-golm.mpg.de/03-research/researchGroups/01-dept1/Root_Metabolism/smp/index.html) (Fig. 6.11). AMDIS can only deal with a single sample. XCMS adds a function of bulk sample peak-matching apart from dealing with a single sample, so it is most widely used. The package was first developed for LC-MS data preprocessing and was then applied to other data processing. Because its source code is completely open, it can be modified to meet actual needs. However, such code is complex and requires more energy investment in the early stage, and modification is difficult. Besides, the package is developed on the basis of R platform, so it does not have a graphical interface, which causes difficulty in software upgrades and application. Based on Java, TagFinder is the software specifically developed for metabolomics data preprocessing and peak

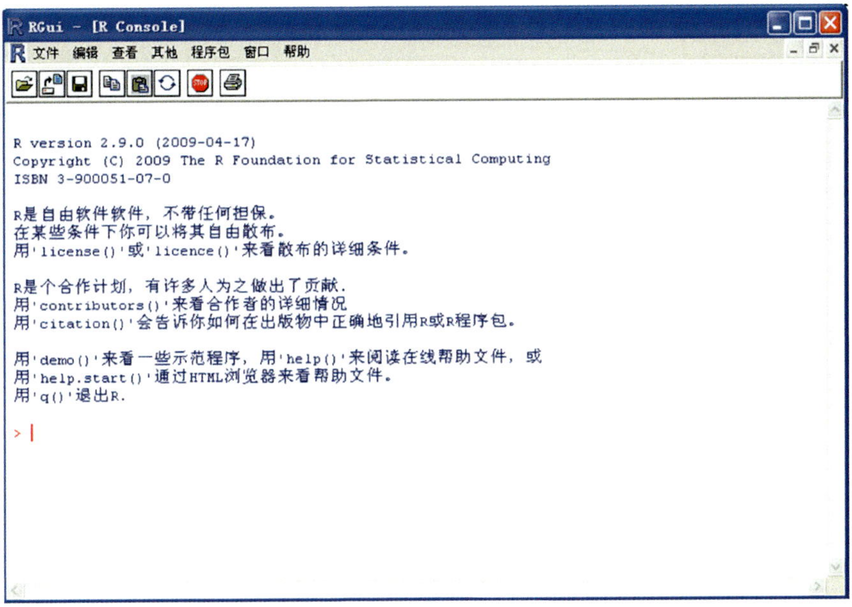

Fig. 6.10 *R* platform main interface

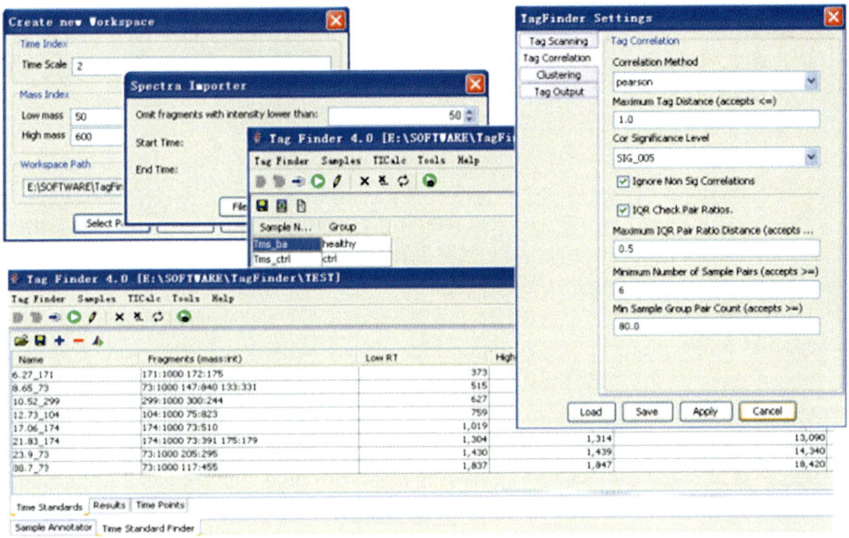

Fig. 6.11 TagFinder key interface

extraction. This software and the accompanying set of small software are functional and easy to operate and can be downloaded and used free of charge, so its number of users and the influence have been increased rapidly.

(3) home-developed software such as HDA developed by Umea (Umeå) University in Sweden, which is based on gas chromatography–time-of-flight mass spectrometry detector (GC-TOFMS) by Leco Company (Jonsson et al. 2004). The software is based on MATLAB, so the functions and interfaces can be changed according to actual needs, modifications and applications are very convenient. The source code has not been made public.

Depending on sample processing capacity, peak preprocessing and extraction software (packages) can be divided into two categories: Software that can only handle a single sample, such as Chromatof, AMDIS, and Turbomass; and the ones capable to deal with multiple samples, such as MassLynx, XCMS, TagFinder, HDA, and so on.

6.2.5 Multi-platform Data Integration

Although analysis and separation technology in metabolomics have enjoyed sound development, currently there is still no such analytic technique that can accurately describe all possible compounds in the samples under test. In view of the inadequacy of such analysis mean, more and more researchers obtain information from a variety of analytical techniques (GC, LC-MS, or NMR) and/or a number of biological samples (such as blood, urine, or tissue), hoping to take advantage of the different derivatives and methods for analysis of information, as well as the complement function among them to expand sights (Nicholson et al. 1999; Dunn and Ellis 2005; Qiu et al. 2009, 2010). Of course, this brings great challenge for subsequent processing and analysis. There are several reasons for this situation: First, the inherent differences in biological samples together with different analytical separation techniques and instrumentation differences inevitably produce significant differences in the set of data to be analyzed. Second, the theory and technical development of information fusion are still immature, the application in the field of bioinformatics is still in its initial stages; also, there are rarely previous metabolomics data processing examples for reference. Third, there are few research groups that have multiple mature metabolomics analysis platforms, neither is there exists a common database resource available for public download, so access to multiple sources of data is of certain difficulties.

Currently, for multiple platforms data, most researchers at home and abroad will process and analyze metabolomics data from a variety of biological samples separately on different platforms, identify differences in metabolites, respectively, and engage such results in general discussions. From the point of information fusion, such approach belongs to the decision-making level of integration. Through integration, one can capture changes in metabolism of the detected samples and also take advantage of the overlapping material detected by different platforms to make substance validation. However, this metabolite level of integration and validation uses only a very small proportion of all information; there is still plenty of available information that has not yet been exploited. In order to maximize the use of measurable information, some researchers gradually started the information fusion on a variety of raw data or peak data during the preprocessing and peak extraction stage. Although being complex, diverse, flexible, and difficult, this data layer or feature level of fusion has great potential for further development.

6.3 Pattern Recognition Based on Statistics

Pattern recognition is an important part of informatics and artificial intelligence, and it refers to the process of describing, identifying, classifying, and interpreting things or phenomena via processing and analysis of all kinds of (numeric, literal, and logically related) information that represents an object or a phenomenon. According to the difference of basic theories, pattern recognition methods are mainly divided into statistical methods and syntactic methods, and they are widely used in character and speech recognition, remote sensing, medical diagnosis, and so on. Pattern recognition of metabolomics information is mainly a statistical pattern recognition (including single- and multi-dimensional statistical pattern recognitions), which is early applied and more mature. Single- and multi-dimensional pattern recognitions are just mathematical approaches, and we recommend to look at the data from different perspectives in order to obtain more reliable results.

6.3.1 Single-Dimensional Statistical Pattern Recognition

Statistics divide variation in quantitative characteristics and its influencing factors into two categories: One is random changes caused by random factors; the other is systemic change by controlled factors. Commonly used in metabolomics, single-dimensional analysis methods such as *T* test, variance analysis, and Mann–Whitney examination are such methods and process that decompose the amount changes reflected by the data and take significance test under some significant level, so as to determine whether such changes are due to random changes caused by random factors or systemic changes introduced by controlled factors. Each method often assumes that samples are independent of each other, and data population meets the normal distribution.

6.3.1.1 *T* test and *U* test

T test and *U* test are hypothesis tests with statistic *T* and *U* and also are the most commonly used methods to compare the differences between the means of two groups (for example, the differences between comparison drug treatment group and the placebo treatment group). When samples are of large numbers and normally distributed, it is recommended to take U test. However, when the sample size is small (i.e., sample size is 10), T test is suggested as long as variables in each group are normally distributed and variances of both groups are not significantly different. Generally, T test is used in metabolomics to make preliminary judgment of two samples (both normal population) to check whether they have significant difference.

T test has the following basic procedures.

Suppose $X_1, X_2, \ldots, X_n$ are samples from normal population $N(\mu_1, \sigma^2)$, $Y_1, Y_2, \ldots, Y_n$ are that from normal population $N(\mu_2, \sigma^2)$, samples of both population are independent and their variances are equal (to verify this condition please refer to variance analysis section). Also assume that $\overline{X}, \overline{Y}$ is the overall sample mean, respectively, S_1^2, S_2^2 are sample variances, μ_1, μ_2, σ are unknown. The issue of determining whether there is significant difference between two samples equally means the question of calculating the rejection region of the assumption $H0$: $\mu_1 - \mu_2 = \delta$, $H0$: $\mu_1 - \mu_2 >$ (δ is a known constant that controls the size of differences). Significance level is denoted by α. Applying the T test and estimate statistics

$$t = \frac{(\overline{X} - \overline{Y}) - \delta}{S_w \sqrt{\frac{1}{n_1} + \frac{1}{n_2}}} \tag{6.17}$$

where

$$S_w^2 = \frac{(n_1 - 1)S_1^2 + (n_2 - 1)S_2^2}{n_1 + n_2 - 2} \tag{6.18}$$

When $H0$ is true, $t \sim t(n_1 + n_2 - 2)$, so its rejection region is

$$t = \frac{|(\overline{x} - \overline{y}) - \delta|}{S_W \sqrt{\frac{1}{n_1} + \frac{1}{n_2}}} \geq t_{\alpha/2}(n_1 + n_2 - 2) \tag{6.19}$$

According to the t value and T distribution table, one can tell whether the two populations are different (deny $H0$).

6.3.1.2 Variance Analysis

Analysis of variance (ANOVA) was first proposed by statistician R.A. Fisher from the UK in the year 1923 (Xu and Shao 2004; Kong and Zhang 2009; Shi et al. 2010). In memory of Fisher, it is named by F, so the variance analysis is also known as F test. ANOVA is a method to test whether differences between means of multiple sets of sample have statistical meaning. Strictly speaking, the ANOVA is not researching the variance; rather, it is the study of variation among data. In metabolomics, it is aided by this method in testing whether there exists difference

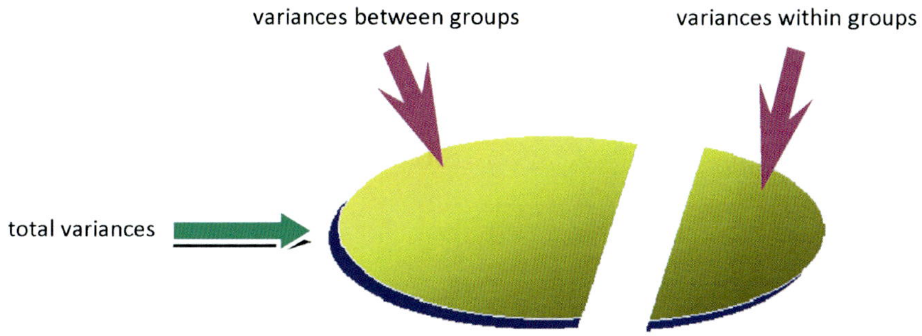

Fig. 6.12 Composition of total sample variations

between sample groups. Variance analysis includes single factor, two factors, multiple factors, and covariance analysis. Take the case study of the most commonly used single factor variance analysis as an example.

The basic idea of single factor variance analysis is as follows: according to their source of variation, divide the total variation of all samples into the ones which are within the group and others belongs to that of between groups (Fig. 6.12), and then make comparisons, evaluate whether the changes arising from some factor have statistically significant or not. The total variation (SS总) is group a, which represents the difference of overall mean (Y) of all measured values [$Y_{ij}(i = 1, 2, \ldots, a;$ $j = 1, 2, \ldots n_i)$] from n_i measurements, differences in variances between groups (SS组间) mean the differences between the mean of each group.

[$Y_i(i = 1, 2, \ldots a)$] and the overall mean (y), and the variance differences within group (SS组内) mean the differences between each measured value per group (Y_{ij}) and the group mean (Y_i). Usually, one can use the sum of squares (SS) to represent the size of variation. The mathematical expressions of three variations are as follows.

Total variation:

$$\begin{aligned} SS_{\text{total}} &= \sum_{i=1}^{a}\sum_{j=1}^{n_i}\left(Y_{ij}-\overline{Y}\right)^2 = \sum_{i=1}^{a}\sum_{j=1}^{n_i} Y_{ij}^2 - C \\ &= \sum_{i,j}^{N} Y_{ij}^2 - C = (N-1)S^2 \end{aligned} \tag{6.20}$$

where N denotes the sample size; S the variance; C the correction factor,

$$C = \frac{\left(\sum_{i=1}^{a}\sum_{j=1}^{n_i} Y_{ij}\right)^2}{N} = \frac{\left(\sum_{i,j}^{N} Y_{ij}\right)^2}{N} \tag{6.21}$$

degrees of freedom

$$V_{\text{total}} = N - 1 \tag{6.22}$$

variation between groups:

$$\text{SS}_{\text{between}} = \sum_{i=1}^{a} n_i \left(\overline{Y}_i - \overline{Y}\right)^2 = \sum_{i=1}^{a} \frac{\left(\sum_{j=1}^{n_i} Y_{ij}\right)^2}{n_i} - C \tag{6.23}$$

degrees of freedom

$$\text{V}_{\text{between}} = \alpha - 1(\alpha\,\text{is the number of groups})$$

$\text{SS}_{\text{between}}$ reflects the degree of variation of the mean of each group, and it is the result of the combined effect of random errors and influencing factors.

Variation within group:

$$\begin{aligned}\text{SS}_{\text{within}} &= \sum_{i=1}^{a} \sum_{j=1}^{n_i} \left(Y_{ij} - \overline{Y}_i\right)^2 \\ &= \sum_{i=1}^{a} (n_i - 1) S_i^2\end{aligned} \tag{6.24}$$

degrees of freedom

$$\text{V}_{\text{within}} = \text{N} - \alpha \tag{6.25}$$

$\text{SS}_{\text{within}}$ is represented by quadratic sum of the differences between each measured value Y_{ij} in various groups and the corresponding group mean, and it reflects the influence of random error.

The following relationship exists among the three variations:

$$\text{SS}_{\text{total}} = \text{SS}_{\text{between}} + \text{SS}_{\text{within}} \tag{6.26}$$

Degree of variation not only relates to the value of sum of squares, but it also is relevant to its degree of freedom. Because the degrees of freedom of various parts are not equal, their sum of squares cannot be compared directly. To deal with that, one needs to divide sum of squares by the corresponding degrees of freedom, such ratio is defined as the mean square, which in short, is called MS. Mean squares between groups and within group are calculated as:

$$\text{MS}_{\text{between}} = \frac{\text{SS}}{v} \quad \text{MS}_{\text{within}} = \frac{\text{SS}}{v} \tag{6.27}$$

The ratio of mean square between groups and that within group is defined as a statistic.

$$F = \frac{MS_{\text{between}}}{MS_{\text{within}}} \quad (6.28)$$

As F becomes closer to 1, the differences between and within groups become almost unanimous, which means that all the samples come from the same population, and they have no significant differences. The larger F is, more significant differences exist between groups.

There are two prerequisites for variance analysis:

① samples come from random, independent normal population.
② population variances of sample groups are equal.

Before use, it is necessary to confirm that the data set meets these two conditions; otherwise credibility of the results will be greatly reduced. Normal randomness can be validated by D method, W method, or the Chi-squared test, and homogeneity of variances can be tested by Bartlett or Levene test. Practically, a simplified method is to check whether the ratio of maximum and minimum variance is less than 3, if so, one can preliminary speak of variance homogeneity.

If the data set does not satisfy the above two prerequisites, there is need to make proper adjustments via transforms. Commonly used transforms include logarithmic transformation and arcsine square root transformation.

6.3.1.3 Mann–Whitney U test

The methods mentioned above belong to parameter tests. This type of tests has certain requirements on the distribution of data set population. In theory, the data to be processed should be subject to the normal distribution. However, the actual distribution is often unknown. Therefore, nonparameter test methods which have low requirements on the data set properties can be effective supplementary to parameters tests. Mann–Whitney U test (Mann–Whitney U test, also known as Mann–Whitney–Wilcoxon (MWW), Wilcoxon rank-sum or Wilcoxon–Mann–Whitney) checks the nonparametric rank S of two independent samples. When the distributions of two population are the same (not necessarily be normal), the work is to check whether the two have significant differences, in other words, whether two samples are from the same population. When both populations are of normal distribution, such method is the T test. However, when it is not possible to determine whether the two populations are the same, Z- or Wald-Wolfowits tests is recommended to examine the full picture of the data.

Unlike the T test and ANOVA, this method looks at the central location of both groups, the median, rather than the mean. In addition, the object of the method must be ordered data. As a result, the data need to be sorted before being used. Briefly, the steps are listed below (for the case of two groups).

① the sample groups are ordered and numbered by the numerical size and are recorded as vectors R_1 and R_2. Calculate the sum of numbers ΣR_1 and ΣR_2. The group with larger sum corresponds to R_2.

②

$$\text{calculate statistic} \quad U = N_1 N_2 + \frac{N_2(N_2+1)}{2} - \sum R_2 \tag{6.29}$$

where N_1 and N_2 denote the number of samples in the two groups.

③ take simple transformation to U

$$Z = \frac{U - m}{\sigma} \tag{6.30}$$

where $m = \frac{N_1 N_2}{2}$, $\sigma = \sqrt{\frac{N_1 N_2 (N_1 + N_2 + 1)}{N}}$ (N is the number of samples).

④ refer to the distribution table to determine whether the two groups are significantly different or not.

6.3.1.4 Simple Correlation Analysis

Generally, it is considered that the concept of correlation is proposed by Francis Galton during the year 1877–1878. But Karl Pearson is the one who actually systematized this theory. It is because of Pearson's excellent work that makes the theory to shine and obtain a wide range of applications. In honor of his contribution, correlation coefficient used in simple correlation analysis (Pearson correlation) is also called Pearson correlation coefficient.

When the scatters of two continuous variables X and Y in the scatter diagram show a trend of straight line, it can be considered that they have linear correlation trend, also known as simple correlation trend. Pearson correlation coefficient, also known as coefficient of production-moment correlation, is a commonly used indicator of the degree of linear dependence. Correlation coefficient is in fact the standardized covariance, where SSXY is the covariance of X and Y, $SSXX$ and $SSYY$, respectively, represents variance of X and Y. The value range of r is $[-1, 1]$. The value reflects the degree of correlation; positive and negative represent correlation directions.

In metabolomics, one can use the Pearson coefficient to establish certain links between biological and/or physical parameters and the metabolites of samples, also, it can be used to mark the relevance between parameters and metabolites, hence to provide the basis for redundancy analysis and metabolic pathways study. On this basis, there developed canonical correlation analysis (CCA) which is used for multivariate correlation analysis. That part will be introduced in the section of multi-dimensional statistical pattern recognition.

6.3.1.5 Receiver Operating Characteristic Curve

Receiver operating characteristic (ROC) curve, also known as sensitivity curve, is a widely used statistical method. In the field of medicine and botany, usually one takes the covered area under the curve as the evaluation indicator to assist researchers compare a variety of test results and to find out the best diagnosis or classification (Bradley 1997).

For two sets of plant species diagnosed by a gold standard, take a new classification analysis, and then, the results can be summarized as several types in Table 6.1.

$$\text{True positive (sensitivity)} = 100\,\%\,{}^{*}\,\mathrm{a}/(\mathrm{a}+\mathrm{c}) \quad (6.31)$$

$$\text{True negative (specificity)} = 100\,\%\,{}^{*}\,\mathrm{d}/(\mathrm{b}+\mathrm{d}) \quad (6.32)$$

$$\text{Fault positive (False rate)} = 100\,\%\,{}^{*}\,\mathrm{b}/(\mathrm{b}+\mathrm{d}) \quad (6.33)$$

$$\text{Fault negative (Missing rate)} = 100\,\%\,{}^{*}\,\mathrm{c}/(\mathrm{a}+\mathrm{c}) \quad (6.34)$$

Take "1-specificity" as horizontal axis, "sensitivity" as vertical axis, portray the line which links all data points whose coordinate is (1-specificity and sensitivity) in rectangular coordinate system, the resulting curve is the ROC curve. According to principle of ROC curve, an excellent analysis experiment should have the ROC curve with a shape of from the lower left vertical ascents to the top line, and then extends to the upper right in the horizontal direction. The case when ROC curve distributes along the diagonal direction represents that classification is random, the probability of correctness and error is 50 % respectively, which points out that current analysis method is null and void. If two curves do not intersect, the merits of the two experiments can be compared based on their manifestations. The curve that stays outer and farther away from the diagonal covers larger area. Its corresponding experiment has higher sensitivity and specificity than that of the inner curve which is closer to the diagonal.

In metabolomics, one can also use the area under the ROC curve to represent the effectiveness of a metabolite or metabolites group in distinguishing two kinds of plants.

In addition to what is introduced in this book, commonly used one-dimensional statistical testing methods also include Spearman Factor, Fisher accurate test, Kruskal–Wallis test, which is used for two or more groups of samples and so on.

Table 6.1 Diagnosis of plant varieties unit

Experiments	Plant 1 (number)	Plant 2 (number)	Total (number)
Positive	a	b	$a + b$
Negative	c	d	$c + d$
Total	$a + c$	$b + d$	$a + b + c + d$

6.3.2 *Multi-Dimensional Statistical Pattern Recognition*

In metabolomics studies, it is always necessary to observe several variables (peaks), which reflect the samples, and to collect data from all aspects in providing a more comprehensive analysis hence to search for patterns. Multivariable and large amount of samples will no doubt provide a wealth of information for scientific research, but also to some extent increase the workload of data acquisition. What is more important is that in most cases, the possible correlation between variables will increase the complexity of analysis. Although data will be preprocessed, all kinds of disturbances are still present and cannot be ignored. Multi-dimensional statistical pattern recognition with high fault tolerance naturally becomes a suitable tool for metabolomics analysis. According to the nature of the problem and the solution, multi-dimensional statistical pattern recognition can be divided into two categories: supervised classification and unsupervised classification.

6.3.2.1 Unsupervised Classification

Unsupervised pattern recognition method only looks at the data itself, it checks the holistic nature of the data and the internal variables association without any external guidance (Ebbels and Cavill 2009).

(1) *Principal component analysis and factor analysis* because of the correlation between various variables, it is possible to extract all kinds of information from the variables via less composite indicators. Principal components analysis and factor analysis are such a method of dimension reduction. Principal components analysis is a special case of factor analysis (Xu et al. 2001; Zhang 2004; Hendriks et al. 2007; Trygg et al. 2007).
Also known as K–L (Karhunen–Loeve) transformation, principal components analysis explains the variance and covariance structure of a set of variables according to their linear combinations, so as to achieve the purpose of data compression and interpretation (Jolliffe 2002). The less principal components have been chosen, the better result of dimension reduction there will be. However, more components indicate a better representation of the features of original data set. As a result, the selection standard is the ratio of the sum of the length of the spindle, which represents the chosen principal components and the total length of the spindle. Typically, selected total length covering approximately 85 % of total spindle length is an ideal case. In fact, this is only a general standard, it is necessary to make choices according to actual situation.
The mathematical formulation of principal component analysis is as follows. The idea of factor analysis is almost similar, so we will not repeat it.

Assume there are n biological samples; each sample has p variables, so there forms an $n \times p$ data matrix

$$X = \begin{bmatrix} x_{11} & x_{12} & \cdots & x_{1p} \\ x_{21} & x_{22} & \cdots & x_{2p} \\ \vdots & \vdots & & \vdots \\ x_{n1} & x_{n2} & \cdots & x_{np} \end{bmatrix}$$

When p is comparatively large, testing issues in p-dimensional space is troublesome. In order to overcome this difficulty, dimension reduction is necessary. That is, with a smaller number of indicators replacing the original variable indicators, variable information reflected by indicators is not reduced; in the meantime, the indicators are independent from each other.

Definition: take $x_1, x_2, \ldots, x_p$ as the original variable indicator, $z_1, z_2, \ldots, z_m (m \leq p)$ the new variable indicator

$$\begin{cases} z_1 = l_{11}x_1 + l_{12}x_2 + \cdots + l_{1p}x_p \\ z_2 = l_{21}x_1 + l_{22}x_2 + \cdots + l_{2p}x_p \\ \cdots\cdots\cdots\cdots \\ z_m = 1_{m1}x_1 + l_{m2}x_2 + \cdots + l_{mp}x_p \end{cases}$$

Establish principle for coefficient l_{ij} is:

① z_i and z_j $(i \neq j;\ i, j = 1, 2, \ldots, m)$ are completely independent from each other;

② z_1 represents the one with biggest variance in all the linear combinations of $x_1, x_2, \ldots, x_p$, z_2 is the one with biggest variance in all the linear combinations of $x_1, x_2, \ldots, x_p$ that are uncorrelated with z_1, the rest can be done in the same manner so that finally z_m is the one with biggest variance in all the linear combinations of $x_1, x_2, \ldots, x_p$ that are uncorrelated with $z_1, z_2, \ldots, z_{m-1}$. New variable indicators $z_1, z_2, \ldots, z_m$ are known respectively as the first, second,…, the mth principal component of original variable indicator $x_1, x_2, \ldots, x_p$.

It can be seen from the above analysis that principal component analysis is to determine the load $l_{ij}(i = 1, 2, \ldots, m;\ j = 1, 2, \ldots, p)$ which the original variables $x_j(j = 1, 2, \ldots, p)$ place on the principal components $z_i(i = 1, 2, \ldots, m)$. It can be mathematically proved that they are the eigenvectors corresponding to the m larger eigenvalues of the correlation matrix.

The matrix representation of principal component analysis is $X = ZL'$ (Fig. 6.13a). Through projection matrix L', the matrix X is projected onto the m-dimensional subspace, and then, one can get the coordinates Z in the subspace. Columns in Z are often called score vectors, rows in L' are also known as load vectors, both of them are orthogonal vectors. From another viewpoint, principal components can be regarded as the projection which the original data X places on the score matrix Z, that is $Z = XL$ (Fig. 6.13b).

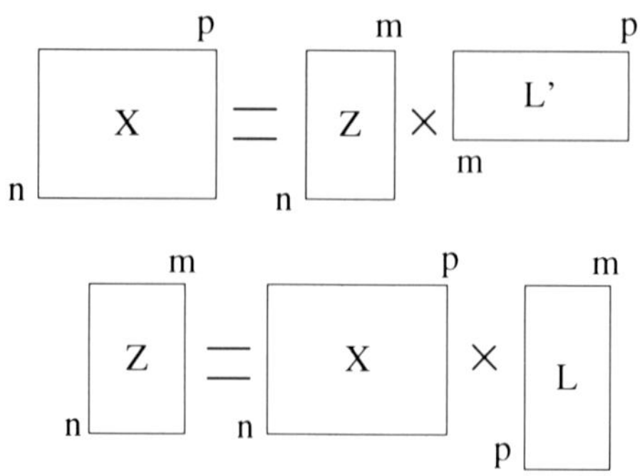

Fig. 6.13 Matrix representation of PCA process

As most variances can be described by 1–3 principal components, the corresponding score vector diagram can be used to check the differences between sample groups. Through the load vector or its transformation, one could obtain variables that contribute more to the sample group separation, hence provide the basis for subsequent biological analysis.
Principal components analysis is calculated as follows.

① compute the correlation coefficient matrix

$$R = \begin{bmatrix} r_{11} & r_{12} & \cdots & r_{1p} \\ r_{21} & r_{22} & \cdots & r_{2p} \\ \vdots & \vdots & & \vdots \\ r_{p1} & r_{p2} & \cdots & r_{pp} \end{bmatrix}$$

$r_{ij}(i, j = 1, 2, \ldots, p)$ is the correlation coefficient of the original variable x_i and x_j, so $r_{ij} = r_{ji}$, it is calculated as

$$r_{ij} = \frac{\sum_{k=1}^{n} (x_{ki} - \overline{x_i})(x_{kj} - \overline{x_j})}{\sqrt{\sum_{k=1}^{n} (x_{ki} - \overline{x_i})^2 \sum_{k=1}^{n} (x_{kj} - \overline{x_j})^2}} \tag{6.35}$$

② calculate eigenvalues and eigenvector

a. Solve characteristic equation $|\lambda_i - R| = 0$ (where I is the identity matrix), usually the Jacobi method is applied to generate eigenvalues, and then sort them in order of size, namely, $\lambda_i \geq \lambda_2 \geq \cdots \geq \lambda p \geq 0$.
b. Compute the eigenvector e_i $(i = 1, 2, \ldots, p)$ corresponding to the eigenvalues, respectively, additional requirement is $\|e_i\| = 1$, equally means $\sum_{j=1}^{p} e_{ij}^2 = 1$, where e_{ij} represents the jth component of vector.
c. Calculate the contribution rate and cumulative contribution rate of principal components. The contribution rate of pth principal component is:

$$\frac{\lambda_i}{\sum_{k=1}^{p} \lambda_k} \quad (i = 1, 2, \ldots, p)$$

Cumulative contribution rate of all *P* principal components is:

$$\frac{\sum_{k=1}^{i} \lambda_k}{\sum_{k=1}^{p} \lambda_k} \quad (i = 1, 2, \ldots p)$$

Generally, one may take the first, second, ..., *m*th ($m \leq p$) principal components corresponding to eigenvalues of λ_i, λ_2, ..., λ_m whose cumulative contribution rate is 85–95 %.

d. Compute principal components load

$$l_{ij} = p(z_i, x_j) = \sqrt{\lambda_i} e_{ij} \quad (i, j = 1, 2, \ldots, p) \tag{6.36}$$

e. Obtain the score of each principal component

$$z = \begin{bmatrix} z_{11} & z_{12} & \cdots & z_{1m} \\ z_{21} & z_{22} & \cdots & z_{2m} \\ \vdots & \vdots & & \vdots \\ z_{n1} & z_{n2} & \cdots & z_{nm} \end{bmatrix}$$

Usually, one takes the location of each sample in the two-dimensional space (score figure), which is constituted by the scores of various samples' first and second principal components to visualize the differences between them. The location of each variable in the two-dimensional space (load map) which is comprised of the load of various variables' first and second principal components shows the correlation between variables and the degree of contribution to classification.

It can be seen that principal component analysis (including factor analysis) is dependent on the original variables, so it can only reflect information conveyed by the original variables. The selection of original variables is thus very important. If the original variables are inherently independent, it is difficult to summarize many independent variables using few comprehensive variables, and then, the dimension reduction might fail. Higher correlated data indicate better dimension reduction.

(2) *Cluster analysis* the above-described pattern recognition method represented by principal component analysis and factor analysis, mainly focuses on the changes in the whole data set. By projecting the original eigenvector to a smaller subspace, one can achieve the purpose of dimensionality reduction and redundancy elimination. After dimensionality reduction, the lost feature information is minimum, which not only secures recognition performance,

but also significantly reduces the calculation costs of subsequent phases. However, although the distinction between categories can be achieved to some extent, also it achieves extraction of important variables, its essence is extracting or building the best descriptive characteristics instead of optimal classification features. Sometimes for more complex problems, or when there is further need to look at whether some samples will gather in a certain area or space, this method is no longer ideal. Cluster analysis is a common type of unsupervised pattern recognition method based on similarity. The method directly sets classification as its purpose, establishes the intrinsic link between samples according to the similarity of given data, and then divides the sample into the corresponding category. Number of categories can be either known or unknown. The capability of this method is limited in extracting important variables for subsequent biochemical interpretation.

① *Cluster basis* Cluster basis is the measure of the differences or similarities. It mainly relies on the distances and similarity coefficients.

(a) Difference measurement. The distance d between two points X and Z is as follows.

$$d_{XZ} = \left[\sum_{i=1}^{n} (x_i - z_i)^p\right]^{1/p} \tag{6.37}$$

Each different p value produces a different distance representation method. Common representations include the continental distance ($p = 2$), the Minkowski distance ($p = 1$), and the absolute distance ($p \to \infty$).

(b) Similarity measure

a. The similarity coefficient S for two vectors X and Z is the cosine of the angle between them. More similar the two vectors are, the smaller angle between them and the greater cosine value it has.

$$S_{XZ} = \frac{X^T Z}{\|X\|\|Z\|} = \cos\theta \tag{6.38}$$

b. Similarity coefficient S for two samples x_i and x_j:

$$S_{ij} = \frac{\sum_{k=1}^{n} \min\{x_{ik}, x_{jk}\}}{\sum_{k=1}^{n} \max\{x_{ik}, x_{jk}\}} \leq 1 \tag{6.39}$$

c. When the number of samples is N, similarity coefficient r between two variables is:

$$r_{ij} = \frac{\sum_{k=1}^{N}\left(x_{ki} - \overline{x_i}\right)\left(x_{kj} - \overline{x_j}\right)}{\sqrt{\left[\sum_{k=1}^{N}\left(x_{ki} - \overline{x_i}\right)^2\right]\left[\sum_{k=1}^{N}\left(x_{kj} - \overline{x_j}\right)\right]^2}} \tag{6.40}$$

② *Basic ideas and typical methods of clustering* Different similarity metrics method and different clustering criteria form different cluster analysis. The uniform goal is to maximize the proximity between samples within classes and to minimize that between classes and samples.
The basic idea is summarized below.

(a) *n* samples, each point form one class;
(b) Define difference or similarity and its amendments between points, points and point sets;
(c) Merge the two closest or most similar categories together, and the number of classes is lessened by 1;
(d) Merge all samples into one category;
(e) By setting a threshold to decide the termination of clustering.
Large-scale cluster analysis can be classified into three series: sequence clustering, hierarchical clustering, and model clustering. Sequence clustering method checks one by one the shortest distances between the sample and each class and decides whether it belongs to the class or a new class according to predefined thresholds. Such method is fast, but the results are highly related to the sample sequence and the given threshold values. Hierarchical clustering uses agglomerative or divisive algorithm to break down the object hierarchically, forming a hierarchical nested cluster (pedigree or tree cluster). In general, single-linkage cluster analysis is suitable for strips or *S*-shaped class, complete linkage, median, and average-linkage cluster analysis apply to oval-shaped classes. The fatal flaw of this series method is that once an emergence or division is complete, it cannot be revoked; in other words, such method cannot do correctness to wrong decisions. Furthermore, termination conditions of hierarchy must be specified. Model clustering method is most widely used; it divides the data set according to the evaluation function of the model and optimization.
Clustering methods that are commonly used in metabolomics include *k*-means (minimum sum of squared distance clustering) and iterative self-organizing data analysis technology algorithm (ISODATA). They are both model clustering methods. Effect of *k*-means clustering is very good when the classes are intensive, and differences between classes are significant (such as spherical aggregation). In addition, the algorithm has comparatively low complexity, so it is efficient for the handling of large data sets. However, the results are highly associated with the initial centroid and are sensitive to "noise" and

outlier data, small amounts of such data have greater influence on the mean. Moreover, it requires in advance given clustering category number, so usually there will be several attempts to determine the appropriate initial parameters before using the method. What is similar to k-means clustering is k-center point clustering. This method avoids the sensitivity of k-means clustering to "noise" and that of a few isolated points; it changes the mean (centroid) of each object of a class into the center point, with improved classification effect in exchange for increased operation costs. ISODATA method is an interactive adaptive method based on experience, and it has wide range of applications and a lot of improved versions. In addition, this method requires seven human-defined parameters, so it is more dependent on experience.

(3) CCA in one-dimensional statistics, studies on linear relationship between two random variables can refer to the Pearson simple correlation coefficient. If one wants to study the relationship between two sets of variables, such as metabolic information of some kind of plant (all observable metabolites available for identification or all peaks tested by instrument) and other information (species, place of origin, cultivation practices, multiple parameters of soil irrigation, a number of parameters of environmental conditions, height, width, color, and so on), metabolic group, and clinical biochemistry information of a disease, then directly using multivariate analysis methods is easier than examining the correlation coefficients between each two variables (Zhang and Dong 2004).

In 1936, Hotelling first introduced CCA to be used in the analysis of the correlation between two variables, borrowing the idea of principal component analysis. The basic idea is: for two sets of variables $X(x_1, x_2, \ldots, x_n)$ and $Y(y_1, y_2, \ldots, y_m)$, separately build linear combinations $u = a'X$ and $v = b'Y$ of each variable, and then replace the set of original variables by the two new compositional variables (called canonical variables), finally analyze the correlation of U and v to represent the correlation between these two sets of variables. Principle of constructing a canonical variable is to select the linear combinations of a and b so that the Pearson correlation coefficient $r(U, V)$ of two canonical variables is of maximum. This is different from the principal of constructing principal components which maximizes the extraction of variables volatility information in the principal component analysis. CCA can be regarded as a Pearson correlation analysis between canonical variables, and it is an extension to Pearson correlation analysis between two variables and factor analysis between a variable and a set of variables. When both variable groups have only one variable, the CCA is simple correlation analysis.

Similar to principal components analysis, if the Pearson correlation coefficient of the first pair of canonical variables is not 1, then the variances of both groups have not been fully explained. Such condition produces a second pair of canonical variables to represent the secondary correlation between the two sets of variables. This process will continue until all variances have been interpreted or the number of pairs of canonical variables equals to that of the variables of the group with less variables. Pearson correlation coefficient between each pair of canonical variables reaches its maximum. Pearson correlation coefficient of canonical variables that belong to different pair is zero. Here are some important indicators and terminologies of canonical correlation analysis.

① Rc (canonical correlation coefficient): Pearson correlation coefficient between two canonical variables. Because of the conversion introduced by standardization of linear combination coefficient, this value cannot be negative, it ranges from [0, 1].

② Rc^2 (squared canonical correlation): Represent the percentage of variance shared by a pair of canonical variables.

③ Standardized Canonical Coefficients: Coefficient of each variable in linear combinations after data standardization when establishing canonical variables. The coefficient, to some extent, can reflect the degree of importance for each variable in each group. Multicollinearity may exist (one of a set of variables can be expressed as a linear combination of a number of other variables), it can also be regarded as the reference index of important variable filtering.

④ Structure Coefficients: Pearson correlation coefficient between a single primitive variable and a canonical variable. Compared with canonical coefficient of standardization, it avoids the multicollinearity problem and in the meantime indicates the degree of importance for each variable. The structure coefficient between a single primitive variable and a set of canonical variables in the same group or that between the variable and a set of canonical variables in another group is respectively referred to canonical loadings and cross loadings. In metabolomics study, structure coefficient is often used to help filter important variables or metabolites.

⑤ Redundancy. It represents the proportion of a single canonical variable explaining the total variance of original variables in the same group or the proportion of that cross-explaining total variance of another set of original variables. It expresses the total ability of explanation or prediction of such canonical variable has toward the original variables in its own group or another group.

CCA is a complex, multi-dimensional analysis method that has a lot of parameters and results. In practice, one should take special scrutiny on its application condition and results interpretation. Several recommendations are as follows.

First of all, when analyzing the degree of correlation between variables, it is best to first sort out the hierarchy of variables by qualitative analysis, determine the impact of several other variables on a particular variable, and then apply the simple correlation analysis. If the variables are of large numbers and tangled, it is recommended to divide them into two categories. Besides, if variables have complex structure, say they render the mesh structure, a better choice is the canonical correlation analysis. However, this is just the first step of data analysis. After having discovered the basic laws contained in data, it is best to swap with other multidimensional statistical analysis model to carry out more accurate and in-depth analysis. Secondly, both sets of variables need to be preliminary judged before the canonical correlation analysis. This aims for analyzing whether the two have a two-way relationship or a one-way causation. This is important for rational interpretation of the results. For cases of one-way causation, there is need to interpret some selected portion of the results data. Finally, when the results are analyzed, one must pay attention to focuses and relationship between major results and minor ones. Canonical correlation coefficient and structure coefficient are most important factors. Typically, only one or two pairs of canonical variables are chosen to have their results interpreted, and then, a typical structure chart can be drawn.

6.3.2.2 Supervised Method

Data grouping can be achieved by both the clustering and projecting high-dimensional data onto low-dimensional space. These methods do not require objects in known categories for instructions. Because of the complexity and high specificity of biological metabolites test data, results of unsupervised method are often less than ideal and hard to explain. On the other hand, in practice, one often has prior knowledge of whether some samples belong to the same class. Therefore, researchers attempted to integrate the priori class information into pattern recognition, and hence, there created the supervised approach of pattern recognition. Through the integrated use of category guide information and the set of variables after preprocessing, classification results of training samples or modeling have improved significantly (Duda 2001; McKinney 2006; Bao et al. 2009). Widely used supervised methods in metabolomics include partial least square projection to latent structure (PLS), linear discriminant analysis (LDA), support vector machine (SVM) as well as random forest (RF), and so on (George et al. 1999; Li and O'Shaughnessy 2007; Mahadevan et al. 2008).

(1) PLS and OPLS partial least squares method is a new type of multivariate statistical data analysis for multiple independent variables on multiple dependent variables, and it was first proposed in 1983 by the S. Wold and C. Albano. In recent decades, it have been developed rapidly in terms of theory, methods, and applications, and it is able to simultaneously achieve regression modeling (multiple linear regression), data structure simplification (principal component analysis), and the correlation analysis between two variable sets (CCA). It was a leap of multivariate statistical data analysis.

① PLS similar to the matrix diagram in principal component analysis, Fig. 6.14 demonstrates the basic process of PLS. Because of the introduction of guide variable Y, PLS decomposes not only the sample variable X, but also the guide variable Y, then obtains respectively the sub-matrices

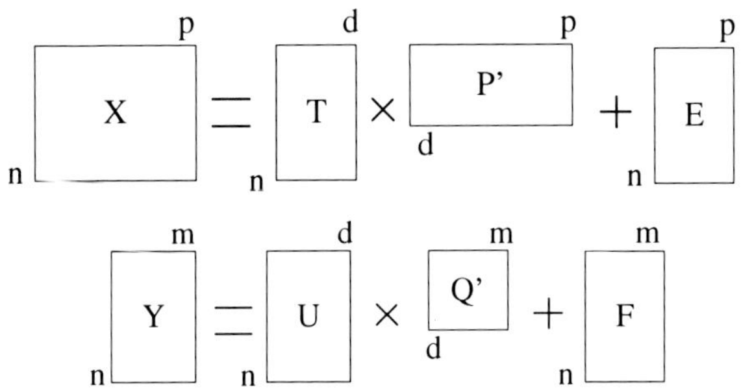

Fig. 6.14 Matrix denotation of PLS

T and *U*, the load matrices *P* and *Q*. *E* and *F* respectively represents the residual matrix of *X* and *Y*. Score matrices are still used to study difference between groups and the load matrix for selection of critical variables.

Partial least squares and principal component analysis share similar ideas and calculation processes, so we will not go into details here. The difference lies in the extraction method of scoring matrices. In short, the weight matrix produced during the process of principal component analysis reflects the covariance between variable *X*, but that of partial least squares is a reflection of covariance between variable *X* and the guide variable *Y* (Xu et al. 2001).

A standard algorithm for partial least squares is nonlinear iterative partial least square (NIPALS). There are many variables in this algorithm and some get standardized, so there are several modified versions. The following algorithm is considered to be the most effective form of NIPALSs. NIPALS iterative procedure is as follows.

a. The first column of the real matrix *Y* is assigned to its score vector *u*, as the initial vector,

$$u = y_1 \tag{6.41}$$

b. Calculate the weight of matrix *X*

$$w' = \frac{u'X}{u'u} \tag{6.42}$$

c. normalize the weight vector

$$w' = \frac{w'}{(w'w)^{1/2}} \tag{6.43}$$

d. approximate score vector of matrix X

$$t = Xw' \tag{6.44}$$

e. calculate the load matrix of matrix Y

$$q' = \frac{t'Y}{t't} \tag{6.45}$$

f. generate score vector of Y

$$u = \frac{Yq}{q'q} \tag{6.46}$$

Compare the difference between two u introduced by two iterations, if it is less than a predefined threshold, one could say it converges, and then, the iterative process stops. Otherwise, repeat the above a–f steps of the iterative process.

g. Compute scalar b to be used for internal connection,

$$b = \frac{u't}{t't} \tag{6.47}$$

h. calculate the load matrix of matrix X,

$$p' = \frac{t'X}{t't} \tag{6.48}$$

i. calculate residual errors

$$E = X - tp \quad \text{and} \quad F = Y - btq \tag{6.49}$$

j. establish the relationship between X and Y,

$$\begin{aligned} Y &= XB \\ B &= WQ' \end{aligned} \tag{6.50}$$

The relationship between X and Y can be used to calculate Y values and to predict respective categories of new samples.

② OPLS. Because of the complexity, uncertainty, and ambiguity of biological samples, data redundancy removal is usually applied before the implementation of PLS. Orthogonal PLS (OPLS) is one of the widely used and effective supervised pattern recognition methods. This method in advance takes correlation analysis of its each column (each variable) with the guide vector Y on the basis of PLS. If the variable has small

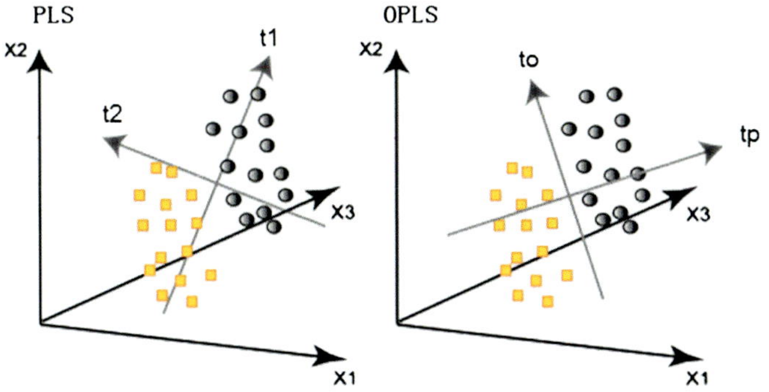

Fig. 6.15 PLS and OPLS relationship diagram

correlation with *Y*, or the classification contribution of its orthogonal variable under the guidance of *Y* is considered to be ignored, the variable will not be used in modeling. Variables having large correlation with *Y* will be retained. Therefore, the original data set *X* is decomposed into orthogonal (T_oP_o') and non-orthogonal (T_pP_p) parts.

$$X = T_oP_o' + T_pP_p' + E \tag{6.51}$$

The graphical interpretation of the method is presented in Fig. 6.15. This is equivalent to rotating axis of PLS, so that the one-dimensional projection on *tp* is able to distinguish two types of samples. This method applies to data sets with interference and poor stability.

(2) *LDA* The general formulation of distinguishing problem is: assume there are k population G_1, G_2 …, G_k, it is known that sample *X* is from one of these k population, but the specific one which it comes from remains unknown. Discriminant analysis is to determine (according to discriminant function for each class) which class sample x belongs to, on the basis of known knowledge (obtained from past experience or from the sampling from the k population) and some observation indicators of samples to be determined. There are various types of discriminant analysis, including distance discriminant analysis, Bayes discriminant, as well as Fisher discriminant. Among them, Fisher discriminant is the LDA we want to focus on [based on class K–L (Karhunen–Loeve) transform].

LDA and principal component analysis are very similar in form, but there are essential differences in principle. Principal component analysis is concerned with finding the most efficient direction to express the information of original data, while LDA lays emphasis on seeking the best way to distinguish data of different types. LDA method allows the ratio of distance between and within classes to reach the maximum, so new data will receive the maximum distinction after LDA transforms. This has similarities with clustering methods (Ueki et al. 2006).

① *basic idea of LDA* It is known that in one-dimensional ANOVA, the basic idea of judgment "whether there is a significant difference between the mean values of several population" is to decompose the total sum of squared deviation from the mean into sum of squares "between groups" and that "within group," then take the ratio between the two to measure the size of the difference. When introduce this idea into multiple dimension case, we can get linear discriminant method.

Suppose there are k p-dimensional population $G_1, G_2, \ldots, G_k$, the corresponding mean vector and covariance matrix is $\mu_1, \mu_2, \ldots, \mu_k, v_1 > 0, v_2 > 0, \ldots, v_k > 0$. Randomly given a sample X, considering its linear function $a(X) = a'X$, the mean and variance of $a(X)$ under the condition that X is from G_i are as follows:

$$e_i = E(a(X)|G_i) = a'\mu_i, \quad i = 1, 2, \ldots k \tag{6.52}$$

$$v_i = D(a(X)|G_i) = a'v_i a \quad i = 1, 2, \ldots, k \tag{6.53}$$

Let $B_0 = \sum_{l=1}^{k}\left(e_i - \frac{1}{k}\sum_{j=1}^{k} e_j\right)^2$, $E_0 = \sum_{l=1}^{K} v_i$, then B_0 is equivalent to the sum of squares in the one-dimensional statistical ANOVA between groups, E_0 equivalent to that within the group. We hope B_0 should be as high as possible, while E_0 as small as possible to ensure a strong discriminant capability of the function $a(X) = a'X$. Using the idea of variance analysis, it means to select such that $\Phi(a) = \frac{B_0}{E_0}$ reaches its maximum. $\Phi(a)$ is called discriminant efficiency. The obtained a is used as a linear discriminant function, one could further generate the linear discriminant function. This is the basic idea of linear discriminant method.

② *solution to linear discriminant function* Applying Lagrange multiplier method used to compute conditional extreme values, one can derive the Fisher discriminant function theorems.

Theorem 1 The coefficient vector a in Fisher linear discriminant function $a(X) = a'X$ is the regularized eigenvector a_1 corresponding to the largest eigenvalue λ_1 of matrix $E - 1B$ and determine the corresponding discriminant efficiency is $\phi(a_1) = \lambda_1$, where $E = \sum_{i=1}^{k} v_i$ (v_i is the covariance matrix of population G_i), $B = M'(I_k - \frac{1}{k}J_k)M = \sum_{i=1}^{k}(\mu_i - \bar{\mu})(\mu_i - \bar{\mu})'$.

③ *Judging rules* According to theorem 1, linear discriminant function is $a(X) = a'X$, where a is the eigenvector corresponding to the largest eigenvalue λ_1 of matrix $E - 1B$. For a sample X to be classified, calculate the distance between $a(X)$ and $a(\mu i)$,

$$\mathrm{d}(a(X), a(\mu_i)) = \left|a_1'X - a_1'\mu_i\right|, \quad i = 1, 2, \ldots k \tag{6.54}$$

the judging rule $D = (D_1, D_2, \ldots D_k)$ is

$$D_j = \{X : |a_1'X - a_1'\mu_j|\} = \min_{j \le i \le k}\{a_1'X - a_1'\mu_i\}, \quad j = 1, 2, \ldots k \quad (6.55)$$

If sample $X \in D_j$, then $X \in G_j$.

In some cases, merely one linear discriminant function does not differentiate each population well because there is too much lost information during the process of projecting p-dimensional vector to a one-dimensional vector space. At this point, the eigenvector a_2 corresponding to the second largest eigenvalue λ_2 of matrix $E - 1B$ could be used to establish a second linear discriminant functions $a(X) = a_2'X$. If that is still not enough, one can turn to the third eigenvector corresponding to the third largest eigenvalue to establish the linear discriminant functions $a(X) = a_3'X$, and so on. Finally, the first s discriminant function $a_1'X$, $a_2'X$, …, $a_s'X$ can be set via the s eigenvectors corresponding to s nonzero eigenvalues. This is equivalent to compressing the original p indexes into s composite indicators and then compresses these s indicators to the overall distance, so that judgment of the classification of the samples according to discriminant rules is able to be made.

It can be seen that the idea in the LDA of using s composite indicators instead of p indexes is the same with that of the principal component analysis. The thinking of using distance as the discriminant rule is similar with clustering. Using the ratio of differences between and within classes is similar with ANOVA.

For judging objectives that can be divided into m classes, it is required to substitute variables into the discriminant function of each class and obtain discriminant scores. As a result, the object belongs to the class with largest score. In general, variables with larger coefficients in the discriminant function also have greater contribution to classification and can be used as key variables to be further analyzed.

On the basis of linear discriminant, there derives several other discriminant methods such as the ones based on DA (quadratic and logit).

(3) *SVM* pattern recognition methods introduced above are mainly linear decomposition methods. Apart from clustering methods, they do not apply to nonlinear classification cases. However, nonlinear case is hard to avoid in practical applications. SVM is a new generation of learning algorithm developed from statistical learning theory of VC dimension theory and structure risk minimization principle; it can achieve both linear and nonlinear classification. It has nice applications in the field of bioinformatics, text classification, handwriting recognition, and image processing, but is at a tentative stage in metabolomics data processing. The method seeks the best compromise between complexity of the limited sample information in the model (learning precision of specific training samples) and its learning ability (the ability to identify any samples without errors), in hope of obtaining the best generalization ability (Guyon et al. 2002; Statnikov et al. 2008).

A couple of major advantages for SVM method are as follows:

① It is specifically applied to the case of limited samples; the goal is to generate optimal solutions with existing information rather than the optimal value when the sample size tends to infinity.

② Algorithm would eventually be translated into a quadratic optimization problem. In theory, the result would be the global optimal point, and hence, it settles the unavoidable problem of local extreme problem in neural network method.

③ Algorithm transforms real problem via nonlinear transition to a high-dimensional feature space and constructs a linear discriminant function in the high-dimensional space to achieve nonlinear discriminant function in the original space. The special nature ensures that the machine has good generalization ability, and in the meantime, it neatly solves the dimension problem, its algorithm complexity is independent of sample dimension. Kernel function is the core of SVM method. Provided different definitions of kernel functions, it can be achieved by many existing learning algorithms such as polynomial approximation, Bayes classifiers and radial basic function method, and multilayer perceptron network. Kernel function of SVMs in general includes nuclear kernel, Gaussian radial basis kernel, exponential radial basis kernel, multiple hidden layers nuclear perception, Fourier series kernel, spline kernel, *B* spline kernel, and so on. Results generated by different kernel functions in classification have slight differences. The formula of three most commonly used Kernel functions are as follows:

$$\text{Polynomial} \quad K(x,y) = (xy+1)^m \tag{6.56}$$

$$\text{Gauss(radial)} \quad K(x,y) = \exp\left(-\gamma\|x-y\|^2\right) \tag{6.57}$$

$$\text{Two-layer neural network} \quad K(x,y) = \frac{1}{1+\exp(v(x\cdot y)-c)} \tag{6.58}$$

Main idea of SVM is to transfer the input information through the kernel function and project it to a higher-dimensional space where it can be modeled and classified using a variety of methods and then return it back to the low-dimensional space which gives the classification results (Suykens and Vandewalle 1999). The simplest case that cannot be distinguished in one dimension (information) is able to be successfully separated by increasing the number of dimensions (aspect and information of classification). This feature is the key to achieve nonlinear classification. However, it is this feature that increases the difficulty in extracting key variables after classification.

Weights, thresholds, as well as cross-validation (CV) can be used to extract key variables. Recursive feature elimination is a widely used variable selection method in genomics and proteomics data analysis. This method repeats operations of "classify, variables sort, delete last *x* % variables in line" until the reduced number of variables satisfy a predetermined value *t*. *x* and *t* are generally set by experience. The rest *t* variables and their sorting indicate their levels of contribution to classification.

Although SVM can be superior to traditional methods in dealing with the nonlinear classification of small sample, it is still in its developing stage, whose many respects are not perfect, for example: Many theories only have theoretical significance and are yet to be implemented in a practical algorithm; some theoretical explanations of the SVM algorithm are also less than perfect (for example, the structural risk minimization principle cannot rigorously prove why SVM has good generalization ability); besides, there is no common methodology for the VC-dimensional analysis of a practical learning machine; SVM method has no theoretical basis on how to select the appropriate kernel function depending on the specific issue. This methods is widely used in data processing of genome, proteome, and metabolomics (Guyon et al. 2002; Duan et al. 2003; McKinney 2006; Mahadevan et al. 2008; Ebbels and Cavill 2009).

(4) *RF* Random forest was first proposed by Leo Breiman and Adele Cutler and was named by the Bell Labs in 1995. It is a classifier consisting of several decision trees. Its output is a voting decision by the outputs of all trees. This relatively young and simple machine-learning method has been implemented successfully in various fields of engineering and genetic screening (Guyon et al. 2002; Diaz-Uriarte and de Andres 2006; Chen et al. 2007; Wu et al. 2008; Hanselmann et al. 2009; Jiang et al. 2009; Menze et al. 2009; Katz et al. 2010). Although there have been no reports yet, some scholars both at home and abroad have applied it to metabolomics data classification and metabolite screening. RF has some main features: Number of variables has little effects on the accuracy rate of the results but has significant impact on the learning speed; there are several available standard methods to assess the importance of variables but results exist some contradictions; randomness makes forest possess high fault tolerance, even if there is large fraction of data missing, it is still able to maintain accuracy; it has a certain compensation mechanism for groups of data set having unbalanced number of samples.

Since decision tree is the subject of forest, it is necessary to first introduce the decision tree. A decision tree is a hierarchical data structure made up of nodes and directed edges to realize divide-and-conquer strategy. The tree contains three nodes: root node, intermediate nodes, and leaf nodes. The root node is a collection of all training data (purple oval in Fig. 6.16). Each intermediate node represents a split problem (green oval in Fig. 6.16), and it will take block

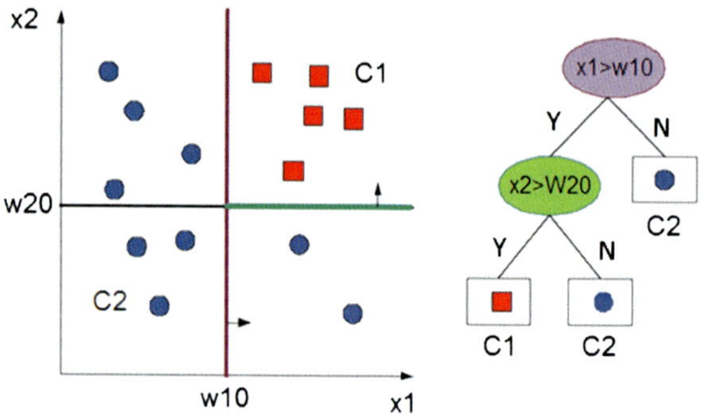

Fig. 6.16 Data sets diagram and the corresponding decision tree structure

operation to its arriving samples in accordance to the size value (cut-off point *w*) of a particular variable *x* (such as metabolite or peak intensities). Leaf nodes are data collections with the classification labeling (white box in Fig. 6.16). A path from the root node of the decision tree to one of the leaf nodes forms a criterion rule (tree structure chart in Fig. 6.16).

Algorithm of decision tree uses the top-down algorithm: Each intermediate node selects the variable *x* and block threshold *w* with best classification results, divides the data arriving at the node into blocks, and then repeats the process until the tree can classify all of the training data accurately. The core of decision tree algorithm lies at choosing superior split variable *x* and the split threshold *w*. There are bunches of optional standard, for example, information gain, information gain ratio, Gini indexes, and so on. Corresponding to different methods of variable selection, there are different decision trees algorithms (such as ID3, C4.5, CART).

The rational combination of a large number of decision trees is the RF. Its building process is to step-by-step determine the number of trees, training samples, and variables to be used in each tree. The formation of the *k*th tree in the forest is based on the general data training sets and the random vector Θ_k. The new random vector Θ_k needs to be independent of the generated random vectors Θ_1, …, Θ_{k-1}, and they should be subject to the same distribution. Training samples for a single decision tree are initially formed (Bagging sampling method) by randomly taking *N* samples from all the training samples with replacement. Repeating such sampling process *M* times could respectively generate learning samples for *M* decision trees. The variable *x* used in single decision tree construction is also randomly selected. Suppose there are *Q* variables in the sample, usually each tree is randomly assigned with $q < Q$ (usually take *q* as the square root of *Q*) variables with replacement. When selecting the split threshold *w* of each node in the tree, it is usually not to compare all available thresholds, but rather compare the ones randomly selected from the whole numbers, and then choose those whose classification results are better. Such process of building forest can increase the degree of difference between each tree, thereby enhancing the generalization error of the entire forest. After the formation of the forest, it can be seen that for a new sample to be classified, each tree will draw its corresponding classification conclusion. The final result is decided by vote of all trees. Compared with the other group and classification techniques, RF is less prone to overfitting phenomenon in the case of a huge number of trees. It can be proved that its upper bound of generalization error is less than $\rho(1 - s^2)/s^2$, where ρ is the average correlation coefficient between trees, s is the classification efficiency of a single tree.

It can be seen that constructing RFs has two main features: first, selected samples and variables are random; second, each tree is established with only a proportion of samples and variables. Construction of decision tree is also a relatively simple step-by-step binary classification model. Therefore, this method belongs to a completely different system from those on the basis of PCA pattern recognition. For less-than-ideal analysis results that processed by conventional methods, RF can always get satisfactory results (Diaz-Uriarte and de Andres 2006; Amaratunga et al. 2008; Statnikov et al. 2008; Jiang et al. 2009; Scott et al. 2010).

6.3.2.3 Model Evaluation Method

There are lots of supervised multi-dimensional statistical methods. The basis of model selection and integration is to determine whether or not the model successfully established and the credibility of the model results. Apart from complexity, speed and the classification accuracy toward unknown samples (including specificity and sensitivity), there is also need to consider the sensitivity evaluation of singular samples and variables, whether or not the number of variables and irrelevant variables have significant influence, whether there is overfitting, and if it dependents on the input order of variables and samples. In addition to the use of representative samples and randomly changing the order samples and variables for multiple tests, commonly used model evaluation methods also include CV and permutation method (Kim 2009; Magdon-Ismail and Mertsalov 2010). CV method often uses a fraction of samples (such as 1/3, 1/5, 1/7, 10, and 30 %.) as the test set while other samples as training set, to take multiple trainings and tests, and then use the comprehensive results from plenty of tests to represent the model performance. Such method could simultaneously evaluate the classification accuracy, predictive performance, and stability of the model, and hence, it is the most commonly used basic evaluation method. Permutation method also uses the results from multiple tests and is applied to study whether or not the model is overfitting. In each test, there is need to randomly change the grouping information of guiding variables, theoretically, predictive capability of the model established on the basis of error guiding variables should be lower than that of the original model. If its predictive ability is almost the same with that of the original model, or even higher, then the original model exists certain overfitting, besides, its classification accuracy of the training set tends to be significantly higher than the results of the test set. In this case, it is necessary to properly increase the number of training samples or increase its representativeness and coverage rate on group identities.

6.3.3 Software Tools

There are a lot of software to be used for single- and multi-dimensional statistical analysis. This book merely names a few of the most commonly used ones in metabolomics data analysis.

- SPSS: the famous software for statistical analysis, it has functions of the above-mentioned *T* test, M-W test, ANOVA, Pearson correlation analysis (PCA), cluster analysis, factor analysis, ROC, CCA, and LDA methods, and it can also provide appropriate graphical results.
- SIMCA: specific software for statistical analysis in metabolomics developed by umeå University. It has been updated to 12.0 version now. It is able to carry out part of data preprocessing and has modeling and evaluation functions of PCA, PLS, OPLS, permutation, and CV, also it provides a variety of two- and three-dimensional graphical results.
- RF: specific software developed by Salford in USA.
- MATLAB: famous matrix calculation software, powerful, flexible, and widely used. It is capable to achieve all the algorithms and functions mentioned in this book. However, this software requires some basis of programming, so it is not suitable for beginners.

6.3.4 Differences of Metabolite Screening Based on Single- and Multi-dimensional Analysis

Single- and multi-dimensional approaches have their own advantages, disadvantages, and applicability. Due to the different angle of investigation and data characteristics, single- and multi-dimensional approaches may generate conflicting results on the measurement of metabolite difference significance. It is recommended to take into account the various attributes of the method and give preference to substances recommended by all methods to ensure the credibility of the results. For the test results of a variety of units, typically it is required that *P* value be less than 0.05 or 0.1, areas under the curve in Pearson and ROC be greater than 0.6. For multi-dimensional model results, SIMCA series methods (PLS and OPLS) require a VIP value greater than 1. RF, SVM, and LDA methods sort the importance of variables according to their corresponding weights or coefficients in the process of classification and select as much as possible the materials rank near the top of the order. In addition, filtering can also be achieved by referring to the extent of changes, biochemical meaning, and its metabolic pathways.

References

Amaratunga D, Cabrera J, Lee YS. Enriched random forests. Bioinformatics 2008;24:2010–14.

Bao YQ, Zhao T, Wang XY, Qiu YP, Su MM, Jia WP, Jia W. Metabonomic variations in the drug-treated type 2 diabetes mellitus patients and healthy volunteers. J Proteome Res. 2009;8:1623–30.

Bradley AP. The use of the area under the roc curve in the evaluation of machine learning algorithms. Pattern Recognit. 1997;30:1145–59.

Chen X, Liu CT, Zhang M, Zhang H. A forest-based approach to identifying gene and gene-gene interactions. Proc Natl Acad Sci USA. 2007;104:19199–203.

Dettmer K, Aronov PA, Hammock BD. Mass spectrometry-based metabolomics. Mass Spectrom Rev. 2007;26:51–78.

Diaz-Uriarte R, de res SA. Gene selection and classification of microarray data using random forest. BMC Bioinf. 2006;7:1–9.

Duan K, Keerthi SS, PooA N. Evaluation of simple performance measures for tuning SVM hyperparameters. Neurocomputing. 2003;51:41–59.

Duda RO HP, Stork DG. Pattern classification. New York: Wiley; 2001.

Dunn WB, Ellis DI. Metabolomics: current analytical platforms and methodologies. Trends Anal Chem. 2005;24:285–94.

Ebbels TMD, Cavill R. Bioinformatic methods in NMR-based metabolic profiling. Prog Nucl Magn Reson Spectrosc. 2009;55:361–74.

Frans MK, Ivana B, Elwin RV, Renger HJ. Analytical error reduction using single point calibration for accurate and precise metabolomic phenotyping. J Proteome Res. 2009;8:5132–41.

Garrido M, Rius F, Larrechi M. Multivariate curve resolution-alternating least squares (MCR-ALS) applied to spectroscopic data from monitoring chemical reactions processes. Anal Bioanal Chem. 2008;390:2059–66.

George AM, Wynne WC, Carol S. A critical look at partial least squares modeling. MIS Quart. 1999;33:171–6.

Guyon I, Weston J, Barnhill S, Vapnik V. Gene selection for cancer classification using support vector machines. Mach Learn. 2002;46:389–422.

Hanselmann M, Kothe U, Kirchner M, Renard BY, Amstalden ER, Glunde K, Heeren RMA, Hamprecht FA. Toward sigital staining using Imaging mass spectrometry and random forests. J Proteome Res. 2009;8:3558–67.

Hendriks MM, Smit S, Akkermans WL, Reijmers TH, Eilers PH, Hoefsloot HC, Rubingh CM, de Koster CG, Aerts JM, Smilde AK. How to distinguish healthy from diseased? Classification strategy for mass spectrometry-based clinical proteomics. Proteomics. 2007;7:3672–80.

Jiang R, Tang W, Wu X, Fu W. A random forest approach to the detection of epistatic interactions in case-control studies. BMC Bioinf. 2009;10(Suppl 1):65–76.

Jonsson P, Johansson AI, Gullberg J, Trygg J, Grung B, Marklund S, Sjöström M, Antti H, Moritz T. High-throughput data analysis for detecting and identifying differences between samples in GC/MS-based metabolomic analyses. Anal Chem. 2005;77:5635–42.

Jolliffe IT. Principal component analysis. New York: Springer; 2002.

Jonsson P, Gullberg J, Nordström A, Kusano M, Kowalczyk M, Sjöström M, Moritz T. A strategy for identifying differences in large series of metabolomic samples analyzed by GCMS. Anal Chem. 2004;76:1738–45.

Katz JD, Mamyrova G, Guzhva O, Furmark L. Random forests classification analysis for the assessment of diagnostic skill. Am J Med Qual. 2010;25:149–53.

Kim JH. Estimating classification error rate: repeated cross-validation, repeated hold-out and bootstrap. Comput Stat Data Anal. 2009;53:3735–45.

Kong XR, Zhang XG. The tutorial examples of statistical software SPSS in medical applications. Beijing: Tsinghua University Press; 2009.

Li XB, O'Shaughnessy D. Clustering-based two-dimensional linear discriminate analysis for speech recognition. In: Interspeech: 8th An Con ISCA, vol. 4. 2007; p. 1949–52.

Magdon-Ismail M, Mertsalov K. A permutation approach to validate. In: Proceedings of SIAM SDM, Columbus, Ohio, USA. 2010; pp. 882–93.

Mahadevan S, Shah SL, Marrie TJ, Slupsky CM. Analysis of metabolomic data using support vector machines. Anal Chem. 2008;80:7562–70.

McKinney BARD, Ritchie MD, Moore JH. Machine learning for detecting gene-gene interactions: a review. Bioinformatics. 2006;5:77–88.

Menze BH, Kelm BM, Masuch R, Himmelreich U, Bachert P, Petrich W, Hamprecht FA. A comparison of random forest and its Gini importance with standard chemometric methods for the feature selection and classification of spectral data. BMC Bioinf. 2009;10:213.

Ni Y, Su MM, Lin JC, Wang XY, Qiu YP, Zhao AH, Chen TL, Jia W. Metabolic profiling reveals disorder of amino acid metabolism in four brain regions from a rat model of chronic unpredictable mild stress. FEBS Lett. 2008;582:2627–36.

Nicholson JK, Lindon JC, Holmes E. Metabonomics: understanding the metabolic responses of living systems to pathophysiological stimuli via multivariate statistical analysis of biological NMR spectroscopic data. Xenobiotica. 1999;29:1181–90.

Qiu Y, Su M, Liu Y, Chen M, Gu J, Zhang J, Jia W. Application of ethyl chloroformate derivatization for gas chromatography-mass spectrometry based metabonomic profiling. Anal Chim Acta. 2007;583:277–83.

Qiu Y, Cai G, Su M, Chen T, Zheng X, Xu Y, Ni Y, Zhao A, Xu LX, Cai S, Jia W. Serum metabolite profiling of human colorectal cancer using GC-TOFMS and UPLC-QTOFMS. J Proteome Res. 2009;8:4844–50.

Qiu Y, Cai G, Su M, Chen T, Liu Y, Xu Y, Ni Y, Zhao A, Cai S, Xu LX, Jia W. Urinary metabonomic study on colorectal cancer. J Proteome Res. 2010;9:1627–34.

Richard GB. Chemometrics: data analysis for the laboratory and chemical plant (part I). Bristol: Wiley; 2003.

Scott IM, Vermeer CP, Liakata M, Corol DI, Ward JL, Lin W, Johnson HE, Whitehead L, Kular B, Baker JM, Walsh S, Dave A, Larson TR, Graham IA, Wang TL, King RD, Draper J, Beale MH. Enhancement of plant metabolite fingerprinting by machine learning. Plant Physiol. 2010;153:1506–20.

Shi YG, Li B, Tian GY. Chemometrics methods and MATLAB application. Beijing: China Petrochemical Press; 2010.

Statnikov A, Wang L, Aliferis CF. A comprehensive comparison of random forests and support vector machines for microarray-based cancer classification. BMC Bioinf. 2008;9 (319–328):160.

Suykens JAK, Vandewalle J. Least squares support vector machine classifiers. Neural Process Lett. 1999;9:293–300.

Trygg J, Gullberg J, Hohansson AI, Jonsson P, Moritz T. Plant Metabolomics: Biotechnology in Agriculture and Forestry 2006;57:117–28.

Trygg J, Holmes E, Lundstedt T. Chemometrics in metabonomics. J Proteome Res. 2007;6:469–79.

Ueki K, Hayashida T, Kobayashi T. Two-dimensional heteroscedastic linear discriminant analysis for age-group classification. In: 18th conference on pattern recognition, vol. 2, p. 585–588.

Wu XY, Wu ZY, Li K. Identification of differential gene expression for microarray data using recursive random forest. Chin Med J. 2008;121:2492–6.

Xie G, Zheng X, Qi X, Cao Y, Chi Y, Su M, Ni Y, Qiu Y, Liu Y, Li H, Zhao A, Jia W. Metabonomic evaluation of melamine-induced acute renal toxicity in rats. J Proteome Res. 2010;9:125–33.

Xu L, Shao XG. Chemometrics methods. Beijing: Science Press; 2004.

Xu QS, Liang YZ, Shen HL. Generalized PLS regression. J Chemomet. 2001;15:135–48.

Zhang WT, Dong W. Advanced textbook for SPSS statistical analysis. Beijing: Higher Education Press; 2004.

Zhang WF. The study of speaker recognition based on principal component analysis and linear discriminant analysis. Master dissertation, Zhejing University, 2004.

Chapter 7
Metabolite Qualitative Methods and the Introduction of Metabolomics Database

Li-Xin Duan and Xiaoquan Qi

7.1 Introduction

Metabolomics aims at the unbiased and high-throughput analysis of all the small molecular metabolites and their dynamic changes in an organism or a tissue or even a single cell. It has crossed the first decade far from the proposing of the metabolomics concept. There were great development either in analysis techniques or in metabolomics applications. Metabolomics is penetrating into the new technology development, multi-platforms integration, data processing, and quantitative and qualitative analysis of metabolites, wherein the qualitative analysis of metabolites is one of the important and difficult problems. Metabolomics generates large and complex data, the vast majority of which are derived from metabolites, others including the background noise of the instrument, chromatography column bleed, and experimental exogenous contamination signals. Qualitative analysis of metabolites is the transition of complex data into metabolite information with biological significance, so as to better illustrate metabolic changes in organism. It is of important theoretical and practical significance for the qualitative analysis of metabolites associated with functional genes and biological traits, especially for differential metabolites such as biomarkers. With the emergence of high-resolution and high-precision mass spectrometer and the development of multi-platforms integration technology, both the detection capabilities of metabolomics analysis and the qualitative level of metabolite structures are improved. At the same time, the intelligent and software-based structural analysis of organic compounds and the accumulation of professional database provide feasibility for the rapid identification of metabolites.

L.-X. Duan (✉) · X. Qi
Institute of Botany, Chinese Academy of Sciences, Beijing 100093, China
e-mail: nlizn@ibcas.ac.cn

X. Qi
e-mail: xqi@ibcas.ac.cn

X. Qi et al. (eds.), *Plant Metabolomics*, DOI 10.1007/978-94-017-9291-2_7

The complex and diverse structures and large number of metabolites make the identification of metabolites very challenging. Plants synthesized thousands of metabolites to adapt to different external environments during the long-term evolution. It is estimated that there are about 0.2–1 million metabolites in plants (Dixon and Strack 2003), including phenylpropanoids, quinones, flavonoids, tannins, terpenes, steroids and their glycosides, alkaloids, and other secondary metabolites, each species known to have thousands of, even tens of thousands of or more, metabolites. There are about 10,000 species of triterpenes, which are synthesized from 2,3-oxidosqualene or squalene cyclized by oxidosqualene cyclase (OSC) or squalene cyclase (SC) to 200 skeleton structures with a variety of different types, and then through oxidation, glycosylation, and other modification to generate complex triterpenes metabolites with different structures (Xu et al. 2004). The first complete study of OSC-encoding gene products in plant kingdom is in the model plant Arabidopsis, in which the 13 OSC-encoding genes and the gene products are identified. The triterpenes profile in Arabidopsis, see Fig. 7.1. The structures of these products are complex and diverse. If the subsequent biological modifications are added, more intermediates and end products of sterols and terpenoids will be produced. The structures of these products are similar, and isomer and chiral isomers may exist. Their polarity and mass spectra are quite similar, so it is not easy to be identified. And most of them have no commercially available standards nor recorded by mass spectra database (such as the NIST and WILEY database).

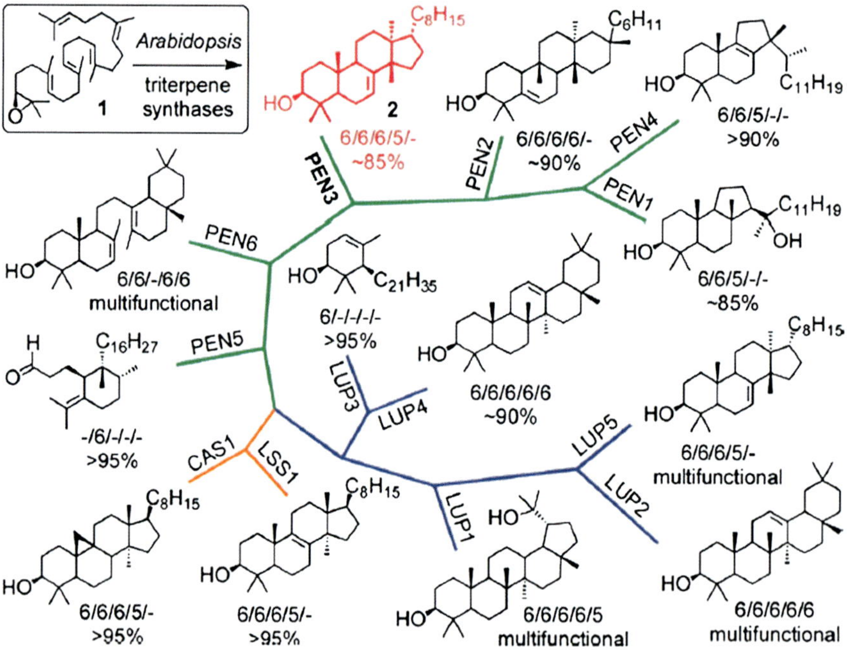

Fig. 7.1 Structures of metabolite products of triterpenes synthase in Arabidopsis. Abbreviations represent genes (Reprinted from Morlacchi et al. (2009). Copyright © 2009 American Chemical Society. Reprinted with permission)

Traditional phytochemistry methods are not suitable for high-throughput and trace samples, as well as simultaneously qualitative analysis of hundreds of metabolites because they generally use tens of kilograms plants to extract, separate, purify, and identify pure compounds. This process is very lengthy, time-consuming, and labor intensive. The identification of compounds generally includes ultraviolet spectra (UV), infrared spectra (IR), nuclear magnetic resonance (NMR), and mass spectrum (MS). The identification of new structure often need the jointly application of above four methods, and single-crystal X-ray diffraction is used to determine the atomic and molecular structure of a crystal compound. Plant metabolomics analysis only need tens to hundreds of milligrams or less for metabolic profiling analysis. Therefore, traditional structure identification methods are difficult to play a role in metabolomics qualitative analysis. Without doubt, authentic substances prepared by phytochemistry technology can be used as reference substance and the structural identification data and spectra can apply to metabolite qualitative analysis.

MS and NMR are the two technical platforms of metabolomics, and each has its advantages and disadvantages. For NMR-based metabolite identification methods, refer to chapter IV. MS is one of the conventional means of metabolite qualitative analysis, but not the most reliable method for structure confirmation. In particular, it is difficult to discriminate isomers by MS, such as glucose and fructose, which have very similar mass spectra. In addition, there are disadvantages in metabolite identification relying on MS in metabolomics analysis in the following aspects: Firstly, the sample that generally does not or rarely goes through pre-treatment purification process is a very "dirty" sample, because non-targeted metabolomics enquiries keep all the components as much as possible. So this will result in co-eluting peaks and matrix interference in the chromatographic separation process and cause the impurity of mass spectra; secondly, the partial deconvolution and impure spectra produced from the deconvolution process affect metabolite identification accuracy; thirdly, currently suitable metabolomics database is needed. Existing databases such as NIST, EPA, NIH, and WILEY contain a large number of synthetic substances and contain only limited proportion of natural products. And the commercial natural products' standards are also far from satisfying the need for qualitative analysis of metabolomics. Thus, it is difficult to confirm all the structures of metabolites in a short period. Moreover, different laboratories use different qualitative methods and standards. The metabolomic qualitative analysis of compounds has become a bottleneck. In 2005, Metabolomics Standards Initiative (MSI) was established aiming to promote the standardization of metabolomics research report. In 2007, according to MSI, the metabolite qualitative annotation was divided into four levels. At present, this standard is being gradually improved (Sumner et al. 2007).

The first level: Identified compounds. Typically, non-novel metabolites have been previously characterized, identified, and reported at a rigorous level in the literature. Thus, non-novel metabolites not being identified for the first time are often identified based upon the co-characterization with authentic samples. The following minimum standards are proposed.

① Obtain reference substances and align identification data. A minimum of two independent and orthogonal data relative to an authentic compound analyzed under identical experimental conditions are proposed as necessary to validate non-novel metabolite identifications (e.g., retention time/index and MS, retention time and NMR spectrum, accurate mass and tandem MS, accurate mass and isotope pattern, full ^{1}H, and/or ^{13}C NMR, 2-D NMR spectra). The use of literature values reported for authentic samples by other laboratories result in level 2 identifications.

② Compare with spectra similarity. If spectral aligning and comparing are utilized in the identification process, then the authentic spectra used for the spectral alignment should be described appropriately or libraries made publicly available. Concerning the non-availability of commercialized libraries (NIST, Wiley, etc.), authors should document and provide the spectral evidence to validate the metabolite identifications. If the authors choose not to provide the experimental evidence to support the identifications, then the identifications should be reported as "putative identifications."

③ Provide additional identification data. Based upon the identification data of ① or ②, provide additional identification data or additional stereo configuration data. Additional data consistent with best chemical practices might include selective solvent extraction, retention time, *m/z*, photodiode array spectra, maximum absorption wavelength λ_{max} and maximum molar extinction coefficient ε_{max}, chemical derivatization, isotope labeling, and 2D-NMR and IR spectra.

First identification of the novel compounds needs sufficient structural identification data. Requirements such as that proposed by American Chemistry Society (http://pubs.acs.org/journals/jacst/) can be adopted, including the extraction, separation and purification methods, elemental analysis results, precise molecular weight, fragment ion mass spectra, NMR data (^{1}H, ^{13}C, 2D-NMR), and other spectral data, such as infrared spectroscopy, ultraviolet spectrum, and chemical derivatization.

The second level: Putatively annotated compounds, e.g., without chemical reference standards, based upon physicochemical properties, and/or spectral similarity with spectral libraries.

The third level: Putatively characterized compound classes, e.g., based upon characteristic physicochemical properties of a chemical class of compounds or by spectral similarity to known compounds of a chemical class.

The fourth level: Unknown compounds. Beyond above three levels, although unidentified or unclassified, these metabolites can still be differentiated and quantified based upon spectral data.

7.2 Common Organics Mass Spectrometry Fragment Rules

Plant metabolites have various and complex structures, but they are composed of basic structural units in different ways, e.g., fatty acids, phenols, and polyketides such as benzoquinone are composed of C_2 unit (acetate unit); terpenes and sterols are composed of C_5 units (isopentenyl units); phenylpropanoids such as coumarin and lignin are composed of C_6 units; alkaloids are composed of amino acid units. Compounds with similar structures generally have similar MS fragmentation regularity. The knowledge of common compounds' spectra is helpful for the determination of compound class and the deduction of structure or substituent group.

7.2.1 Hydrocarbons

(1) Straight-chain hydrocarbons: Usually, molecular ion peak can be seen in the spectra, but the intensity decreases with the growth of the hydrocarbon chain. Fragment ion peaks with odd mass stay in clusters (C_nH_{2n+1}), e.g., *m/z* 29, 43, 57, 71, 85, 99… (Fig. 7.2) at intervals of *m/z* 14 (CH_2) units. Behind C_nH_{2n+1} fragment ion, there is generally a small peak with the *m/z* 1 smaller than the former, i.e., C_nH_{2n} at *m/z* 28, 42, 56, 70, 84, 98…. Usually, fragment ions with the highest intensity are C_3 (*m/z* 43) and C_4 (*m/z* 57), and others decrease as smooth curve down till the M-C_2H_5. The spectra of straight-chain hydrocarbons with more than 8 carbon atoms are similar with each other, the only difference lies in the molecular ion peaks.

(2) Branched hydrocarbons: Weak molecular ion peak. Branched hydrocarbon presents advantage fragmentation at the branch point, and the intensity of the

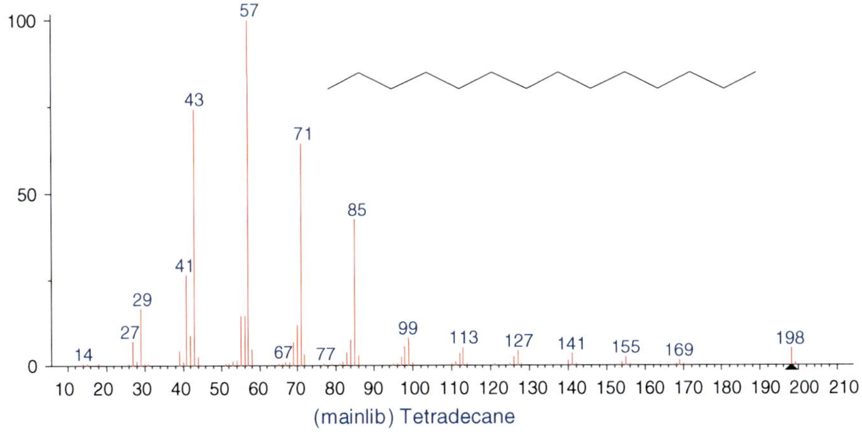

Fig. 7.2 The mass spectra of tetradecane

generated fragment ion peak is greater than the molecular ion peak. The location of the branch can be determined from the spectra of branched hydrocarbons.

7.2.2 *Olefin Compounds*

Compounds containing double bonds in the structure often conduct β-cleavage, Mclafferty rearrangement, and retro-Diels–Alder fragmentation.

7.2.3 *Aromatic Compounds*

Due to the presence of the benzene ring, a strong molecular ion peak can be seen in the aromatic compound. The molecular ion peak of benzene at *m/z* 87 is the base peak, and the molecular ion peak of polycyclic aromatic compounds is a base peak. β-cleavage, Mclafferty rearrangement, α-cleavage, hydrogen rearrangement, and retro-Diels–Alder fragmentation often occur. *m/z* 39, 51, 65, 77, 78, 91, and 92 are the characteristic ions of aromatic compounds. The molecular ion peak of phenolic compounds is strong. CO will be lost when phenolic compound subjects to cleavage.

7.2.4 *Alcohol Compounds*

The molecular ion peak of alcohol compound is usually small. General cleavage patterns include dehydration and α-cleavage. Dehydration peak (M-18) is very obvious.

7.2.5 *Ethers*

The molecular ion peak of aliphatic ether is small, but the molecular ion peak of aromatic ether is obvious. Aliphatic ethers conduct mainly α-cleavage, generating strong peaks at *m/z* 45, 59, 73, 87.... Toluene ether conducts β-cleavage, generating a fragment at *m/z* 93 and then loss CO, producing a stable ion at *m/z* 65.

7.2.6 *Aldehyde Compounds*

The molecular ion peaks of aliphatic aldehydes are obvious. But the molecular ion peaks of aliphatic aldehydes with carbon number more than 4 are obviously decreased with the increase of carbon number. The molecular ion peaks of aromatic aldehydes are stable. The M-1 peak obtained by α-cleavage is a characteristic peak of aldehyde, the intensity is strong, and sometimes, it is stronger than the molecular ion peak. The α-cleavage in the C_1–C_3 aliphatic aldehydes causes the loss of R • free radical and the formation of $HC{\equiv}O^+$ ion peak at *m/z* 29, which is a strong peak. Aliphatic aldehyde with longer carbon chain conducts α-cleavage and the loss of HCO • free radical, forming M-29 R^+ ion peak. The C4 above aliphatic aldehydes can also conduct McLafferty rearrangement, obtaining a base peak at *m/z* 44.

7.2.7 Ketones

Similar to aldehydes, ketones often conduct α-cleavage on both sides of the ketone carbonyl group, which comply with the rule of losing the maximum alkyl. McLafferty rearrangement often occurs at the presence of γ-H.

7.2.8 Acids

The molecular ion peak of branched monobasic acid is small, and the molecular ion peak of aromatic acid is big. The base peak at *m/z* 60 produced by McLafferty rearrangement is a characteristic ion of branched monobasic acid ($C \geq 4$). The α-cleavage causes the loss of R • radical, forming an ion peak at *m/z* 45.

7.2.9 Esters

Esters can conduct α-cleavage, losing R • radicals or • OR radicals and generating ion peaks at *m/z* $59 + n \times 14$ and $29 + n \times 14$. The characteristic ion of fatty acid methyl ester is *m/z* 74.

7.2.10 Amine Compounds

The molecular ion peaks of aliphatic amines are small, and the molecular ion peaks of aromatic amines are big. The molecular ion peaks of monoamines are odd. α-cleavage is the most important fragmentation pattern of amine compounds, with priority loss of the largest alkyl.

7.2.11 Compounds Containing Sulfur and Halogen

Compounds containing sulfur and halogen can be easily determined according to the intensity ratio of M/M + 2 on the mass spectra (Yao 1996).

7.3 GC-MS Metabolites Qualitative Methods

GC-MS is one of the important platforms widely used in metabolomics studies. Chapter II describes in detail the principles, methods, and applications of GC-MS. GC-MS is characterized by hard ionization, i.e., the compound is bombarded and crushed by electron beams with certain energy, generating charged ions (fragment ions) with different mass. The structural information of the compounds is included in the quality and intensity distribution of these fragment ions, i.e., the often said mass spectra. The above process is stable, and thus, a mass spectra library of standard substances can be established. The identification process of a compound is also a process of manual interpretation of mass spectra or intelligent software mass spectra library search. In addition, the chromatographic behavior of metabolites, i.e., retention time, is associated with structure, which can be converted into retention index as an auxiliary parameter for identification. Common GC-MS qualitative analysis steps are as follows:

① Mass spectrometry data can be preprocessed to obtain pure mass spectra. The total ion chromatogram (TIC) produced by GC-MS analysis contains matrix background (signal noise), column bleed, and co-elution peaks. So the raw data produced by GC-MS is not suitable for direct metabolites identification and need to go through data process such as noise filtering, smoothing, peak detection, and deconvolution of overlapping peaks to obtain pure mass spectra and then to conduct qualitative analysis such as library search. For example, Fig. 7.3a is the original MS signal at 27.063 min chromatographic peak point, Fig. 7.3b is the background spectra signal near the peak of Fig. 7.3a, c is the pure mass spectra obtained by Fig. 7.3a, b. It can be seen through the example that after the background subtraction to exclude the interference of impurities signal, the pure mass spectra of compound is quite different to the original mass spectra. Figure 7.3d is the candidate substance hexachlorobiphenyl retrieved by the mass spectra library search with similarity of 99 %.
Deconvolution is a mathematical algorithm process, which is very important to the split of co-elution peaks. Assumed that the ratio of the fragment ion peak intensity of the pure compounds does not change with concentration, deconvolution is the process of extracting a single peak from a complex overlapping peak. Deconvolution results will directly affect the identification of metabolites. At present, deconvolution algorithms of different software are quite different, e.g., the deconvolution results of the free AMDIS software and the commercial Chroma TOF software of LECO company are different due to the difference of deconvolution algorithm. In addition, the MS acquisition rate, the concentration of the compound, the sample matrix, the peak smoothing effect, and other factors will also affect the results of deconvolution. Deconvolution also produces false-positive or false-negative results (Lu et al. 2008). Figure 7.4 is a diagram of the LECO Chroma TOF software deconvolution effect, (a) is a TIC of GC-TOF-MS at 23.9–24.0 min, and (b) is the analysis chromatogram after deconvolution, which contains three substances (peak

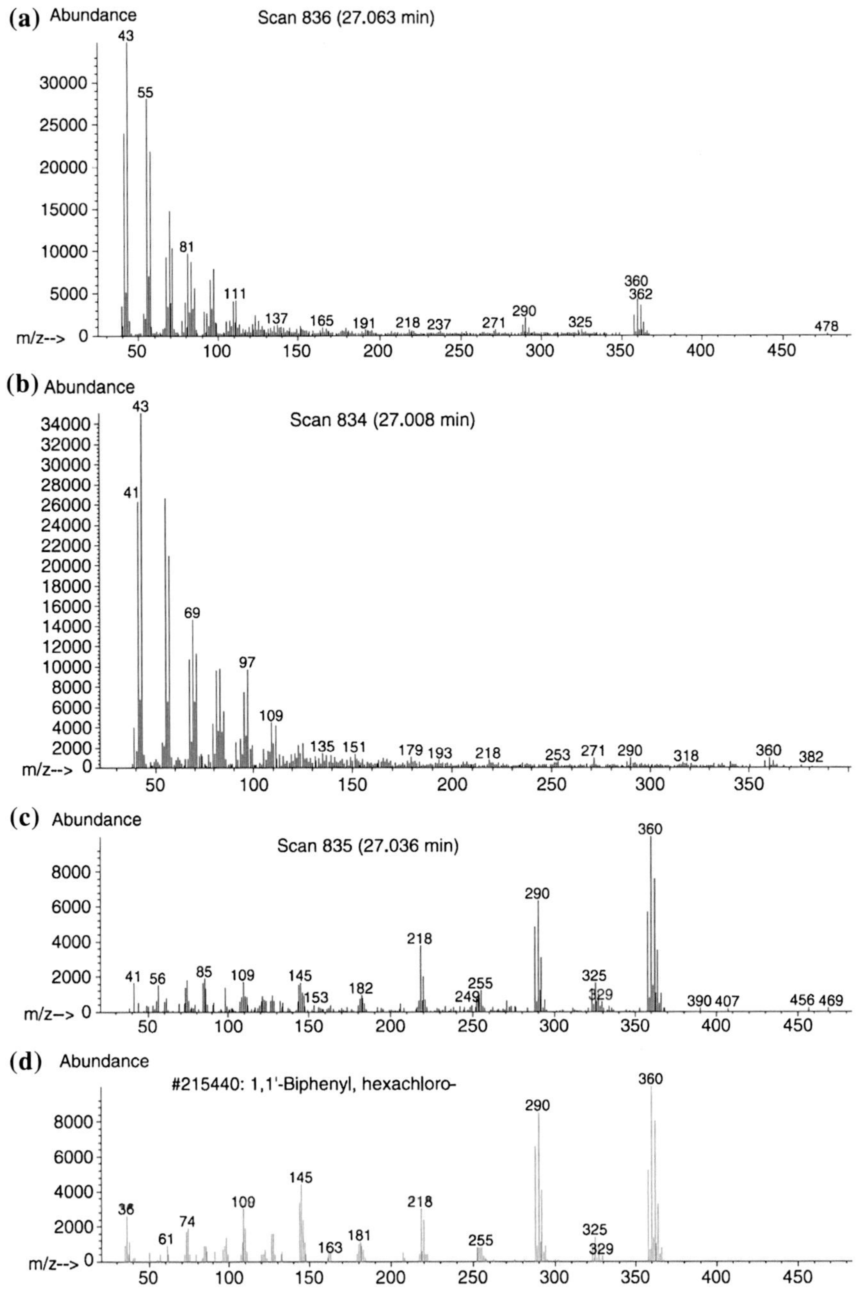

Fig. 7.3 Date pre-procession for mass spectrum comparison. **a** The raw mass spectrum of a peak at 27.063 min; **b** the background MS signal near the peak; **c** the pure mass spectrum for the peak deducted the background; **d** the candidate similar compound from library (Reprinted from Ekman et al. (2008). Copyright © 2009 John Wiley & Sons, Inc. Reprinted with permission)

349, peak 350, and peak 351) marked with different colors, respectively, with characteristic ions of 92, 81, and 202, respectively. Figure 7.4c, d, e is mass spectra of the three substances in Fig. 7.4b.

② Mass spectra library search. Computer-aided retrieval method is generally used in GC-MS qualitative analysis, namely computer automatically aligns the mass spectra of components with the mass spectra in library. As a general guide, a match factor 900 or greater is an excellent match; 800–900, a good match; 700–800, a fair match. Less than 600 is a very poor match. For isomers, homologs, and compounds with relatively similar structural characteristics, due to the little differences in their mass spectra, the structure confirmation must be done combining other qualitative methods. Compared with standard reference spectra, for mass spectra obtained from GC-TOF-MS, the ion abundance of high *m/z* is generally low, while the ion abundance of low *m/z* is high, which has an impact in the spectra library searching.

③ Compound identification combined with retention index. The mass spectra of isomers and homologs are closely similar, so it is difficult to differentiate by GC-MS and other qualitative parameters are needed. Thanks to the high separation capacity of gas chromatography for isomers, homologs, and structurally similar compounds, the use of retention time for qualitative is also an effective method. But as retention time is closely associated with instrument, chromatographic column, and operating conditions such as column temperature, injection volume, flow rate, and column length, the retention time alone is not suitable for qualitative comparison of data between different laboratories. The retention behavior of the compounds on the column by its very nature is the interaction between the compounds and the stationary phase. When the stationary phase is unchanged, the size of this interaction directly related with the topology and geometrical and electrical characteristics of compounds. On the basis of compound structure and chromatographic separation principle, quantitative structure–retention relationships (QSRR) theory was proposed that the functional relationship between the structure and retention time can be used in predicting and analyzing the chromatographic behavior of compounds (including unknown compounds).

Retention index or Kovats index (RI or KI) was proposed by Kovats in 1958, which is one of the most widely used GC qualitative parameters. At present, retention indices of thousands of compounds can be retrieved. For homologs with similar structures (only different at CH_2), such as *n*-paraffins or saturated fatty acids methyl ester homologs, the retention times present a characteristic of equidistant distribution (it presents a linear function between retention time and the number of carbon atoms) in constant temperature analysis (Fig. 7.5). This class of compounds has a good QSRR and can be used as a reference to measure other substances' retention capacity. When adding these standard substances into the analytes as a reference, the retention time of the analytes can be converted into retention values (retention index) relative to *n*-alkane or saturated fatty acid methyl ester. In general, the retention index of analyte can be

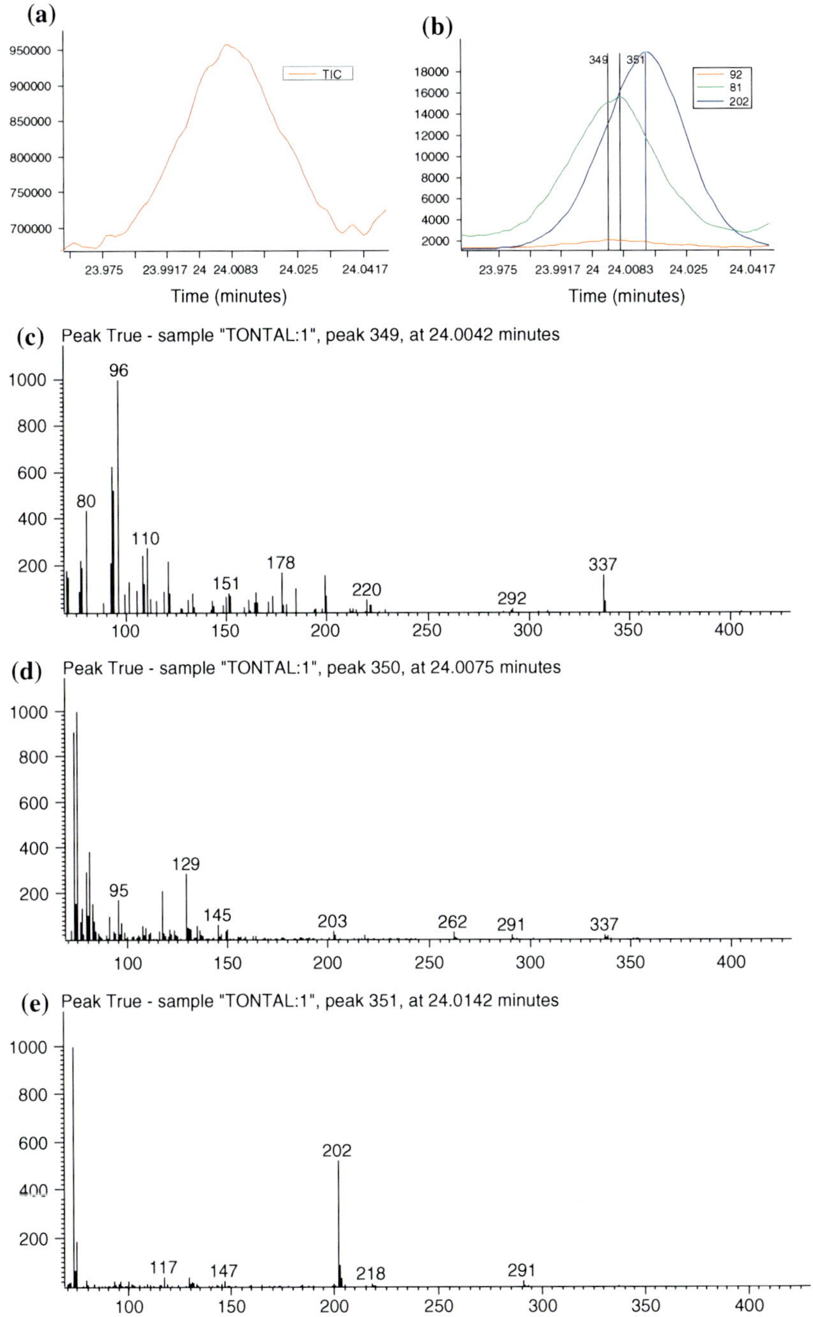

Fig. 7.4 Peak deconvolution from raw TIC. **a** TIC peak having co-eluted signal at 23.9–24.0 min; **b** the deconvoluted compound plot, the TIC peak includes three overlapped peaks, namely peak 349, 350 and 351; **c, d, e** the deconvoluted mass spectrum for peak 349, 350 and 351 respectively

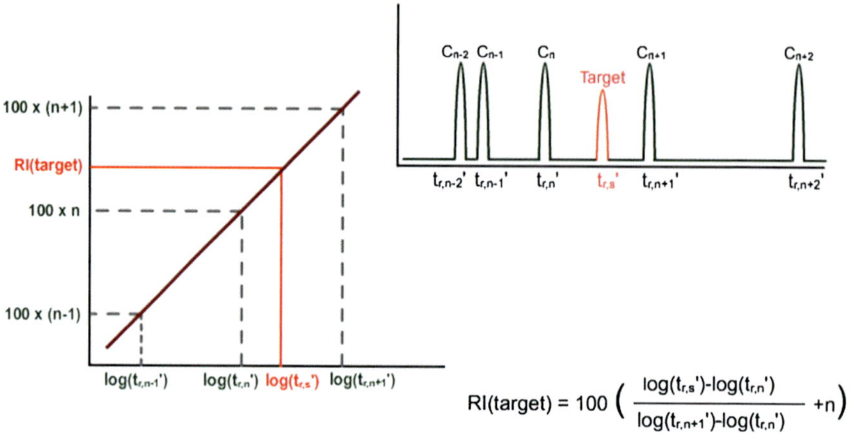

Fig. 7.5 The principle of retention index calculation using *n*-alkanes as the standard

calibrated by the retention indices of two *n*-alkanes or saturated fatty acid methyl esters before and after analyte. In order to facilitate the calculation, provide the retention indices of *n*-paraffins as 100 times the carbon atoms of the hydrocarbon, e.g., the RI of *n*-hexane is 600, the RI of *n*-heptane is 700, and the RI of *n*-pentadecane is 1,500. Then, the retention time of analyte is converted to retention index by certain formula. In the case of the same chromatographic column and column temperature, the retention index is only related with the structure of the compound while not related with other experimental conditions, with good accuracy and reproducibility. As long as GC can separate structurally similar substances, the retention indices of these substances are different. Combined with mass spectrometry, the accuracy of structure identification will be significantly improved. Kovats proposed that the retention index formula is only suitable for thermostatic analysis. For complex mixture with a wide range of boiling points, temperature programmed analysis is generally used. When using temperature programmed analysis, the determination of retention index of analyte is different and needs to be corrected.

After calculation, Van Den Dool et al. defined the temperature programming retention index formula in 1963 as the following:

$$I_X^T = 100n + 100[(T_{Rx} - T_{Rn})/(T_{R(n+1)} - T_{Rn})]$$

wherein T_R is the retention temperature; x denotes the compound to be analyzed; and n and $n + 1$ represent the number of carbon atoms of the *n*-paraffins. For single linear temperature programming, T_R can be replaced by corresponding retention time t_R. Generally speaking, the determination of retention temperature is more tedious than the determination of retention time. As retention temperature and retention time are highly correlated, use corresponding retention time to calculate retention index instead of retention temperature of above formula.

Joachim Kopka of Max-Planck Institute of Molecular Plant Physiology proposed the joint identification based on retention index and mass spectra of metabolites in metabolomics studies (Wagner et al. 2003) and established Golm Metabolom Database (Kopka et al. 2005).

For example: a case of analysis perilla essential oils by GC-MS, two peaks at retention time 14.90 and 15.68 are both similar to α-caryophyllene or β-caryophyllene, and the similarities are from 0.95 to 0.97, respectively (Liang et al. 2008).The attribution of these two peaks cannot be determined only by mass spectrometry. The author used n-alkanes as retention index reference, using a linear temperature program to calculate the retention indices of the two peaks as, respectively, 1,429 and 1,460. By literature searching found that the reported retention index of β-caryophyllene is about 1,430, and the reported retention index of α-caryophyllene is about 1,460, which, respectively, consistent with experimental results. For compounds with similar matching degree and difficult to be identified, a method of canceling base peak to calculate mass spectra similarity is used to eliminate the significant contribution of the base peak to similarity, thereby more accurately reflect the difference between the fingerprints of different compounds and improve identification accuracy. For example, the mass spectra similarity of a volatile compound A without base peak canceling to the mass spectra of β-phellandrene and 3-carene in NIST02 database are 0.9725–0.9562, respectively, both are large enough (Fig. 7.6). After base peak canceling, the similarity to β-phellandrene and 3-carene are 0.9608 and 0.945, respectively. The retention index of the separated substance with a DB-5 capillary column is 1029. According to literatures, the retention indices of β-phellandrene and 3-carene are 1,030 and 1,011, respectively. Thus, the interested substance was identified as β-phellandrene (Su et al. 2009).

④ Manual interpretation mass spectra, predicting possible structure. For metabolites with low similarity in library search, it may be because the library does not contain the substance and manual inspecting of mass spectra is needed. The structure of unknown metabolite was predicted by mass spectra interpretation, which is a challenging work, and needs good experience and knowledge. The following are some empirical methods typically used in determining the structure of unknown compounds.

a. Inspecting molecular ion. Molecular ion peak is the basis for predicting molecular weight. In theory, the m/z of molecular ion should be the largest

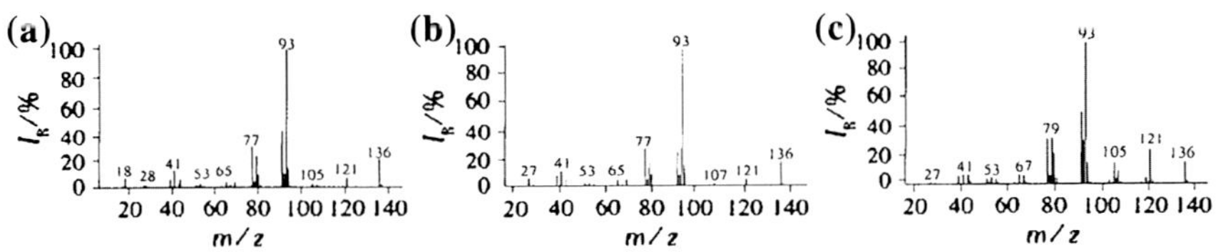

Fig. 7.6 A case of combining retention index to identify peak. Mass spectrum of unknown compound A from a flavor (**a**), mass spectrum of β-phellandrene (**b**), and mass spectrum of 3-carene (**c**). They three mass spectra are very similar, but the retention index of compound A is more close to β-phellandrene

in the mass spectra. If the molecular ion is instable, the relative abundance of molecular ion peak is very low or simply does not appear. Sometimes, molecular ions impact with other ions or gas molecules and may generate an ion with higher mass. The high mass ion peak generated by the incorporation of impurities will also constitute a difficulty for the determination of molecular ion peak. When estimating a molecular ion peak, you should consider if the mass difference between the molecular ion peak to be estimated and other fragment ion peak is reasonable. When cleavage occurs in molecular ion, the mass of the lost free radicals or neutral small molecules has certain regularity, such as M-1, M-15, and M-18, and the mass loss of 3–14, 21–26, 37, 38, 50–53, 65, and 66 generally does not appear. In addition, it should be considered that if the molecular ion is comply with the nitrogen rule. If a compound contains an even number of nitrogen atoms or no nitrogen atom, the molecular weight is even; if it contains an odd number of nitrogen atoms, the molecular weight is odd. Ion peaks do not comply with the nitrogen rule cannot be molecular ion peak. For a compound with molecular ion peak does not appear, a softer ionization mode can be used, such as chemical ionization (CI).

b. Analyzing the isotope pattern and analyzing element composition through isotope pattern. Plant natural products generally contain elements of C, H, O, N, P, S, Cl, Si, etc., which all have natural isotopes, and the abundance ratios are constant. The light isotope with the largest abundance is represented by A, and the heavy isotopes with 1 unit and 2 units heavier are represented by A + 1 and A + 2, respectively. Table 7.1 shows the common element isotopes and their relative abundance. For small molecular

Table 7.1 Common element isotopes of natural products and their abundance①

Element	Isotope A		Isotope A + 1		Isotope A + 2		Element type④
	Mass	Abundance (%)	Mass	Abundance③ (%)	Mass	Abundance③ (%)	
H	1	100	2	0.015	-	-	"A"
C	12	100	13	1.1②	-	-	"A + 1"
N	14	100	15	0.37	-	-	"A + 1"
O	16	100	17	0.04	18	0.2	"A + 2"
Si	28	100	29	5.1	30	3.4	"A + 2"
S	32	100	33	0.79	34	4.4	"A + 2"
Cl	35	100	-	-	37	32	"A + 2"
Br	79	100	-	-	81	97.3	"A + 2"
I	127	100	-	-	-	-	"A"
F	19	100	-	-	-	-	"A"
P	31	100	-	-	-	-	"A"

① Wapstra and Audi, 1986
② 1.1 ± 0.02, depending on the source
③ Relative abundance (relative to the highest concentration of isotopes)
④ "A": elements with only one natural abundance; "A + 1": elements with two isotopes, in which the mass of the second isotope is 1 unit heavier than the isotope with the highest abundance; "A + 2": elements containing isotopes with abundance 2 units heavier than the isotope with the highest abundance. Due to the low content of ^{2}H in nature, the impact to the heavy isotope peaks can be ignored, so usually take ^{2}H as the A element

compounds containing only C, H, O, and N and with small molecular weight, the relative abundance of the isotope ion peaks can be used to estimate the molecular formula of sample. For compounds containing S, Cl, Br, and Si, the relative abundance of M + 2 significantly increased. The relative abundance distribution of heavy isotopes can be calculated by the following binomial expansion formula:

$$(a + b)^n = a^n + na^{n-1}b + n(n-1)a^{n-2}b^2/2! + n(n-1)(n-2)a^{n-3}b^3/3! + \cdots.$$

wherein a is relative abundance of light isotope; b is relative abundance of heavy isotope; and n is number of isotope atoms.

For compounds containing a lot of heavy isotope elements, check Beynod table to determine the composition and formula of the compound. The higher the molecular weight, the higher the requirements of the mass spectrometer for the measurement accuracy of isotope abundance.

c. Analyzing fragment ions and interpretating the fragmentation pathway. It is a complex and tedious work to manually determine the structure of compound from mass spectra. There is a detailed monograph in this aspect by McLafferty and Turecek (2003). Here, we will not describe in detail.

⑤ Comparing with standard substances. Final confirmation of metabolite structure can be done by buying standard substance, comparing retention time or retention index, and mass spectra under the same experimental conditions. However, currently, standards for metabolites, especially for secondary metabolites in the market, are often non-available.

Other identification methods further comprising the high-resolution GC-MS, MS/MS, inferring compound structure using derivatization reagent MTBSTA (Fiehn et al. 2000), and MSTFA-d_9 (Herebian et al. 2005).

7.4 Metabolites Identification Methods Using LC-MS

The steps of compound structure identification using LC-MS are similar with that using GC-MS. Comparing to the huge number of library for GC-MS, there is no "standard mass spectra library" for LC-MS. Thus, the metabolite identification is more dependent on the standard substance. Soft ionization modes, such as electrospray ionization (ESI) and atmospheric pressure chemical ionization (APCI) technology, are usually employed in LC-MS. Generally, pseudo-molecular ion peaks can be obtained, and the fragment ion peaks are rarely obtained. The high-resolution molecular ions (calculating accurate molecular weight and predicting molecular formula) are the most important clues for LC-MS identification. Combined with isotope pattern, possible molecular formula can be inferred. Candidate substances can be obtained by looking up database. From MS/MS or MS^n cleavage fragments information, estimate possible groups or skeletons in the molecule, then infer the cleavage pathway of the candidate substance for confirmation, and finally align and compare it with the standard substance.

7.4.1 Be Familiar with Analyte Information

Before the identification of an unknown metabolite, it is necessary to fully understand the source, species, tissue site, known metabolic pathways of the analyte, and known metabolite structures in the species and family that the analyte belongs. Gene annotation, co-expression, and other biological information can be used to infer possible structural modification. A diode array detector can be connected to LC-MS to observe ultraviolet absorption of metabolites and predetermine the structure type and possible groups of the compound. Further, according to the sample nature to select ionization mode, e.g., amines, quaternary ammonium salts, and heteroatom-containing compounds such as carbamate are suitable for ESI ionization; weakly and medium polar small molecules such as fatty acids, phthalic acids, and heteroatom-containing compounds such as carbamate and urethane are suitable for APCI ionization; basic compounds are suitable for positive ion mode; acidic compounds are suitable for negative ion mode; metabolites of unknown nature need to be detected both in positive and negative ion modes. Of course, ion peaks of some compounds can be detected both in positive and negative ion modes (Wang and Zhang 2006).

7.4.2 Data Preprocessing

LC-MS data need to go through data processing such as de-noising and alignment to remove background ions derived from LC-MS. Common background ions are shown in Table 7.2.

7.4.3 Inferring Metabolite Molecular Formula

The high-resolution mass spectrometry is a powerful approach of LC-MS in the identification of compounds. A compound is composed of certain proportions of different elements which contain different numbers of protons, neutrons, and electrons. Provides the mass of *C* element as 12.000u, the mass of other elements is the ratios relative to *C* element, e.g., $^{1}H = 1.007825u$, $^{14}N = 14.003070u$, and $^{16}O = 15.994910u$. For low-resolution mass spectrometry, masses with the same nominal mass may have many elemental permutations, e.g., compounds with the same relative mass 28 may be CO, N_2, or C_2H_4. But as long as the mass accuracy of the mass spectra is high enough, the element composition of compound will be able

Table 7.2 Common background ions derived from LC-MS

Background ion	Source	Ion composition
m/z 102	H + acetonitrile + acetic acid	$C_4H_7NO_2H^+$, 102.0549
m/z 102	Triethylamine	$(C_2H_5)_3NH^+$, 102.1283
m/z 149	Phthalic anhydride in the channel	$C_8H_4O_3H^+$, 149.0233
m/z 288, 316	Characteristic ions from 2-mm centrifuge tube	
m/z 279	Dibutyl phthalate in the channel	$C_{16}H_{22}O_4H^+$, 279.1591
m/z 384	Ions generated from light stabilizer of the bottle	
m/z 391	Dioctyl phthalate in the channel	$C_{24}H_{38}O_4H^+$, 391.2843
m/z 413	Dioctyl phthalate +Na	$C_{24}H_{38}O_4Na^+$, 413.2668
m/z 538	Acetic acid + O + Fe (injection spray pipe)	$Fe_3O(O_2CCH_3)_6$, 537.8793

to be inferred, e.g., CO (27.9949), N_2 (28.0061), and C_2H_4 (28.0313). Current mass spectrometers such as Fourier transform ion cyclotron resonance mass spectrometry (FT-ICR-MS) are able to achieve a resolution of more than 20 million, mass accuracy less than 1 ppm. The resolution of the latest quadrupole time-of-flight mass spectrometry (Q-TOF-MS) is 0.6 million. However, even if the mass accuracy can reach 0.1 ppm (currently, such a mass spectrometer is non-available), it is not enough to infer the element composition simply rely on high mass accuracy alone. If the isotope abundance patterns are supplemented, some error molecular formulas can be significantly excluded. Kind and Fiehn (Kind and Fiehn 2006) proposed that even if the mass accuracy is 3 ppm, the result of inferring molecular formula by the combination of isotope patterns (2 % measurement error of isotope abundance) is superior to the results simply by common mass spectrometer with mass accuracy of 0.1 ppm (Table 7.3). Most of the mass spectrometer manufacturers provide the isotope abundance filter function.

Table 7.3 The numbers of possible molecular formulas in different mass accuracy and isotope abundance patterns

Relative molecular weight	No isotope abundance information					2 % Isotope abundance accuracy	5 % Isotope abundance accuracy
	10	5	3	1	0.1	3	5
150	2	1	1	1	1	1	1
200	3	22	2	1	1	1	1
300	24	11	7	2	1	1	6
400	78	37	23	7	1	2	13
500	266	115	64	21	2	3	33
600	505	257	155	50	5	4	36
700	1,046	538	321	108	10	10	97
800	1,964	973	599	200	20	13	111
900	3,447	1,712	1,045	354	32	18	196

Unit: ppm

Nitrogen rule and the unsaturation degree can be used to determine the molecular composition. For the calculation of unsaturation degree, initially determine the numbers of double bonds or rings. The unsaturation degree = C + Si − 1/2 (H + F + Cl + Br + I) + 1/2(N + P) + 1, wherein O and S do not have to be counted. This formula is based on the lowest valence state of elements and is applicable to the vast majority of compounds. But when N and P are in the pentavalence state (covalent bond) while not in the trivalence state, and S is in the tetravalence or hexavalence state but not in the bivalence state, the unsaturation degree calculated by above formula is not accurate.

Tobias Kind and Oliver Fiehn proposed the seven golden rules in inferring molecular formula by high-resolution mass spectrometry (Kind and Fiehn 2007):

(1) Limiting the number of elements in the formula of unknown peaks. The absolute number of elements can be obtained from the element mass divided by molecular weight. For example, the maximum number of C atoms in a molecule with the mass of 1,000 Da does not exceed 1,000/12 = 83. The numbers of N, P, and S in natural products of plant are far less than those in peptides.
(2) Determining the molecular composition according to LEWIS and SENIOR rules;
(3) Filtering isotope abundance mode;
(4) Checking H/C proportion;
(5) Checking heteroatom proportion;
(6) Checking the multi-element H, N, O, P, and S probability;
(7) Checking the TMS substituted position for GC-MS.

7.4.4 *Searching Database, Looking for Candidate Substances*

By above precise molecular weight, isotope abundance mode, nitrogen rule, and unsaturation degree to infer element composition of unknown peaks and generate molecular formula. Then, search possible structures in various databases such as METLIN, ChemSpider, KEGG, MassBank, and PubChem by molecular formula.

7.4.5 *Inferring Compound Structure Combined with Tandem Mass Spectrometry*

In ESI or MALDI, spectra often only contain the ionized molecule with very little fragmentation data and consequently, the spectra are of little use for structural characterisation. In the cases when structural data are required, tandem mass spectrometry (MS/MS) can be routinely employed by Q-TOF-MS. MS^n is a technique that allows the re-fragmentation of product ions (fragment ions from MS/MS) to determine possible fragments in the molecule and then combined with molecular weight to infer possible candidate substance.

7.4.6 Aligning and Comparing with Standard Substances

The final confirmation of metabolite structures need to be done by aligning and comparing commercial available standard substances. For a compound with new structure and no standard substance can be obtained, it is necessary to purify and enrich it and identify the structure using NMR.

Example: Christoph Bőttcher (Bőttcher et al. 2008) using LC-Q-TOF-MS analyzed the metabolic profiles of Arabidopsis wild type (Ler), chalcone synthase mutant (*transparent testa4, tt4*), and chalcone isomerase (*tt5*) mutant, wherein the synthesis of all the flavonoids is blocked in *tt4* mutant, the synthesis of flavone and Dihydroflavonol is blocked in *tt5* mutant, and several new compounds accumulate in *tt4* and *tt5* mutants. Seventy-five compounds were structurally characterized from the three Arabidopsis materials, of which 40 compounds are found for the first time from Arabidopsis. According to the four annotation levels defined by MSI (Sumner et al. 2007), 21 metabolites were identified at the first level, 22 metabolites were identified at the second level, and 11 metabolites were identified at the third level. The identified metabolites are mainly flavonoids (glycosides), flavonols (glycosides), chalcone (glycosides), phenolic choline esters (glycosides), sinapine, spermine (glycosides), and glucosinolate breakdown products. For the identification of metabolites, firstly identify (pseudo-) molecular ions, cluster ions, in-source fragment ions as well as the determination of their charge states. Ions detected in narrow retention time windows and exhibiting high chromatogram correlation were grouped. In the second step, purposely select molecular ions to conduct collision cleavage, based on accurate mass obtained from LC-Q-TOF-MS and ESI-FT-ICR-MS, isotope abundance mode, fragment ions and neutral losses to search database, lookup literatures on related metabolite mass spectra, and inferred the compound structures (Fig. 7.7). The authors also confirmed the structures of other substances by standard purchase, chemical synthesis, and pure substances isolation.

Figure 7.7a represents metabolic profiles of wild type and *tt4* mutant by LC/ESI (+)-TOF-MS and found that there exists differential peaks at 7.0 min, *m/z* 386.183, 387.188, 388.189, and 165.058 are co-elution ions, which have similar peak shapes. After analyzing, we found that *m/z* 386.183 is a molecular ion, *m/z* 387 and *m/z* 388 are isotope ions, and *m/z* 165 may be an in-source collision-induced dissociation. These ions belong to the same compound. Ten potential molecular formulas were obtained by elemental composition analysis [constrains: elements ^{1}H, ^{12}C, and ^{16}O without limitation; $^{14}N \leq 5$, $^{32}S \leq 2$; even electron state, double bond equivalents (DBE, unsaturation degree) ≥ -2, molecular weight difference of less than 15 ppm]. Figure 7.7b shows that using ultra-high-resolution instrument ESI (+)-FT-ICR-MS determined the accurate molecular weight as 386.18029. Using the same limiting

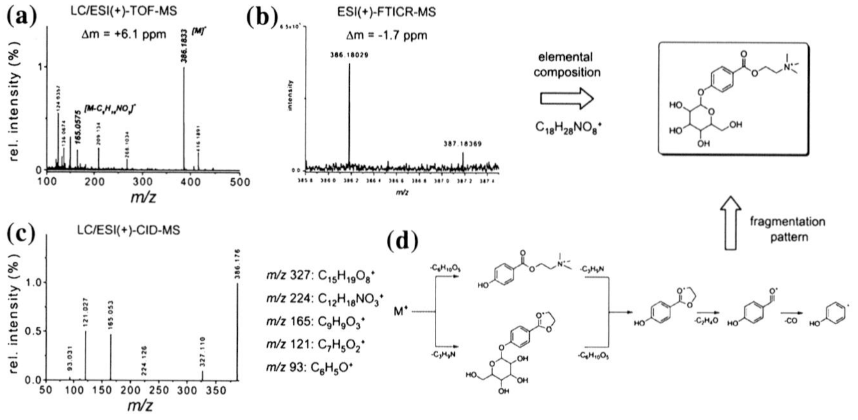

Fig. 7.7 The structural resolution process of a new compound T21 in Arabidopsis

conditions, the single molecular formula $C_{18}H_{28}NO_8^+$ (< 2 ppm) was determined. Figure 7.7c shows the MS/MS analysis of molecular ions with the collision energy of 15–25 eV. The elemental compositions of each fragment ion and neutral loss were calculated, and the molecular information of fragment ions was obtained. Figure 7.7d shows that combining biosynthetic origin information, the characteristic fragment ion *m/z* 121.017 was identified as 4-hydroxybenzoyl, and neutral loss of *m/z* 59.07 is C_3H_9N and neutral loss *m/z* 162.05 is hexose $C_6H_{10}O_5$. And the possible molecular structure and cleavage process of T21 were finally inferred. To further validate the structural assignment, the authors chemically synthesized the aglycone of 4-hydroxybenzoyl. The CID mass spectra of synthetic 4-hydroxybenzoylcholine shows fragment ions at *m/z* 165, 121, and 93. Thus, the unknown compound was identified as 4-hexosyloxybenzoylcholine, which is a new compound first identified from Arabidopsis.

7.4.7 The Multi-Platform Cross-Validation Methods in Metabolite Identification

Single identification method is inadequate to cope with the complex plant metabolic system and a large number of metabolites. Xu proposed a new micro-preparation method for discovery and identification using biomarkers, including LC-MS fingerprint analysis and multivariate data analysis to find markers. Compounds can be identified by combining with FT-MS to determine accurate mass, micro-preparation, MS/MS fragmentation information, GC-MS and retention indices, literature search, and isotope compounds synthesis (Xu 2008).

7.4.8 The Application of Stable Isotope Labeling Technique in the Identification of Plant Metabolites

Stable isotope labeling technique has wide applications in proteomics and biochemistry, plays an important role in revealing secrets of in vivo and intracellular physical and chemical processes and clarifying the material basis of life activities, and is a very effective means. In recent years, stable isotope labeling and mass spectrometry techniques can be used in labeling DNAs, proteins, and metabolites through a certain culture system, by incorporating reagent rich in heavy isotopes (^{13}C, ^{2}H, ^{18}O, and ^{15}N) into a cell, tissue, plant, or animal. Natural abundance metabolites and labeled metabolites can be distinguished by mass spectrometry. Plant metabolome can be labeled by $^{13}CO_2$. External contaminants and metabolites of plant can be distinguished by comparing $^{13}CO_2$ and $^{12}CO_2$ metabolic extract. The number of carbon atoms in the compound can be inferred by mass shift. Plant extract labeled with $^{13}CO_2$ may play a role in assisting identification as internal standard. It is a very promising technology.

7.5 Metabolomics Database

Metabolomics is still in the progress of innovation, development, and accumulation. Different laboratories used different analytical instruments to analyze biological samples and generated a lot of metabolomics data. The data exchange and assimilation require a proper database platform, which can store, manage, publish, search, and annotate a variety of data. The establishment of database also helps to join metabolomics and other branch platforms of systems biology.

Wiley Registry of Mass Spectral Data and NIST/EPA/NIH Mass Spectral Database are the two largest commercial and comprehensive GC-MS database with popular usages. The latest integrated version of Wiley 9th Edition/NIST 2008 contains a total of 796,000 mass spectra and 667,000 compounds, of which 746,000 spectra have chemical structures and 2,900,000 compound names. NIST MS/MS library contains 5308 precursor ions. The NIST GC retention index database contains 43,000 retention indices.

The Golm Metabolome Database (GMD) is a metabolomics database developed by scholars in German Max-Planck and can be obtained free of charge, which including GC-MS of derivative metabolites and GC TOF MS spectra library. The current GMD contains about 2000 evaluated mass spectra, including 1089 non-redundant MS tags (MSTs) and 360 identified MSTs provided by both quadrupole and TOF-MS platform. In addition, GMD database also contains MS retention index (MSRI) libraries, which greatly improve the identification of structurally similar compounds. GMD also provides experimental methods and chromatography and mass spectrometry conditions.

MassBank is a mass spectra database jointly established by many Japanese universities and research institutions. It mainly collects data from high-resolution mass spectrometers including ESI-QqTOF-MS/MS, ESI-QqQ-MS/MS, ESI-IT-$(MS)^n$, GC-EI-TOF-MS, LC-ESI-TOF-MS, FAB-CID-EBEB-MS/MS, FAB-MS, FD-MS, CI-MS, and LC-ESI-FT-MS. The reference spectra contain mass spectrometry information of multistage mass spectrometry. So far, it contains more than 24,993 mass spectra including positive and negative ion modes for more than 12,000 primary and secondary metabolites. MassBank allows free Web search, mass spectra comparing and three-dimensional visualization comparing by inputting mass spectra in text format.

The METLIN metabolite database is established by the Biological Mass Spectrometry Center of The SCRIPPS Research Institute (TSRI), which contains LC-MS, MS/MS, and FTMS data of 23,000 endogenous and exogenous metabolites of human, small molecule drugs and drug metabolites, small peptides (8,000 di- and tripeptides), etc., which can be retrieved by mass and molecular formula and structure. METLIN provides free online search.

The Fiehn GC-MS database is established by the Olive Fiehn laboratory and currently contains about 1,050 entries of about 713 common metabolites, including entries of partial derivative metabolites. Every entry includes searchable EI mass spectra and retention indices.

MoTo DB is a metabolite database of tomato fruit obtained from Q-TOF-MS, which contains retention times, accurate masses, UV absorption spectrum, MS/MS fragment ions, and reference literatures of metabolites. The MeT-RO plant and microbial metabolomics database are established by the Rothamsted Institute, which includes GC-MS, LC-MS, and NMR spectra.

The Kyoto Encyclopedia of Genes and Genomes (KEGG) is a database of systematic analyzing gene functions and genomic information. The KEGG integrated metabolic pathways' query is very excellent, including the metabolism of carbohydrates, nucleosides, and amino acids and the biodegradation of organics. It not only provides all possible metabolic pathways, but also comprehensively annotates the enzymes for each step of catalytic reaction, including amino acid sequence and links to PDB library. KEGG is a powerful tool for in vivo metabolism analysis and metabolic network research. KEGG now consists of six separate databases, including gene database (GENES database), pathway database (PATHWAY database), ligand chemical reaction database (NGAND database), sequence similarity database (SSDB), gene expression database (EXPRESSION), and protein molecular relationship database (BRITE). The PATHWAY database contains about 90 reference metabolic pathway patterns, and each reference pathway is a network composed of enzyme or EC number.

The MetaCyc is a sub-database of the BioCyc database, which is a database of metabolic pathways and enzymes. It expounded metabolic pathways of more than 1,600 organisms and contains metabolic pathways, reactions, enzymes, and substrates obtained from a large number of documents and online resources. Over 1200 metabolic pathways, 5500 enzymes, more than 5,100 genes, and 7,700 metabolites are included.

References

Bőttcher C, Roepenack-Lahaye EV, Schmidt J, Schmotz C, Neumann S, Scheel D, Clemens S. Metabolome analysis of biosynthetic mutants reveals a diversity of metabolic changes and allows identification of a large number of new compounds in Arabidopsis. Plant Physiol. 2008;147:2107–20.

Dixon RA, Strack D. Phytochemistry meets genome analysis, and beyond. Phytochemistry. 2003;62:815–6.

Ekman R, Silberring J, Westman-Brinkmalm A, Kraj A. Mass spectrometry instrumentation, interpretation and application. New Jersey: Wiley; 2008.

Fiehn O, Kopka J, Trethewey RN, Willmitzer L. Identification of uncommon plant metabolites based on calculation of elemental compositions using gas chromatography and quadrupole mass spectrometry. Anal Chem. 2000;71:3573–80.

Herebian D, Hanisch B, Marner FJ. Strategies for gathering structural information on unknown peaks in the GC/MS analysis of *Corynebacterium glutamicum* cell extracts. Metabolomics. 2005;1:317–24.

Kind T, Fiehn O. Metabolomic database annotations via query of elemental compositions: mass accuracy is insufficient even at less than 1 ppm. BMC Bioinformatics. 2006;7:234.

Kind T, Fiehn O. Seven golden rules for heuristic filtering of molecular formulas obtained by accurate mass spectrometry. BMC Bioinformatics. 2007;8:105.

Kopka J, Schauer N, Krueger S, Birkemeyer C, Usadel B, Bergmuller E, Dormann P, Weckwerth W, Gibon Y, Stitt M, Willmitzer L, Fernie AR, Steinhauser D. GMD@CSB.DB: the Golm Metabolome database. Bioinformatics. 2005;21:1635–8.

Liang C, Li YW, Zhao CX, Liang YZ. Qualitative analysis of the essential oils of traditional Chinese medicines using GC-MS and retention indices. J Instrum Anal. 2008;27:84–7.

Lu H, Dunn WB, Shen H, Kell DB, Liang Y. Comparative evaluation of software for deconvolution of metabolomics data based on GC-TOF-MS. Trends Anal Chem. 2008;27:215–27.

McLafferty FW, Turecek F. Interpretation of mass spectra. 4th ed. California: University Science Books; 2003.

Morlacchi P, Wilson WK, Xiong Q, Bhaduri A, Sttivend D, Kolesnikova MD, Matsuda SPT. Product profile of PEN3: the last unexamined oxidosqualene cyclase in *Arabidopsis thaliana*. Org Lett. 2009;11:2627–30.

Su Y, Liu SH, Wang CZ, Guo YL. Qualitative analysis of monoterpene isomers by GC-MS with spectral similarity analysis and retention index. J Instrum Anal. 2009;28:525–8.

Sumner LW, Amberg A, Barrett D, Beale MH, Beger R, Daykin CA, Fan TWM, Fiehn O, Goodacre R, Griffin JL, Hankemeier T, Hardy N, Harnly J, Higashi R, Kopka J, Lane AN, Lindon JC, Marriott P, Nicholls AW, Reily MD, Thaden JJ, Viant MR. Proposed minimum reporting standards for chemical analysis. Metabolomics. 2007;3:211–21.

Wagner C, Sefkow M, Kopka J. Construction and application of a mass spectral and retention time index database generated from plant GC/EI-TOF-MS metabolite profiles. Phytochemistry. 2003;62:887–900.

Wang HR, Zhang LT. The atmospheric pressure ionization interface technique in LC-MS and its applications in drug analysis. Chin J Hosp Pharm. 2006;26:1137–9.

Xu GW. Metabolomics-methods and applications. Beijing: Science Press; 2008.

Xu R, Fazio GC, Matsuda SPT. On the origins of triterpenoid skeletal diversity. Phytochemistry. 2004;65:261–91.

Yao XS. Organic compounds spectral analysis. Beijing: China Medical Science Press; 1996.

Chapter 8
Plant Metabolic Network

Shan Lu

8.1 Introduction

The metabolic network of plants is probably one of the most complicate network systems in nature. Traditionally, the metabolic pathways in this network are categorized as primary metabolism and secondary metabolism, with the concept that only primary metabolism is essential for the growth and development of plants. Now, it is clear that some of the secondary metabolites also possess critical functions in plants. Some of the so-called primary metabolites, such as plant growth regulators gibberellins (GA) and abscisic acid (ABA), are also synthesized from secondary pathways.

It is estimated that each plant could produce 5,000–25,000 different metabolites (Trethewey 2004). Besides carbohydrates, fatty acids, and proteins, about more than 100,000 secondary metabolites, such as isoprenoids, phenylpropanoids and alkaloids, have been identified from different plants. The biosynthesis and accumulation of secondary metabolites showed great spatiotemporal specificity and are sometimes unique in certain plant species. Thus, the secondary metabolites are also termed as plant-specialized metabolites (Gang 2005).

In general, the metabolic network is composed of the metabolic pathways and the regulatory machineries, both of which are interconnected for the transportation of metabolites, signals and energy. The successful operation of a metabolic network requires a precise coordination of metabolites, metabolic flux and regulatory signals among different tissues, cells and subcellular compartments. The coordination is essential for not only the production and transportation of metabolites, but also the regulation at various levels (e.g., transcriptional, posttranscriptional and translational)

S. Lu (✉)
School of Life Science, Nanjing University, Nanjing, China
e-mail: shanlu@nju.edu.cn

X. Qi et al. (eds.), *Plant Metabolomics*, DOI 10.1007/978-94-017-9291-2_8

for either a single gene or a group of genes. The entire network is fine-tuned in response to the endogenous cues for growth, differentiation and development, and to the environmental biotic and abiotic stimuli.

8.2 Plant Metabolic Network and Its Regulation

8.2.1 Compartmentation

In plants, some metabolic processes occur specifically in one or more subcellular compartments. For example, the biosynthesis of starch from ADP-glucose only takes place in plastids (Kirchberger et al. 2007), whereas that of folate requires the cooperation of three compartments, i.e., cytosol, plastids and mitochondria (Diaz de la Garza et al. 2007) (Fig. 8.1).

A well-studied case of metabolic compartmentation is the biosynthesis of isoprenoids in plants. It was commonly accepted that all plant isoprenoids were biosynthesized from the mevalonate (MVA) pathway in cytosol. However, a series of studies from 1990s have uncovered that higher plants utilize a methylerythritol phosphate (MEP) pathway for the biosynthesis of monoterpenoids, diterpenoids and carotenoids in plastids, and the cytosolic MVA pathway only accounts for sesquiterpenoids, triterpenoids, etc. (Rodríguez-Concepción et al. 2002; Zhao et al. 2013).

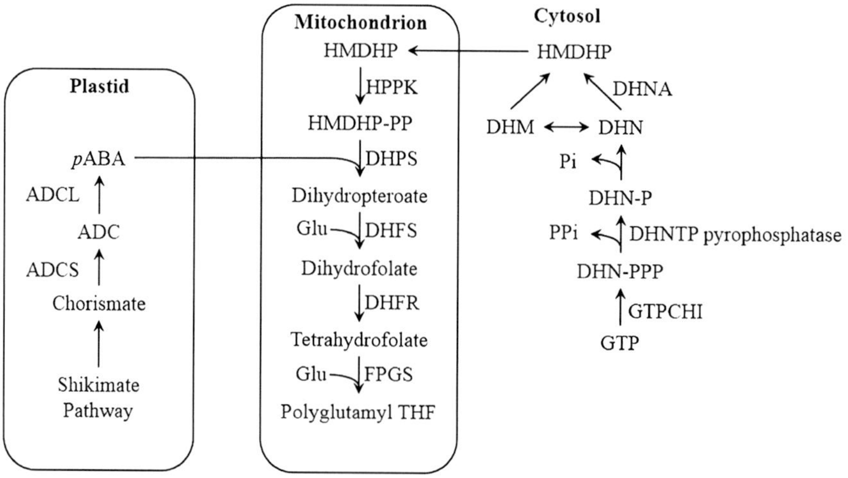

Fig. 8.1 The folate biosynthetic pathway and compartmentation. Abbreviations for metabolites and enzymes are as follows: *ADC* aminodeoxychorismate; *pABA* *p*-aminobenzoate; *HMDHP* 6-hydroxymethyl-dihydropterin; *Glu* glutamate; *THF* tetrahydrofolate; *DHM* dihydromonapterin; *DHN* dihydroneopterin; *GTP* guanosine triphosphate; and *ADCS* aminodeoxychorismate synthase; *ADCL* aminodeoxychorismate lyase; *HPPK* HMDHP pyrophosphokinase; *DHPS* dihydropteroate synthase; *DHFS* dihydrofolate (DHF) synthase; *DHFR* DHF reductase; *FPGS* folylpolyglutamate synthase; *DHNA* DHN aldolase; *GTPCHI* GTP cyclohydrolase I (Redraw according to Hossain et al. 2004)

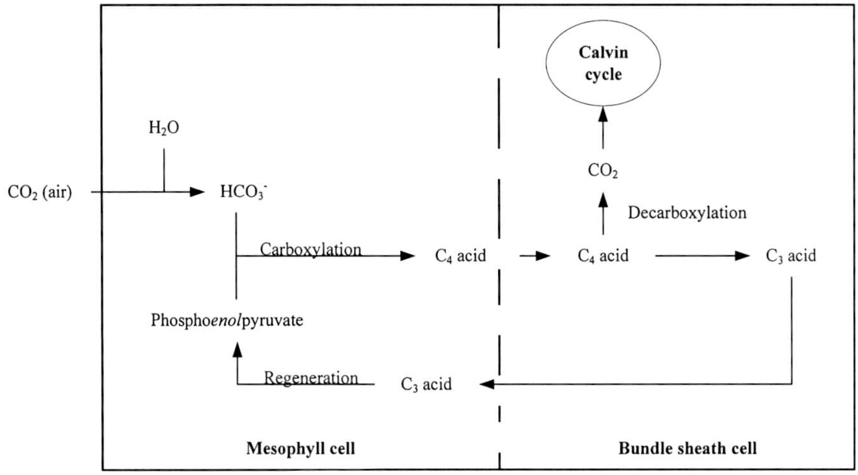

Fig. 8.2 Photosynthesis in a C4 plant (Redraw according to Buchanan et al. 2000)

In addition to subcellular organelles, some of plant metabolic pathways are only active in specific types of cells. In C4 plants, phosphoenolpyruvate (PEP) is carboxylated into oxaloacetate (OAA) and then converted to malate or aspartate in mesophyll cells. The C4 acid is transported to bundle sheath cells and is decarboxylated there to release CO_2 for Calvin cycle, whereas the C3 pyruvate or alanine left is transported back to mesophyll for the next cycle (Fig. 8.2).

For metabolic pathways which share common substrates or metabolic intermediates, compartmentation provides an option for plants to operate these pathways in separate locations, without unexpected competition. The sharing of metabolites brings the puzzles of studying the transportation machineries and their regulatory mechanisms. The identification of transporters on the membrane of plastid or mitochondria would make great contribution to bridge our knowledge of metabolic subnetworks in separate compartments.

The existence of common metabolites in different compartments also infers the possible presence of related isoenzymes and/or dual-targeting enzymes. The catalytic characteristics, substrate affinity and regulatory mechanism of these enzymes might vary with their corresponding compartments, and their coding genes might not express in the same pattern either.

8.2.2 Metabolon and Metabolic Channeling

In plant metabolic network, multiple enzymes on a same pathway could form a structural-functional complex, by protein–protein interaction, which is termed as metabolon. The structure of metabolon facilitates the delivery of metabolic intermediates between the catalytic sites of consecutive enzymes through an inside channel. Such a compact structure of metabolon not only enables a swift metabolite traffic, but also raises the concentrations of both enzymes and metabolites locally to make the enzymatic reactions more efficient. On the other hand, a close channel also prevents either the leaking of metabolic intermediates which could be harmful in cytosol, or the possible interference of the metabolic process in the channel by outside components. The phenylpropanoid metabolic pathway in plants has been well studied for metabolon and metabolic channeling (Winkel 2004). Considering the complexity of an intracellular environment, the regulation of plant metabolism by metabolon and metabolic channeling has two significant advantages. The first is that a plant cell can maintain a relative stable condition for carrying out catalytic processes. And the second is that, when necessary, plant can either change the structure of a metabolon for modulating the orientation of metabolic flux, or transport the entire metabolon as a unit to a new location where its product is needed. To switch the metabolon as a whole is probably more efficient than to de novo synthesize each of the enzymes and to reassemble them. This is exampled by the regulation of plant cellulose biosynthesis by the assembly, trafficking and recycling of the cellulose synthase complexes (CSC) (Gardiner et al. 2003; Lei et al. 2012).

In addition to the protein–protein interaction of enzymes close to each other on a metabolic pathway, two enzymes which are not neighboring can also interact to regulate plant metabolism. In Calvin cycle, both phosphoribulokinase (PRK) and glyceraldehyde-3-phosphate dehydrogenase (GAPDH) can form homodimers. With the help of a small protein CP12, these two enzymes can also form heterodimers. The assembly and dissociation of these dimers regulate the activities of both PRK and GAPDH and the activity of Calvin cycle (Fig. 8.3) (Winkel 2004).

8.2.3 Co-expression and Regulon

Gene co-expression has been widely reported in the metabolisms of plant primary metabolites, cell wall components, isoprenoids, phenylpropanoids, etc. A group of genes under the control of a same regulatory mechanism make up a regulon, which is commonly found in bacteria. Mentzen and Wurtele (2008) analyzed microarray information from 963 Affymetrix chips of *Arabidopsis* under different environmental, developmental and genetic conditions, and identified 998 regulons composed by different numbers of genes. The largest regulon contains 1,623 genes involved in spermatogenesis, pollen development and pollen tube growth. Genes on the genomes of plastids and mitochondria were also found to co-express with nuclear genes (Mentzen and Wurtele 2008). It was also reported that genes neighboring to each other on chromosome are often co-expressed (Mentzen and Wurtele 2008).

Wei et al. (2006) analyzed microarray data from 486 Affymetrix chips and revealed that genes on the same pathway have high ratio of co-expression. In those

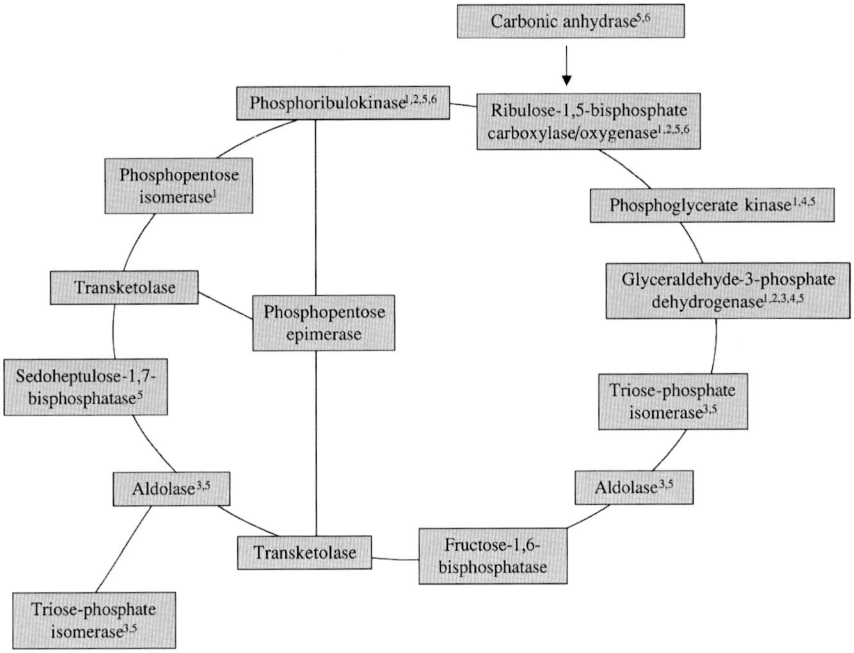

Fig. 8.3 Schematic of the Calvin cycle. *Superscripts* refer to evidence for the presence of some or all of the enzymes in a complex co-purification (*1, 2, 6*), protein interaction (*3, 4*), and co-localization at membranes (*5*) (Winkel 2004) (Reprinted by permission from ANNUAL REVIEW, Copyright © 2004)

metabolic pathways which contain more than 15 genes, they identified more than 100 genes co-expressing with their neighboring genes on the pathway. In all plant metabolic pathways, genes on primary metabolism such as Calvin cycle and Kreb's cycle are highly co-expressed. In addition, genes for carotenoid and chlorophyll metabolism are also highly co-expressed, probably also because these pigments function in photosynthesis. *GUN4* (At3g59400) was found to co-express with many genes in their study, because it both takes part in chlorophyll biosynthesis and functions in the retrograde signaling from the plastid to the nuclear (Larkin et al. 2003). Genes for plastid proteins which are putatively involved in photosynthesis or electron transport also showed high levels of co-expression.

To further analyze the regulatory mechanism of plant metabolism, Jiao et al. (2010) studied the expression of 2,268 genes on 174 pathways. They predicted 91 highly confident transcription factor–gene pairs. Some of the results were confirmed by previous reports. This showed the possibility of using plant metabolic network and co-expression network information to interpret gene regulation network. Web sites such as the *Arabidopsis* Information Resource (TAIR), ATTED-II (http://atted.jp) and Genevestigator (http://www.genevestigator.com) have related information and analytical tools.

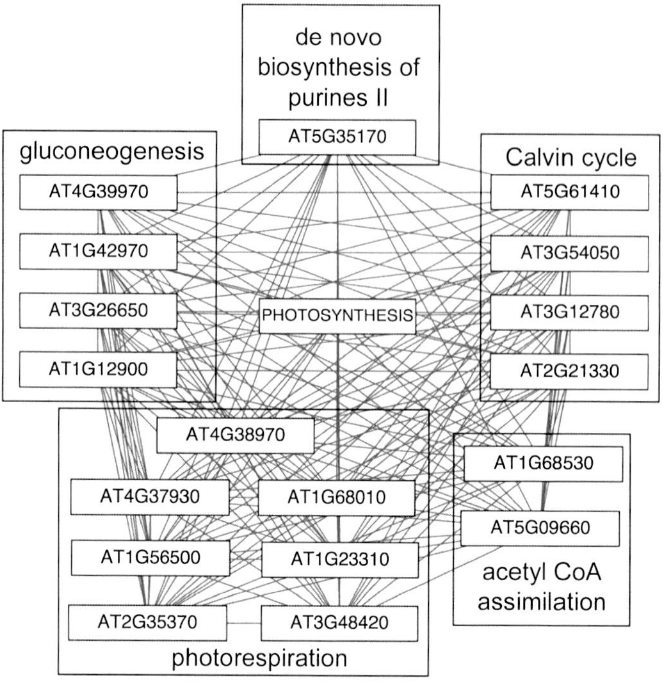

Fig. 8.4 Patterns of co-expression with photosynthesis pathway genes. Genes that are co-expressed (p value <1E−60) with each gene in the photosynthesis light reaction pathway are shown (Wei et al. 2006) (Reprinted by permission from American Society of Plant Biologists, Copyright © 2006)

Moreover, genes on different pathways were also found to co-express. One of the examples is that some photosynthetic genes are co-expressed with those for the metabolism of 3-phosphoglycerate and fructose-1,6-diphosphate (Fig. 8.4).

In 2008, Mentzen et al. (2008) analyzed the co-expression of genes on the metabolic pathways of starch, fatty acid and leucine. In addition to the enzymes directly on the pathways, genes encoding transporters, enzymes for the biosynthesis of cofactors and upstream precursors, and other regulatory molecules were also found to be involved in the co-expression network. Their computation identified a gene encoding a putative protein kinase AtPERK10 (At1g26150), which is tightly co-expressed with starch catabolism. Experimental work proved their bioinformatic postulation. The knockout mutant of this *AtPERK10* gene did accumulate starch as expected. This showed the feasibility of utilizing gene co-expression analysis for searching novel genes for enzymes or regulatory factors in plant metabolic network studies.

In plant metabolic network, some pathways are not running in parallel, and they cross at hub genes. The interconnection of different pathways at a hub gene is one of the reasons for the co-expression of genes belonging to these separate pathways.

Wei et al. (2006) found that genes at each node of the metabolic network may have different number of connections with up- or downstream genes. Most genes have only limited connections, whereas a small number of genes have complicate connection network. In 139 hub genes each with 20 or more connections, 65 % are encoded by single-copy genes. However, for those with fewer than 20 connections, only 37 % are encoded by single-copy genes. This means that, in average, a single-copy gene could have 11.5 connections, and a member of a multigene family usually has only 5 connections. This infers that the evolution of plant metabolism might start from the duplication of enzyme genes. Each of the duplicated genes carried partial of the connections of the ancestral gene and then diversified. Repeated duplication and diversification extended a single pathway into a subnetwork. Some genes retain the expression pattern of their ancestors and thus result in the co-expression of genes on different pathways.

8.3 Examples of Metabolic Network Studies

8.3.1 Simplified Examples of Plant Metabolic Network

Because of the complexity of plant metabolic network, a good strategy in research is to begin with some simplified or specialized systems. Such a system can be unique by either operating a few dominant metabolic pathways or having almost no primary metabolic pathway running. Considering that many of the enzymes are encoded by multicopy genes, it would be helpful if an organism has relatively fewer homolog genes for corresponding enzymes.

8.3.1.1 Algae

The unicellular green algae, such as *Chlamydomonas reinhardtii*, has long been a model organism for plant physiology, cell biology and molecular biology studies. Its nuclear, plastidic and mitochondrial genomes have all been sequenced. Both experimental and bioinformatic analyses found that *C. reinhardtii* has much fewer metabolites and homolog genes for enzymes comparing with higher plants.

On the other hand, cell differentiation increases the complexity in metabolic studies. For an example, leaves of a higher plant contain different types of cells, such as mesophyll, stomata, guard cell, vascular and bundle, which are all different in part of their metabolic network and in gene expression pattern. The result from leaf material actually was from a blend of these cells. The utilization of single-cell alga helps to avoid such a problem and also brings another benefit that all cells can be synchronized easily in their growth. Moreover, some microalgae can grow heterotrophically or mixotrophically with supplied carbon source such as acetate or glucose. This makes the metabolic network even simpler, because photosynthesis, the majority of the metabolic network in a photosynthetic organism, is temporarily suspended.

By feeding algal cells with isotope-labeled carbon source or metabolic intermediates, different metabolites can be simultaneously monitored by NMR conveniently. The discovery of the MEP pathway for plant isoprenoid metabolism was a good example of the advantages of using the algal system. The studies on a green alga *Scenedesmus* and other organisms confirmed the existence of both MEP and MVA pathways in higher plants and found that green algae exclusively utilize the MEP pathway for isoprenoid biosynthesis.

In 2009, Boyle and Morgan (2009) drafted a metabolic network of the primary metabolism in *C. reinhardtii* containing 484 reactions and 458 metabolites in plastids, mitochondria and cytosol. The distribution of metabolites suggested a central role of plastids in primary metabolism, and a key transportation function of cytosol in the network. They also studied the capacity of assimilation of algal cells under autotrophic and heterotrophic conditions. The results showed that each mole of carbon source could generate about 28.9 g of biomass for autotrophic growth, whereas it only generated 15 g of biomass under heterotrophic condition. It showed that about half of the carbon assimilated was probably used for energy supply. This work provided a good reference for further studies in the metabolic network of higher plants.

Using *C. reinhardtii* as a model organism, Manichaikul et al. (2009) described a systems biological strategy to integrate genomic information with methods of the prediction and characterization of novel genes, transcriptome analysis and mutagenesis (Fig. 8.5). The framework proposed for metabolic network construction and refining can also be adopted for higher plant research.

8.3.1.2 Glandular Trichome

Glandular trichomes are specialized cells where isoprenoid and other volatile compounds are usually synthesized and/or stored, and are excellent materials for metabolic network studies, because there are barely no primary metabolic activities in these cells (Wagner et al. 2004). There have been a large amount of studies from the pioneer work on trichome isolation to current omics work (Gershenzon et al. 1992; Dai et al. 2010).

In the past 20 years, peppermint (*Mentha* × *piperita*) has been established as a good system for studying isoprenoid metabolism, especially in its glandular trichomes (Croteau et al. 2005). A high-quality peppermint essential oil consists of a high level of menthol, a moderate level of menthone, and low levels of pulegone and menthofuran. Chemical, biochemical and molecular biological studies have elucidated the biosynthetic pathway for these monoterpene constituents (Fig. 8.6), and showed the involvement of four subcellular compartments, cytosol, endoplasmic reticulum, leucoplast and mitochondrion, in glandular trichomes. Menthofuran was found to be specifically accumulated in glandular trichomes on peppermint leaves, which could reach a concentration as high as 20 mM under low-light

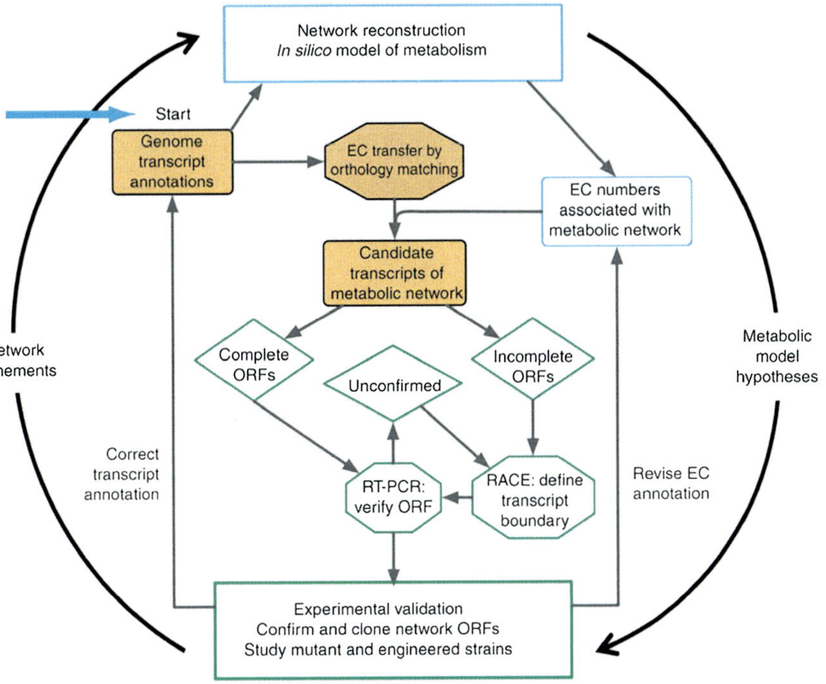

Fig. 8.5 Assessing and improving gene annotation for *Chlamydomonas reinhardtii*: iterative process integrating gene annotation experiments with metabolic network reconstruction and analysis (Reprinted by permission from Macmillan Publishers Ltd: [Nature Methods] (Manichaikul et al. 2009))

condition, comparing with a concentration of 0.4 mM under normal conditions. Rios-Estepa et al. (2008) systematically analyzed the accumulation of monoterpene constituents in peppermint by mathematical modeling and experimental analysis. By calculation, they postulated that menthofuran should have an inhibitory effect on pulegone reductase (PR), the enzyme at the branch point in the metabolic pathway, so that the fluctuation in the levels of each of related metabolites under different conditions could be reasonably explained (Fig. 8.6). In vitro characterization of a recombinant PR proved that menthofuran was indeed a weak competitive inhibitor of PR (Ki = 300 μM). These results supported the systems biology strategy of elucidating plant metabolic regulation by a combination of mathematical modeling of the metabolic network and the quantification of metabolites when a metabolic pathway has been clearly elucidated.

Another example of glandular trichome study is the biosynthesis of artemisinin in *Artemisia annua*. Olsson et al. (2009) analyzed the multicellular glandular trichomes of *A. annua* by laser microdissection (LMD) to separate one pair of apical cells and two pairs of subapical cells from the stalk cells and the mesophyll cells. Transcriptome analysis showed that genes encoding amorpha-4,11-diene synthase

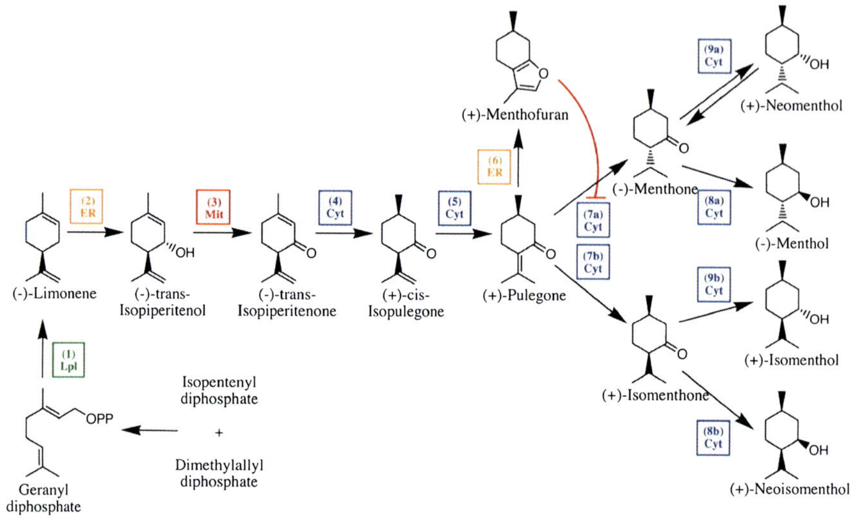

Fig. 8.6 Outline of *p*-menthane monoterpene metabolism in peppermint glandular trichomes (Rios-Estepa et al. 2008). The enzymes are (−)-limonene synthase (*1*); (−)-limonene 3-hydroxylase (*2*); (−)-trans-isopiperitenol dehydrogenase (*3*); (−)-trans-isopiperitenone reductase (*4*); (+)-cis-isopulegone isomerase (*5*); (+)-menthofuran synthase (*6*); (+)-pulegone reductase [(−)-menthone-forming activity] (*7a*); (+)-pulegone reductase [(+)-isomenthone-forming activity] (*7b*); (−)-menthone: (−)-menthol reductase [(−)-menthol-forming activity] (*8a*); (−)-menthone: (−)-menthol reductase [(+)-neoisomenthol-forming activity] (*8b*); (−)-menthone: (+)-neomenthol reductase [(+)-neomenthol-forming activity] (*9a*); (−)-menthone: (+)-neomenthol reductase [(+)-isomenthol-forming activity] (*9b*). The subcellular compartmentations are *Cyt* cytosol; *ER* endoplasmic reticulum; *Lpl* leucoplasts; *Mit* mitochondria (Reprinted from Rios-Estepa et al. (2008). Copyright © 2008 National Academy of Sciences, USA. Reprinted with permission)

(ADS), amorpha-4, 11-dienen hydroxylase (CYP71AV1) and artemisinin aldehyde Δ11(13) reductase (AAD) for artemisinin biosynthesis were all specifically expressed in the apical pair of cells. This suggested that artemisinin is synthesized there and then stored in the subcuticular space (Olsson et al. 2009).

8.3.2 *Metabolic Control Analysis (MCA)*

MAC describes the levels of metabolites and metabolic flux at each enzyme on a specific pathway and their contribution to the metabolic network. The contribution of signaling molecules is also more and more taken into account in recent years. According to the knowledge of a predicted pathway, this can be approached by comparing levels of metabolites under different conditions, such as by sequentially overexpressing enzyme genes, perturbing metabolic process under stress conditions or by using inhibitors, or other methods to analyze. Rios-Estepa and Lange (2007) listed a series of freely accessible data sets, analytical software packages, online tools and mathematical models currently used for different organisms and pathways.

Because photosynthesis is one of the best studied pathways, it is about the earliest example of being studied by MCA. Kruckeberg et al. (1989) studied the influence of cytosolic and plastidic phosphoglucose isomerases on starch biosynthesis. It was revealed that the plastidic phosphoglucose isomerase only showed significant regulatory effect on starch biosynthesis under high light or high CO_2 concentration, and the cytosolic enzyme showed significant control in low light, when it also exerted positive control on the biosynthesis of sucrose.

For carbon assimilation, in addition to sedoheptulose bisphosphatase and Rubisco of the Calvin cycle, other enzymes and triose phosphate transporters on plastid envelope all showed regulatory effects on carbon assimilation. This suggests that it might be impossible to effectively enhance plant photosynthetic capability by simply overexpressing several genes for enzymes of the Calvin cycle. Peterhansel et al. (2008) reported that the components of the photosynthetic electron transport chain and the photorespiration, and the activities of Rubisco, sedoheptulose bisphosphatase and fructose-1,6-diphosphatase should all be taken into account.

Poolman et al. (2009) also built a metabolic model of *Arabidopsis* by MCA with genomic information of the metabolic network of *Arabidopsis* from AraCyc. The analysis on suspension cells growing heterotrophically showed that Rubisco was still active under heterotrophic conditions. This agreed with the finding that Rubisco was active in the recycle of CO_2 in lipid metabolism in oilseed rape. It seems that under non-photosynthetic conditions, the carboxylase activity of Rubisco was cooperated with phosphoglucose isomerase, transaldolase, transketolase, xylulose-5-phosphate epimerase, PRK, etc., to catalyze the conversion of two molecules of 3-phosphoglyceric acid from one molecule of glucose-6-phosphate, and to reduce 2 molecules of NADP with one molecule of ATP (Schwender et al. 2004).

Besides primary metabolism, there are also successful MCA studies on plant-specialized metabolism. By transforming tomato plants with a gene encoding phytoene synthase (PSY), which is a key enzyme for carotenoid metabolism, Fray et al. established a model for analyzing plant carotenoid metabolism and its regulation. The early work by overexpression *PSY* under the control of a constitutive promoter did not result in high accumulation of carotenoid in fruits, but a dwarf phenotype. Metabolic analysis proved a direct competition for the common substrate geranylgeranyl diphosphate (GGPP) between the biosynthetic pathways of carotenoids and GA (Fray and Grierson 1993; Fray et al. 1995). This was improved by using a specific promoter to express *PSY* in the fruits, and using a transit peptide to target the expressed fusion protein into plastids (Fraser et al. 2002). With wild-type and transgenic tomato plants, Rios-Estepa and Lange (2007) calculated the flux control coefficient (FCC) of each of the key enzymes for carotenoid biosynthesis. The result showed that the FCC of PSY was 0.36 in the wild-type tomato plants, but was only 0.15 in the transgenic lines, although its enzymatic activity increased 5- to 10-fold because of the overexpression of its coding gene. This inferred that when *PSY* was overexpressed, the activity of its coding enzyme was less significant in the regulation of metabolic flux than that in the wild-type plants. Other enzymes could also have regulatory functions. The contribution of each regulatory step might not be constant under different conditions (Rios-Estepa and Lange 2007).

8.3.3 Isoprenoid Metabolism of Plants

Isoprenoids are the largest group of plant-specialized metabolites, and their metabolism is also one of the best studied. In plants, isoprenoid metabolism connects the material and energy generated from the photosynthesis and respiration with the production of chlorophylls, carotenoids, sterols, tocopherols, GA, ABA, etc. The common precursors of isoprenoid biosynthesis, isopentenyl diphosphate (IPP) and dimethylallyl diphosphate (DMAPP), come from the MVA pathway in cytosol and from the MEP pathway in plastids. The two pathways utilize different substrate, but both produce the same precursors for the biosynthesis of isoprenoids. A shuttle mechanism has been confirmed to transport metabolic intermediates crossing the plastid membrane. In the MEP pathway, deoxyxylulose 5-phosphate (DXP) was found to be involved in the biosynthesis of pyridoxol and thiamine, in addition to that of IPP. This is the first branching point in isoprenoid metabolism. In *Arabidopsis thaliana*, there are three genes encoding DXP synthase (DXS), but only one gene encoding a DXS reductase (DXR) to direct metabolic flux into the branch for IPP.

GGPP is another important branching point in isoprenoid metabolism. Different groups of diterpenoids were synthesized with GGPP as a common substrate. For example, copalyl diphosphate synthase (CPS) and kaurene synthase (KS) convert GGPP into GA, PSY catalyzes the condensation of two GGPP into phytoene for carotenoids and ABA, GGPP reductase (GGR) reduces GGPP into phytyl diphosphate for chlorophyll and tocopherols, and various diterpene synthases cyclize GGPP into different defensive diterpenoid phytoalexins. Twelve genes were predicted to encode GGPS in *A. thaliana*, with their translation products distributed in plastids, mitochondria, and endoplasmic reticulum (Lange and Ghassemian 2003; Okada et al. 2000). However, the functions of most of the GGPS have not yet been identified.

8.3.4 Potential Metabolic Capability

The analysis of plant metabolites usually only reflects a part of enzymatic activities in the metabolic network. Some of the metabolic capabilities were not revealed for several reasons.

The first reason is that, although a new catalytic activity has been gained during the evolution of one enzyme, the plant cell might not possess a suitable pathway yet to provide a substrate for this new activity. The transgenic work with a linalool synthase gene *LIS* led to different results in separate experiments. The overexpression transformants accumulated 8-hydroxylinalool in tomato (Lewinsohn et al. 2001), linalool glycoside in *Petunia hybrid* (Lücker et al. 2001), and oxygenated derivates in *Dianthus caryophyllus* (Lavy et al. 2002), instead of or in addition to the expected monoterpene linalool itself. These studies suggested the existence of enzymatic activities in different plants for catalyzing further modification of linalool. The metabolic capability was not revealed until linalool, the new substrate from the transgenic linalool synthase, was available.

Another reason comes from gene silencing. In rice, studies showed that genes encoding all enzymes for carotenoid biosynthesis, except only PSY, are expressed in rice endosperm. The absence of *PSY* blocks the metabolic flux from GGPP to carotenoids. It was predicted that the ancestors of rice might possess carotenoid biosynthetic capability in their endosperm (Lewinsohn and Gijzen 2009).

Moreover, the metabolic capacity also relies on the availability of a suitable sink for storage. An analysis on cauliflower found that all enzymes for carotenoid biosynthesis are expressed in the curd. The white phenotype in the wild-type plant curd was probably only because the development from leucoplasts to chromoplasts is blocked (Li et al. 2001). A DnaJ-like zinc finger protein ORANGE stimulates the chromoplastogenesis and results in a massive accumulation of carotenoids in the curd (Lu et al. 2006).

For enzymes with known functions, they could also show variations in their catalytic properties. Geranyl diphosphate synthase (GPPS) catalyzes the condensation from IPP and DMAPP into GPP, which is a substrate for monoterpene biosynthesis. However, in tomato plants, a dwarf phenotype was observed when *GPPS* was suppressed, suggesting a potential function of GPPS in the biosynthesis of GA (diterpenes) (van Schie et al. 2007). It is reported that a long-chain prenyl diphosphate synthase (SlDPS) from tomato could use GPP, fanesyl diphosphate (FPP) and GGPP as its substrates (Jones et al. 2013), and one of the GPPSs (AgGPPS3) in *Abies grandis* posesses 35 % FPP synthase activity in addition to that of GPPS (Burke and Croteau 2002).

8.4 Perspectives

There have been substantial progresses in plant metabolic network studies during the past two decades, benefited from the innovation of new analytical instruments and the accumulative omics information. The structures of more than 100,000 plant metabolites have been elucidated, and the major parts of most of the metabolic pathways have been identified and their genes and enzymes characterized. However, there still exists bottlenecks and challenges.

Novel genes and enzymatic activities Although a large amount of genes encoding enzymes on plant-specialized metabolic pathways have been studied, the cloning and characterization of new genes and the discovery of novel functions of enzymes which are already known will undoubtedly improve our toolbox for future research work, and also shed light on our understanding of the metabolic network of plants. Homolog blast has been and will still be one of the most productive methods for gene searching. This strategy is restricted to those genes sharing reasonable sequence similarities or harboring common structures in their sequences. Co-expression network analysis is an alternative method for identifying novel genes without sequence similarities. However, the prediction of the catalytic specificity is a big challenge, especially for the enzymes that belong to big families such as terpene synthases or P450s. Such identification and characterization work might be tedious but will still last for a long time.

Novel regulatory elements and transcription factors It is now clear that the utilization of constitutive promoter might result in unexpected result in metabolic engineering. Because of the spatiotemporal specificity and the compartmentation of plant metabolic network, specific regulatory elements are necessary to support a predictable modulation of plant metabolism. From another point of view, it is also critical to identify novel transcription factors to globally regulate the expression of genes for enzymes on one or more pathways, and also to coordinate the metabolic network and the developmental scheme.

New regulatory mechanisms Epigenetic regulation was also reported in the regulation of plant metabolic network. For instance, in 2009, Cazzonelli et al. 2009 reported that a gene for carotene isomerase (CRTISO) on carotenoid biosynthetic pathway was regulated at the chromosomal level by SDG8, which also regulates the expression of *FLC* and other flowering genes.

Clustering is another possible regulatory mechanism reported in recent years. By tandemly arranging a set of genes for enzymes on a same metabolic pathway, plant controls the levels of enzymes for consecutive catalytic steps together. Such an operon-like structure was exampled by the studies of Field and Osbourn (2008) on the biosynthetic pathway of triterpene phytoalexin thalianol in *Arabidopsis*. Four genes, for one synthase, two P450s, and one enzyme for further modification, were arranged sequentially in the genome (*At5g47980*, *At5g47990*, *At5g48000*, *At5g48010*), and showed similar expression patterns in different tissues. Evolutionarily, this gene cluster might originate from genome rearrangement and gene duplication and loss. It seems that this event occurred only recently, because such a gene cluster does not exist in other species closely related to *Arabidopsis*. It is possible that such a structure was maintained in *Arabidopsis* genome because of its advantage in regulating the biosynthesis of the defensive triterpene phytoalexin. Large gene clusters for diterpene metabolism were also reported on rice genome. For one of them, 10 genes encoding CPS, KS, and members of the P450 family were arranged in a 245 kb region on chromosome 2. These genes encode enzymes that are sufficient to catalyze a series of reactions from GGPP to 11α-hydroxy-*ent*-cassadiene, the precursor of phytocassane-type diterpene phytoalexins (Swaminathan et al. 2009).

Direct impact from metabolites In addition to the well-known feedback regulation from metabolites, riboswitch is a newly identified mechanism. A riboswitch is a fragment of the mRNA sequence that consists of two parts: an aptamer and an expression platform. The binding of small molecules by the aptamer changes the structure of the expression platform, which, in turn, regulates gene expression. The regulation by a riboswitch differs from other known regulatory mechanisms in that it is self-regulated and relies on no protein component. Most of riboswitches were found in prokaryotic organisms. One of the examples in plants is that for thiamine pyrophosphate (TPP) biosynthesis. A TPP-sensing riboswitch is found in the 3′-untranslated region (UTR) of the thiamin biosynthetic gene (*THIC*) of all plant species examined, including *A. thaliana*, *Oryza sativa*, and *Poa secunda* (Wachter et al. 2007). The binding of TPP contributes to the process and stability of gene transcripts. However, the existence of riboswitches other than that for TPP is still to be confirmed.

In the analysis of plant metabolic network, it is still difficult to quantify those insignificant metabolites which either exist at very low levels, or accumulate at a very short time. In most cases, our knowledge also makes it hard to predict which pathways of the network are perturbed from a specific phenotype, and vice versa. A combination of traditional methods in gene characterization and novel strategies at systems biology level such as functional genomics and genome-wide association studies (GWAS) will help to resolve plant metabolic network and provide a good reference for plant metabolic engineering.

References

Boyle N, Morgan J. Flux balance analysis of primary metabolism in Chlamydomonas reinhardtii. BMC Syst Biol. 2009;3:4.

Buchanan BB, Gruissem W, Jones RL. Biochemistry and molecular biology of plants. Rockville: American Society of Plant Biologists; 2000.

Burke C, Croteau R. Geranyl diphosphate synthase from *Abies grandis*: cDNA isolation, functional expression, and characterization. Arch Biochem Biophys. 2002;405:130–6.

Cazzonelli CI, Cuttriss AJ, Cossetto SB, Pye W, Crisp P, Whelan J, Finnegan EJ, Turnbull C, Pogson BJ. Regulation of carotenoid composition and shoot branching in *Arabidopsis* by a chromatin modifying histone methyltransferase, SDG8. Plant Cell. 2009;21:39–53.

Croteau R, Davis E, Ringer K, Wildung M. (−)-Menthol biosynthesis and molecular genetics. Naturwissenschaften. 2005;92:562–77.

Dai X, Wang G, Yang DS, Tang Y, Broun P, Marks MD, Sumner LW, Dixon RA, Zhao PX. TrichOME: a comparative omics database for plant trichomes. Plant Physiol. 2010;152:44–54.

Diaz de la Garza RI, Gregory JF III, Hanson AD. Folate biofortification of tomato fruit. Proc Natl Acad Sci USA. 2007;104:4218–22.

Field B, Osbourn AE. Metabolic diversification—independent assembly of operon-like gene clusters in different plants. Sci. 2008;320:543–7.

Fraser PD, Romer S, Shipton CA, Mills PB, Kiano JW, Misawa N, Drake RG, Schuch W, Bramley PM. Evaluation of transgenic tomato plants expressing an additional phytoene synthase in a fruit-specific manner. Proc Natl Acad Sci USA. 2002;99:1092–7.

Fray RG, Grierson D. Identification and genetic analysis of normal and mutant phytoene synthase genes of tomato by sequencing, complementation and co-suppression. Plant Mol Biol. 1993;22:589–602.

Fray RG, Wallace A, Fraser PD, Valero D, Hedden P, Bramley PM, Grierson D. Constitutive expression of a fruit phytoene synthase gene in transgenic tomatoes causes dwarfism by redirecting metabolites from the gibberellin pathway. Plant J. 1995;8:693–701.

Gang DR. Evolution of flavors and scents. Annu Rev Plant Biol. 2005;56:301–25.

Gardiner JC, Taylor NG, Turner SR. Control of cellulose synthase complex localization in developing xylem. Plant Cell. 2003;15:1740–8.

Gershenzon J, McCaskill D, Rajaonarivony JIM, Mihaliak C, Karp F, Croteau R. Isolation of secretory cells from plant glandular trichomes and their use in biosynthetic studies of monoterpenes and other gland products. Anal Biochem. 1992;200:130–8.

Hossain T, Rosenberg I, Selhub J, Kishore G, Beachy R, Schubert K. Enhancement of folate in plants through metabolic engineering. Proc Natl Acad Sci USA. 2004;101:5158–63.

Jiao Q, Yang Z, Huang J. Construction of a gene regulatory network for *Arabidopsis* based on metabolic pathway data. Chinese Sci Bull. 2010;55:158–62.

Jones MO, Perez-Fons L, Robertson FP, Bramley PM, Fraser PD. Functional characterization of long-chain prenyl diphosphate synthases from tomato. Biochem J. 2013;449:729–40.

Kirchberger S, Leroch M, Huynen MA, Wahl M, Neuhaus HE, Tjaden J. Molecular and biochemical analysis of the plastidic ADP-glucose transporter (ZmBT1) from *Zea mays*. J Biol Chem. 2007;282:22481–91.

Kruckeberg A, Neuhaus H, Feil R, Gottlieb L, Stitt M. Decreased-activity mutants of phosphoglucose isomerase in the cytosol and chloroplast of *Clarkia xantiana*. Impact on mass-action ratios and fluxes to sucrose and starch, and estimation of flux control coefficients and elasticity coefficients. Biochem J. 1989;261:457–67.

Lange BM, Ghassemian M. Genome organization in *Arabidopsis thaliana*: a survey for genes involved in isoprenoid and chlorophyll metabolism. Plant Mol Biol. 2003;51:925–48.

Larkin RM, Alonso JM, Ecker JR, Chory J. GUN4, a regulator of chlorophyll synthesis and intracellular signaling. Science. 2003;299:902–6.

Lavy M, Zuker A, Lewinsohn E, Larkov O, Ravid U, Vainstein A, Weiss D. Linalool and linalool oxide production in transgenic carnation flowers expressing the *Clarkia breweri* linalool synthase gene. Mol Breed. 2002;9:103–11.

Lei L, Li S, Gu Y. Cellulose synthase complexes: Composition and regulation. Front Plant Sci. 2012;3:75.

Lewinsohn E, Schalechet F, Wilkinson J, Matsui K, Tadmor Y, Nam K-H, Amar O, Lastochkin E, Larkov O, Ravid U, Hiatt W, Gepstein S, Pichersky E. Enhanced levels of the aroma and flavor compound *S*-linalool by metabolic engineering of the terpenoid pathway in tomato. Plant Physiol. 2001;127:1256–65.

Lewinsohn E, Gijzen M. Phytochemical diversity: the sounds of silent metabolism. Plant Sci. 2009;176:161–9.

Li L, Paolillo DJ, Parthasarathy MV, DiMuzio EM, Garvin DF. A novel gene mutation that confers abnormal patterns of β-carotene accumulation in cauliflower (*Brassica oleracea* var. *botrytis*). Plant J. 2001;26:59–67.

Lu S, Van Eck J, Zhou X, Lopez AB, O'Halloran DM, Cosman KM, Conlin BJ, Paolillo DJ, Garvin DF, Vrebalov J, Kochian LV, Küpper H, Earle ED, Cao J, Li L. The cauliflower Or gene encodes a DnaJ cysteine-rich domain-containing protein that mediates high levels of β-carotene accumulation. Plant Cell. 2006;18:3594–605.

Lücker J, Bouwmeester HJ, Schwab W, Blaas J, Van Der Plas LHW, Verhoeven HA. Expression of *Clarkia S*-linalool synthase in transgenic petunia plants results in the accumulation of *S*-linalool-β-D-glucopyranoside. Plant J. 2001;27:315–24.

Manichaikul A, Ghamsari L, Hom EFY, Lin C, Murray RR, Chang RL, Balaji S, Hao T, Shen Y, Chavali AK, Thiele I, Yang X, Fan C, Mello E, Hill DE, Vidal M, Salehi-Ashtiani K, Papin JA. Metabolic network analysis integrated with transcript verification for sequenced genomes. Nat Methods. 2009;6:589–92.

Mentzen W, Peng J, Ransom N, Nikolau B, Wurtele E. Articulation of three core metabolic processes in *Arabidopsis*: fatty acid biosynthesis, leucine catabolism and starch metabolism. BMC Plant Biol. 2008;8:76.

Mentzen W, Wurtele E. Regulon organization of *Arabidopsis*. BMC Plant Biol. 2008;8:99.

Okada K, Saito T, Nakagawa T, Kawamukai M, Kamiya Y. Five geranylgeranyl diphosphate synthases expressed in different organs are localized into three subcellular compartments in *Arabidopsis*. Plant Physiol. 2000;122:1045–56.

Olsson ME, Olofsson LM, Lindahl A-L, Lundgren A, Brodelius M, Brodelius PE. Localization of enzymes of artemisinin biosynthesis to the apical cells of glandular secretory trichomes of *Artemisia annua* L. Phytochem. 2009;70:1123–8.

Peterhansel C, Niessen M, Kebeish RM. Metabolic engineering towards the enhancement of photosynthesis. Photochem Photobiol. 2008;84:1317–23.

Poolman MG, Miguet L, Sweetlove LJ, Fell DA. A genome-scale metabolic model of *Arabidopsis* and some of its properties. Plant Physiol. 2009;151:1570–81.

Rios-Estepa R, Lange BM. Experimental and mathematical approaches to modeling plant metabolic networks. Phytochemistry. 2007;68:2351–74.

Rios-Estepa R, Turner GW, Lee JM, Croteau RB, Lange BM. A systems biology approach identifies the biochemical mechanisms regulating monoterpenoid essential oil composition in peppermint. Proc Natl Acad Sci USA. 2008;105:2818–23.

Rodríguez-Concepción M, Boronat A. Elucidation of the methylerythritol phosphate pathway for isoprenoid biosynthesis in bacteria and plastids. A metabolic milestone achieved through genomics. Plant Physiol. 2002;130:1079–89.

Schwender J, Goffman F, Ohlrogge JB, Shachar-Hill Y. Rubisco without the Calvin cycle improves the carbon efficiency of developing green seeds. Nature. 2004;432:779–82.

Swaminathan S, Morrone D, Wang Q, Fulton DB, Peters RJ. CYP76M7 is an *ent-cassadiene* C11α-hydroxylase defining a second multifunctional diterpenoid biosynthetic gene cluster in rice. Plant Cell. 2009;21:3315–25.

Trethewey RN. Metabolite profiling as an aid to metabolic engineering in plants. Curr Opin Plant Biol. 2004;7:196–201.

van Schie CCN, Ament K, Schmidt A, Lange T, Haring MA, Schuurink RC. Geranyl diphosphate synthase is required for biosynthesis of gibberellins. Plant J. 2007;52:752–62.

Wachter A, Tunc-Ozdemir M, Grove BC, Green PJ, Shintani DK, Breaker RR. Riboswitch control of gene expression in plants by splicing and alternative 3′ end processing of mRNAs. Plant Cell. 2007;19:3437–50.

Wagner GJ, Wang E, Shepherd RW. New approaches for studying and exploiting an old protuberance, the plant trichome. Ann Bot. 2004;93:3–11.

Wei H, Persson S, Mehta T, Srinivasasainagendra V, Chen L, Page GP, Somerville C, Loraine A. Transcriptional coordination of the metabolic network in *Arabidopsis*. Plant Physiol. 2006;142:762–74.

Winkel BSJ. Metabolic channeling in plants. Annu Rev Plant Biol. 2004;55:85–107.

Zhao L, Chang W-C, Xiao Y, Liu H-W, Liu P. Methylerythritol phosphate pathway of isoprenoid biosynthesis. Annu Rev Plant Biol. 2013;82:497–530.

Chapter 9
Applications of LC-MS in Plant Metabolomics

Guo-dong Wang

Although plant metabolomics is still in the development stage, it is developing rapidly. Current international working is on the large-scale metabolomics data accumulation, namely the use of a variety of chemical analytical platforms, especially mass-based GC-MS and LC-MS in plant (mostly concentrate on the model plant Arabidopsis, including different mutants, ecotypes, tissues, and treatments with different conditions) metabolomics analysis. The international plant metabolomics society established a set of standardized procedures on different plant metabolomics analytical tools, including experimental design, data processing, and reporting, and making the metabolomics data sharing between different laboratories possible (Fiehn et al. 2007), which will greatly improve the role of metabolomics in plant research. Plant metabolomics also evolved from simple metabolic analysis to systematically reveal the mechanism of plant cell by combining with other omics techniques.

Different from chemical researchers who mainly engaged in the development of analytical methods, biological researchers engaged in plant secondary metabolic study mostly use metabolomics analysis as a tool for the functional identification of plant genes with the purpose of understanding of the life process of the whole plant. With the rapid development of a variety of modern biological techniques (changing the expression level of target gene by transformation is the most common technique), the function of genes causing "visible phenotypic change" can be rapidly identified by using traditional genetic methods. But in the plant genome, changing in many genes of unknown function cannot cause significant "visible phenotypic change." Fortunately, the chemical difference between transgenic and control plants detected by metabolomic technique further provide clues for functional identification of the gene of interest. Of course, this kind of metabolite-gene link must eventually be confirmed by traditional biological methods. At present, metabolomics technology has been successfully applied in the functional identification of unknown genes in Arabidopsis genome. For example, Hirai et al. positioned

G.-d. Wang (✉)
Institute of Genetics and Developmental Biology, Chinese Academy of Sciences,
Beijing 100101, China
e-mail: gdwang@genetics.ac.cn

X. Qi et al. (eds.), *Plant Metabolomics*, DOI 10.1007/978-94-017-9291-2_9

Arabidopsis network pathway of gene–gene and metabolite–gene under sulfur deficiency using metabolomics and transcriptomics analysis, and determined three genes with wrong functional annotation are the desulfoglucosinolate sulfotransferase involved in the biosynthetic pathway of glucosinolates (Hirai et al. 2005) by the combination of traditional biochemical approach.

Moreover, the combination of different "omics" has become the trend of the study of gene function in plants (Fukushima et al. 2009). In addition to gene function analysis, the joint application of various omics can help us for in-depth understanding of the regulation of metabolic pathways (including hormones) and the coordination and interaction between different metabolic pathways.

In this chapter, we first introduce the sample preparation process in order to highlight its importance in metabolomics analysis and then describe two applications of LC-MS in plant metabolomics.

9.1 Sample Preparation

As mentioned in Chap. 3, the planting conditions and acquisition conditions of experimental material must be maintained as far as possible consistent and the analysis on sample should be as fast as possible after sampling to avoid introducing system error; otherwise, a serious system error often leads to the analysis results do not have any biological significance.

The following experimental procedure is for the analysis of saponin in barrel medic (*Medicago truncatula*) by the use of LC-MS (Huhman and Sumner 2002). For reference, the experimental procedure can be suitably modified according to the reader's requirements:

Step ① Freeze-dry the freshly collected plant material stored in liquid nitrogen (2–3 days). If you cannot immediately proceed to the next step, the dry materials should be kept at −80 °C freezer. The sample taken from the −80 °C freezer will absorb moisture, so it should be reprocessed by freeze-dry; otherwise, weighing accuracy will be affected;

Step ② Shatter the freeze-dried material. Generally, the material is placed in a 4-ml vial which has a flat-head sample spoon in it. Vortex 1–2 min to obtain a very fine powder;

Step ③ Weigh accurately 10 ± 0.6 mg sample and put it into a 4 ml vial;

Step ④ Add 1 ml 80 % methanol and the internal standard (0.018 mg/ml 7-hydroxy coumarin) mixture (the solvent can be adjusted to be mutual soluble with the LC mobile phase);

Step ⑤ Shake mildly overnight in the horizontal shaker;

Step ⑥ Centrifuge at 2,900 g for 30 min (4 °C);

Step ⑦ Transfer the supernatant to a new vial and concentrate with nitrogen to remove organic solvent in the sample;

Step ⑧ Dilute the resulting sample with methanol to a final concentration of 35 %;

Step ⑨ Transfer the sample to a C_{18} solid-phase extraction (SPE) column and rinse with 100 % water and 35 % methanol with two-column bed volume, respectively;

Step ⑩ Elute saponin and other analytical samples with two-column bed volume of methanol; rotary evaporate and concentrate for analysis (if storage is needed, place it in a −20 °C freezer).

Note:

① Usually for the high-throughput omics chemical analysis without preference, the sample from Step ⑥ can be filtered and directly subjected to LC-MS analysis. Step ⑦–⑩ is the use of SPE column for sample pre-treatment, commonly used for the LC analysis of samples with preference. In this example, saponins in the alfalfa root were firstly analyzed and identified. The SPE column with the same packing material as that of analyzing column was used to roughly purify and enrich saponins. SPE is a sample pre-treatment technology developed from the mid-1980s (Berrueta et al. 1995), which is developed by the combination of liquid–solid extraction and LC, mainly through the selective absorption and desorption between sample components and stationary phase to achieve the separation, purification, and enrichment of the chemical of interest, mainly for the purpose of reducing the sample matrix interference and improving detection sensitivity.

② Researchers engaged in biochemical experiments often accustomed to use a pipette and a variety of plastic products, but it should be avoided as far as possible in the preparation of samples for LC-MS analysis; otherwise, the mixed peaks will be introduced and the ionization efficiency of MS will be reduced.

③ The commonly used extraction reagents are chloroform–methanol system, which is suitable for both water-soluble and lipid-soluble metabolites; hot ethanol extraction solution (usually 80 °C 80 % ethanol), which is applicable for the extraction of polar and moderately polar metabolites; and trichloroacetic acid–ether solution, which is applicable for the extraction of acid-stable, water-soluble metabolites. When using methanol-containing extract solvents, it should be noted that esterification easily occurs between methanol and the carboxylic acid group of the metabolites; thus, if the methyl ester group is present in the LC-MS analysis, the source of this methyl group must be carefully determined that if it is plant endogenous or exogenous introduced.

④ Prior to LC-MS analysis, it is preferably filtering the sample with a 0.22-μm pore-size filter membrane, because the insoluble substance will block the column and contaminate the ion source. An alternative cost-reducing method is to centrifuge the sample at high speed (at >12,000 g) for 10 min prior to LC-MS analysis.

⑤ Condition of each laboratory is varied, but the growth of the analytical materials must be strictly controlled consistent, which is an important factor affecting the analysis results.

9.1.1 Applications of LC-MS in Gene Functional Identification

Whole-genome sequencing has been completed for Arabidopsis, a model plant for plant biologists. So far, a very complete range of public resources, including full-length cDNA sequences, various forms of mutant (www.arabidopsis.org), and rich microarray data (www.genevestigator.com and other sites), has been provided, facilitating the process of gene functional identification in Arabidopsis. The most detailed study in the field of Arabidopsis secondary metabolism is the biosynthetic pathway and regulation of flavonoids. Flavonoids are responsible for the testa color of Arabidopsis. If the biosynthetic or transport process of flavonoids is affected, the testa color will become lighter. Based on this phenotype, researchers isolated *tt* (transparent testa) mutant series, cloned, and functionally identified the corresponding genes using forward genetics approach (from phenotype to gene function). Some of these genes are structural genes (*TT3, TT4, TT5, TT6, TT7 FLS1, LDOX,* and *BAN*) encoding enzymes involved in the biosynthesis of flavonoids; some are regulatory genes (*TT1, TT2, TTT8, TTG1, TTG2,* and *TT16*) encoding transcription factors regulating the biosynthetic pathway; some genes (*TT12* and *TT19*) encode proteins involved in the transport and accumulation of pigments (Lepiniec et al. 2006). But the gene mutation downstream the metabolic pathway involving in the structural modification of flavonoids did not cause visible phenotypes like lighter testa color. The functional identification of this class of genes is a great challenge for researchers.

Japanese scientists used LC-MS to comprehensively analyze a variety of flavonoids biosynthesis mutants, combined with transcriptomic analysis, functionally identified several flavonoid glycosyltransferase (Yonekura-Sakakibara et al. 2007, 2008), which is a classic example of metabolomics analysis applications in the identification of gene function in plant secondary metabolites.

9.1.2 Experimental Methods

9.1.2.1 Sample Preparation

Plant growth conditions: biological incubator, 22 °C; 16 h light/8 h darkness; 4 weeks.

Leaf tissue extraction:

1. Take blades grown for 4 weeks with a scissor, accurately weigh the fresh weight, and then add crushing metal beads;

2 Add 5 μl extract solvent (methanol: acetic acid: water = 9:1:10, 0.02 mmol naringenin-7-*O*-glucoside for quantitative analysis) per mg fresh material;
3. Crush the sample in a sample-mixing pulverizer (30 Hz, 5 min);
4. Centrifuge at 12,000 g for 10 min;
5. Take supernatant for LC-MS analysis.

Note that in general, metabolomics analysis requires 4–5 biological replicates for each analytical sample.

9.1.2.2 Instrument Settings: UPLC-PDA-ESI/Q-TOF/MS

1. Column: Phenyl C_{18} (Φ2.1 × 100 mm, 1.7 μm); column temperature: 35 °C;
2. Mobile phase: 0 min, 95 % A + 5 % B; 9 min, 60 % A + 40 % B; 13 min, 95 % A + 5 % B; solvent A is water containing 0.1 % formic acid; solvent B is methanol containing 0.1 % formic acid; the reagents used are HPLC grade;
3. The PDA Detector: UV range is at 210–500 nm;
4. MS condition: electrospray ionization (ESI), positive mode; TOF is used for the detection of the molecular ion peak $[M + H]^+$ of flavonoid glycosides; the detection condition set for the fragment ion peak is desolvation temperature: 450 °C; the nitrogen flow rate: 600 L/h; capillary spray voltage: 3.2 kV; ion source temperature: 150 °C; cone voltage: 35 V.

9.1.3 Experimental Procedure

9.1.3.1 Structure Elucidation and Distribution of Flavonoid Glycosyl Derivatives

For the research of plant secondary metabolism, the first thing is to understand the structure of the metabolite to be studied, then to infer synthetic pathway and predict possible types of enzymes involved in metabolic reactions. The authors first checked the tissue-specificity of total 32 flavonoid glycosides using UPLC-PDA-MS in Arabidopsis and found that the flower tissue contain the highest flavonoid glycosides in types and abundance. Of the 32 flavonoid glycoside components, 12 species (Fig. 9.1, f1–f8, f17, f18, f20, and f24) were previously reported; their chemical structure can be determined by comparing UV and MS information; the structure of other two peaks (f16 and f28) was resolved by comparing the retention time and MS of the standard; the remaining 18 flavonoid glycoside compounds were analyzed by metabolomic comparing wild-type plants with multiple flavonoid synthesis-deficient mutants with known biochemical function (the upstream and downstream relationship of metabolites in the biosynthetic pathway can be obtained), of which 9 species (f14, f15, f19, f23, f25, f27, f29, f30, and f32, all contain the pentose glycosyl group) were initially identified by the combination of

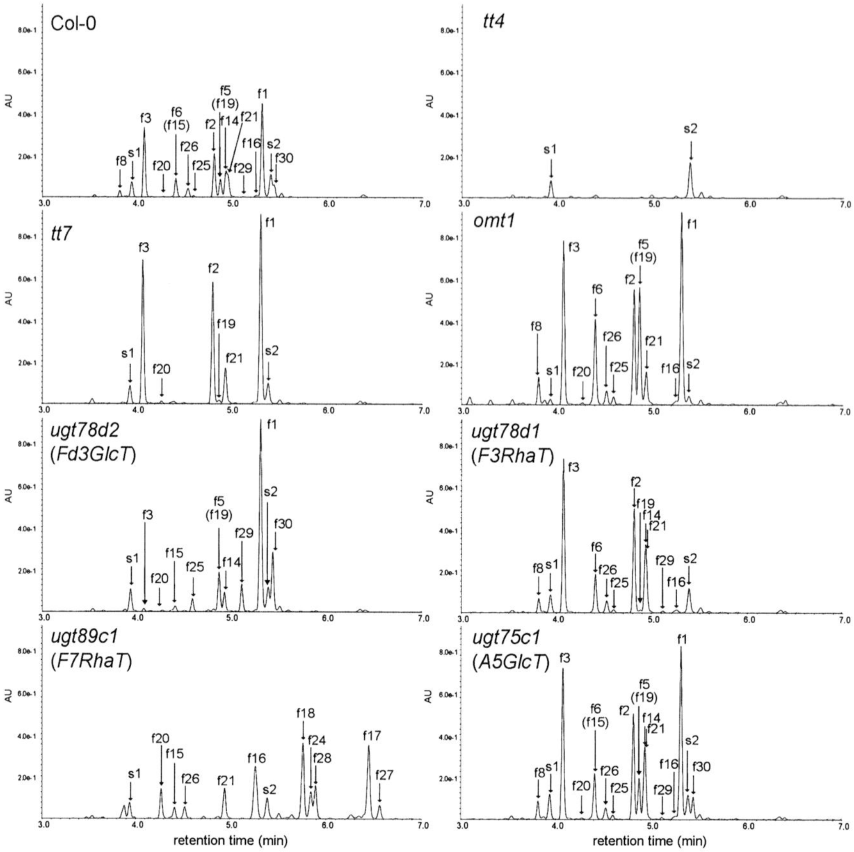

Fig. 9.1 Comparison of flavonoid components in the flower of wild type and a variety of flavonoid synthetic mutants by UPLC-PDA-MS

the biochemical function of each mutant gene and secondary MS information (mostly the glycosyl group fragment loss information, for example, glucose lose 162, rhamnose lose 146, and arabinose lose 132), but the kinds of the pentose (ribose, xylose, arabinose, and lyxose, their molecular weights are the same, just their MS information cannot be distinguished) were not determined (Fig. 9.1); the structures of the remaining 9 flavonoid compounds have not yet been determined, but the possible types of glycosyl group linked to them can be inferred from the MS information. According to the distribution results of 23 structurally identified flavonoid glycoside components in different mutants, it can be inferred that there are at least 4 flavonoid glycosyltransferases in Arabidopsis, which remain to be determined experimentally.

It is noteworthy that it is impossible to elucidate the chemical structure with MS information alone, if there is no mutants resources for chemical comparison. The structural identification of metabolite mostly still depends on the chemical standards usually collected by each laboratory.

Col-0 is the wild-type plant; *tt4* is the mutant of chalcone synthase gene (the structural formula shown in the figure is the parent structure of various identified flavonoid glycoside derivatives); *tt7* is a mutant of flavonoid 3' hydroxylase gene; *omt1* is a mutant of *O*-methyltransferase gene; *ugt78d1, ugt78d2, ugt 89c1,* and *ugt75c1* are mutants of different flavonoid glycosyltransferase gene. f1–f32 are flavonoids with different glycosides; s1 and s2 are phenolic acids. The horizontal coordinate of the chromatogram represents the retention time, and the vertical coordinate represents the UV absorbance at 320 nm (for the detection of flavonoids).

9.1.3.2 Selection of Candidate Genes

The metabolomics analysis of wild-type and various flavonoid synthesis mutants indicates that there are at least four flavonoid glycosyltransferase genes waiting to be functionally identified. But there are 107 genes in Arabidopsis genome annotated as small molecule glycosyltransferase, the finding of the right gene from them is still a huge challenge. Traditional biochemical method is to purify enzyme from plant crude extract by monitoring the enzymatic activity and find corresponding gene according to the amino acid sequence information of the purified enzyme, while the information about rich microarray data of Arabidopsis provides a great convenience for the finding of candidate genes. The gene co-expression analysis of these microarray data is currently a strong assistive tool for the gene function analysis of model plants. Web sites providing such service include Genevestigator, CSB.DB, and ATTED-II. Yonekura-Sakakibara (Yonekura-Sakakibara et al. 2008) used all the reported genes involved in flavonoid biosynthesis and regulation as the query genes to search in the whole genome of Arabidopsis genes related with the entire flavonoids metabolic pathway and found that the gene expression correlation between a functional unknown glycosyltransferase (UGT78D3) and a transcription factor MYB111 regulating flavonoids biosynthetic pathway is very strong (r = 0.572). In addition, both the other two members in glycosyltransferase sub-family UGT78D (D1 and D2) involve in the formation of flavonoid glycosylation derivatives, which support from another side that UGT78D3 may involve in the flavonoid glycosylation modification (the left of Fig. 9.2).

Note that because the transcriptomics co-expression analysis is based on the hypothesis that genes involved in the same metabolic pathway are co-regulated, it should be careful in practice. Firstly, the co-expression analysis ignores the fact that many metabolites only accumulate in living cells of some specific tissues, and the content of different metabolites in the same metabolic pathway is greatly different; secondly, some enzymes may involve in many metabolic pathways with different regulation ways in the organism; lastly, wrong information is probably generated by

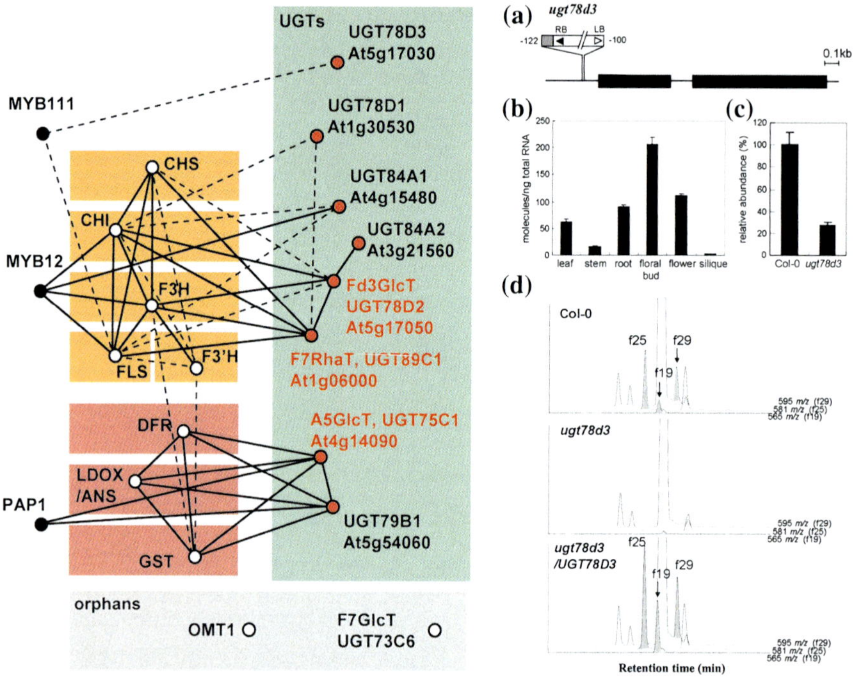

Fig. 9.2 The determination and functional analysis of flavonoids glycosyltransferase candidate genes. *Left* A diagram of correlation between genes involved in flavonoids metabolism and regulation and glycosyltransferase genes. The *solid line* represents correlation coefficient >0.6, and the *dashed line* represents correlation coefficient >0.525. There is no significant correlation between the two functional known genes OMT1 and F7GlcT and other genes involved in flavonoids metabolic pathway. *Right* The chemical analysis of quercetin-3-*O*- arabinose glycosyltransferase (quercetin: UDP-arabinose transferase; UGT78D3) molecule and the deletion or over-expression transgenic plants. In the mutant *ugt78d3*, the products f19 and f29 disappear, and the over-expression of *UGT78D3* in *ugt78d3* can recover its chemical phenotype

the interference between close homologous genes in microarray hybridization. When applying transcriptomics co-expression analysis for the screening of candidate genes, other experimental methods must be applied together.

9.1.3.3 Functional Identification of Candidate Genes

The determination of candidate genes is followed by the functional identification of candidate genes.

In the field of secondary metabolism studies, the functional identification of genes consists of two parts: in vitro biochemical analysis and in vivo functional study (for some non-model plants, it is often unable to carry out in vivo functional verification). In this work, the homozygous mutant *ugt78d3* of the candidate gene was firstly screened, and flavonoids in this mutant and wild type were comparatively analyzed by LC-MS. It was found that the content of flavonoids glycosylation derivatives f19, f25, and f29 significantly decreased in the *ugt78d3* mutant, and the complementary test also confirmed that UGT78D3 is indeed involved in the formation of f19, f25, and f29. The problem is that the LC-MS analysis on the mutant still cannot determine the types of pentose linkage in f19, f25, and f29 (Fig. 9.2, the right). For this reason, the tandem enzyme activity analysis was done with recombinant proteins UGT78D3, UGT89C1 generated from E. coli, and the structures of f19 and f25 were identified as 3-*O*-L-arabinose–7-*O*-L-rhamnose-glycosylated kaempferol and quercetin, respectively. Finally, the same technique was used to reveal that *UGT89C1* (flavonoid 7-*O*-rhamnose glycosyltransferase) and *RHM1* (UDP-rhamnose synthase) also played a role in flavonoids biosynthesis pathway in Arabidopsis.

For those non-model plants which have no genomic information yet, strategy for studying specific metabolic pathway at molecular level is similar to that for the model plant Arabidopsis, i.e., firstly, determine "the interested compound" mostly accumulated in which tissue or organ by chemical analysis, take this tissue or organ such as glandular trichome or flower as a starting material to construct the cDNA library, then carry out large-scale random sequencing. System bioinformatics analyzes the EST sequence, and system functional study genes are involved in the metabolic pathway of "the interested compound." Successful reports include the biosynthesis of alkaloids such as morphine in poppy (*Papaver somniferun*) (Ziegler et al. 2006) and the biosynthesis of bitter acid and prenylated flavonoid in hops (*Humulus lupulus*) (Nagel et al. 2008).

9.2 The Application of "Substantial Equivalence" Detection in Genetically Modified (GM) Plants

In recent years, the rapid development and application of transgenic technology promote the producing of a lot of GM crops with good agricultural traits, for example, the planting of GM soybeans has already accounts for 60 % of the world's soybean acreage. While the GM crop makes a contribution to the alleviation of food shortage and the improvement in food quality, the doubt on the safety of the GM crop in fact has not yet been stopped from the first day of its birth, especially those issues with direct impact on human health. Thus, every country has a set of strict monitoring system to ensure the biological safety of GM crops, one of which is in line with the "substantial equivalence" principle, that is, if some new food or food ingredients are substantially equivalent with existing food or food ingredients, then they are equally safe, which actually is to evaluate food safety using the chemical composition of the final food. The development of metabolomics provides powerful and more comprehensive technical support for the safety evaluation of GM foods. The content analysis of known toxic chemical component in the crop is one of the very important parts. Sterol alkaloid glycosides are the toxic chemical compositions to human health in potato. Two main sterol alkaloid glycosides α-chaconine and α-solanine in the GM potato (exogenous genes were imported into potato to improve insulin content for the treatment of diabetes) were quantitatively compared with that of traditional potato without genetically modification using LC-MS/MS technology by a Germany research group (Catchpole et al. 2005; Zywicki et al. 2005).

9.2.1 Experimental Methods

9.2.1.1 Sample Preparation

Six strains of GM potato and non-GM potato were, respectively, randomly planted in four experimental fields, and potato samples were randomly selected. The collected potatoes were stored in the dark at 4 °C for 2 days, and then in the dark at 10 °C for additional 5 days.

It is worthy to note that the field growing conditions are difficult to control in current experiments, and the analysis of samples is personal sampling. Thus, to ensure statistical significance and reliability of the results, about 40 replicates need to be checked for each strain.

Sample extraction:

1. Take potato samples with the cork drill with 8 mm diameter;
2. Select the potato skins with a thickness of 2 mm and the subsequent potato flesh with a thickness of 2 mm, weigh, and then put them to liquid nitrogen [due to the large difference caused by the sampling method, only samples with weight between (123.2 ± 34.8) mg were used for subsequent analysis];
3. Add 2 ml chloroform–methanol–water (2:5:2, v/v/v) extraction solvent, crush samples with the high-shear dispersing emulsifier at −15 °C, and vortex at 4 °C for 5 min.
4. Centrifuge at 14,000 rpm for 5 min to remove solid impurities and take the supernatant;
5. Dilute potato skins extract 10 times with 1:1 acetonitrile–water, and analyze the potato flesh extract directly.

9.2.1.2 Instrument Settings

1. Column: Reversed-phase Hyperclone C_{18} (Φ2.0 mm × 100 mm, 3 μm); HILIC column, Polyhydroxyethyl (Φ2.1 mm × 100 mm, 3 μm); the column temperature was room temperature;

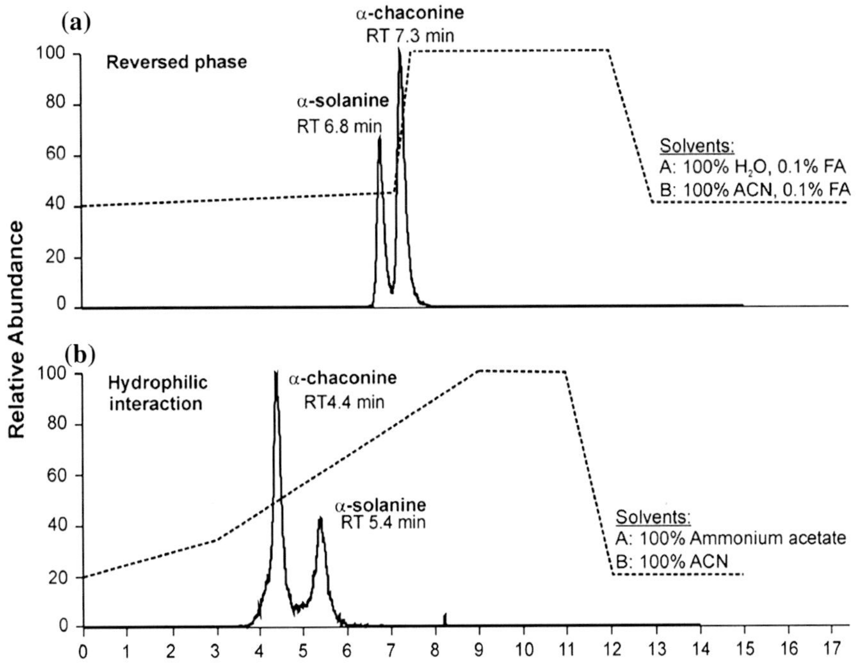

Fig. 9.3 Comparison of separation effect with α-chaconine and α-solanine by RP column and HILIC column

2. The solvent composition and condition of the mobile phase are shown in the dotted lines in Fig. 9.3.
3. MS conditions: electrospray ionization (ESI), positive mode; triple quadrupole (QQQ) for sterol alkaloid glycosides $[M + H]^+$, capillary spray voltage 3.5 kV; ion source temperature 270 °C; collision energy for the detection of fragment ion peak was set to 36 eV.
4. Statistical analysis software: The liquid chromatogram was firstly processed by LCquan (Xcalibur software, version 1.3), MATLAB (version 6.5) for statistical analysis.

9.2.2 Experimental Procedure

Zywicki et al. (2005) firstly used sterol alkaloid glycosides standards α-chaconine and α-solanine to compare the separation and quantitation of RP column and HILIC analytical column and found that both columns have good separation effect for α-chaconine and α-solanine (Fig. 9.3), the difference is that the lifetime of HILIC column was shorter, the peak was wider after analyzing 100 samples, while the RP column still had good separation ability after analyzing 1,000 samples.

The authors also detected and compared the accuracy and precision of the two analysis methods by adding standards in the process of analyzing samples and found that the accuracy and precision of RP column were better than that of HILIC column.

Although the chromatographic conditions (shown in dashed lines, the reversed-phase column mobile phase A is the aqueous phase, B is 100 % acetonitrile, both containing 0.1 % formic acid; HILIC column mobile phase A containing 5.5 mmol/L ammonium acetate) and the compound elution time and sequence were different, both the columns showed good separation effect to the standards. Chromatogram was full ion scanning (total ion current, TIC).

In order to improve the selectivity and sensitivity of the quantitative analysis method, the authors used selected/multiple reaction monitoring (S/MRM) mode instead of total ion chromatogram (TIC) scan mode for quantitative analysis. In this case, the authors, respectively, used m/z 852.4 → 706.4 and m/z 868.4 → 398.4 to quantitatively analyze α-chaconine and α-solanine (Fig. 9.4). The TIC in Fig. 9.4 shows that α-chaconine and α-solanine have been eluted before 9 min, and other substances were eluted after 9 min, but the purpose in this case is just to

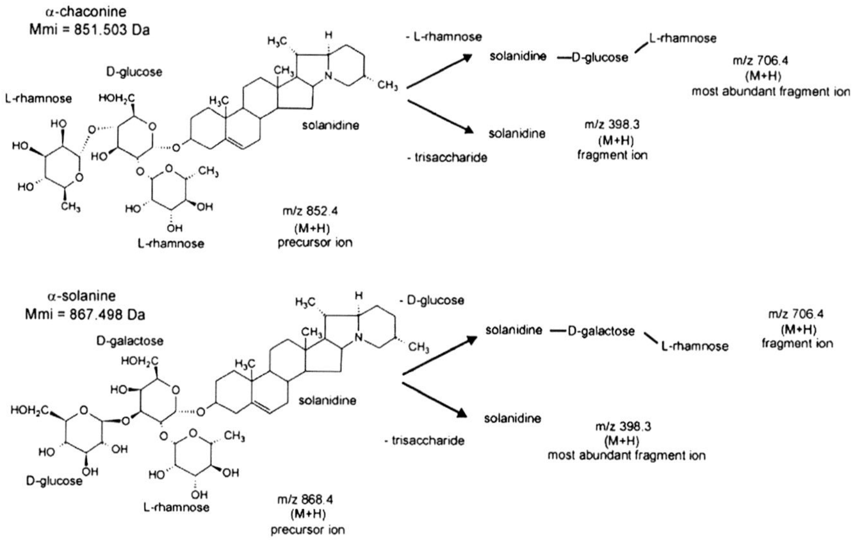

Fig. 9.4 The mass splitting patterns and LC analysis of α-chaconine and α-solanine. The splitting patterns in triple quadrupole MS (ESI ion source) are as follows: α-Chaconine loses a triose molecule with relative molecular weight of 470.1 and generates an ion fragment peak with relative molecular weight of 398.3(868.4 → 398.4, delay time 6.8 min); α-solanine loses a rhamnose molecule with relative molecular weight of 146 and generates an ion fragment peak with relative molecular weight of 706.4(852.4 → 706.4, delay time 7.3 min). The bottom figure shows the comparison of MRM mode and TIC mode in the analysis of potato skins samples, and the results show that the MRM mode has higher selectivity and sensitivity than the TIC mode

quantitatively analyze α-chaconine and α-solanine, other substances will be seen as pollutants producing matrix effect to the ion source of the MS; thus, the authors chose to detect every analysis sample in the range of 2.5–8.5 min (MRM mode).

After analyzing 1,200 samples, the authors finally found that the content of α-chaconine and α-solanine did not increase in the GM potato when compared with wild-type potato (regardless in the skins or in the flesh). Combined with fingerprint results of GC-TOF and flow injection electrospray ionization (FIE-MS) (Catchpole et al. 2005), it was considered that from the aspect of chemical composition, the GM potato has substantial equivalence with traditional non-GM potato at metabolic level, it is safe! Meanwhile, it should be noted that metabolomics is just one of the means for verifying the safety of GM plants, which is a necessary but not sufficient condition. We must hold more cautious attitude to the field release of GM plants and consider carefully the limitations of various techniques, for example, metabolomics platform is not suitable for allergens detection. The introduction of foreign genes which probably affect the allergens formation in a plant host will cause adverse effect to those susceptible people. It is thus necessary to comprehensively apply various different technical means to ensure the safety of GM plants.

References

Berrueta LA, Gallo B, Vicente F. A review of solid phase extraction: basic principles and new developments. Chromatographia. 1995;40:474–83.

Catchpole GS, Beckmann M, Enot DP, Mondhe M, Zywicki B, Taylor J, Hardy N, Smith A, King RD, Kell DB, Fiehn O, Draper J. Hierarchical metabolomics demonstrates substantial compositional similarity between genetically modified and conventional potato crops. Proc Natl Acad Sci USA. 2005;102:14458–14462.

Fiehn O, Robertson D, Griffin J, van der Werf M, Nikolau B, Morrison N, Sumner LW, Goodacre R, Hardy NW, Taylor C, Fostel J, Kristal B, Kaddurah-Daouk R, Mendes P, van Ommen B, Lindon JC, Sansone SA. The metabolomics standards initiative (MSI). Metabolomics. 2007;3:175–8.

Fukushima A, Kusano M, Redestig H, Arita M, Saito K. Integrated omics approaches in plant systems biology. Curr Opin Chem Biol. 2009;13:532–8.

Hirai MY, Klein M, Fujikawa Y, Yano M, Goodenowe DB, Yamazaki Y, Kanaya S, Nakamura Y, Kitayama M, Suzuki H, Sakurai N, Shibata D, Tokuhisa J, Reichelt M, Gershenzon J, Papenbrock J, Saito K. Elucidation of gene-to-gene and metabolite-to-gene networks in Arabidopsis by integration of metabolomics and transcriptomics. J Biol Chem. 2005;280:25590–5.

Huhman DV, Sumner LW. Metabolic profiling of saponin glycosides in *Medicago sativa* and *Medicago truncatula* using HPLC coupled to an electrospray ion-trap mass spectrometer. Phytochemistry. 2002;59:347–60.

Lepiniec L, Debeaujon I, Routaboul JM, Baudry A, Pourcel L, Nesi N, Caboche M. Genetics and biochemistry of seed flavonoids. Annu Rev Plant Biol. 2006;57:405–30.

Nagel J, Culley LK, Lu YP, Liu EW, Matthews PD, Stevens JF, Page JE. EST analysis of hop glandular trichomes identifies an O-methyltransferase that catalyzes the biosynthesis of xanthohumol. Plant Cell. 2008;20:186–200.

Yonekura-Sakakibara K, Tohge T, Niida R, Saito K. Identification of a flavonol 7-O-rhamnosyltransferase gene determining flavonoid pattern in Arabidopsis by transcriptome coexpression analysis and reverse genetics. J Biol Chem. 2007;282:14932–41.

Yonekura-Sakakibara K, Tohge T, Matsuda F, Nakabayashi R, Takayama H, Niida R, Watanabe-Takahashi A, Inoue E, Saito K. Comprehensive flavonol profiling and transcriptome coexpression analysis leading to decoding gene-metabolite correlations in Arabidopsis. Plant Cell. 2008;20:2160–76.

Ziegler J, Voigtlander S, Schmidt J, Kramell R, Miersch O, Ammer C, Gesell A, Kutchan TM. Comparative transcript and alkaloid profiling in Papaver species identifies a short chain dehydrogenase/reductase involved in morphine biosynthesis. Plant J. 2006;48:177–92.

Zywicki B, Catchpole G, Draper J, Fiehn O. Comparison of rapid liquid chromatography-electrospray ionization-tandem mass spectrometry methods for determination of glycoalkaloids in transgenic field-grown potatoes. Anal Biochem. 2005;336:178–86.

Chapter 10
Application of Metabolomics in the Identification of Chinese Herbal Medicine

Li-Xin Duan, Xiaoquan Qi, Min Chen and Lu-qi Huang

10.1 Introduction

Chinese herbal medicine is a special kind of plants with medicinal function, which have been used for thousands of years in China, and gradually forms the unique Chinese medicine system. Chinese medicine is an important traditional industry. The scale of industrial economy has exceeded 100 billion RMB. Due to geographical and historical reasons, the phenomenon of confusing varieties and complex origins as well as the homonyms and synonyms of Chinese herbal medicine is widespread. Just of the 534 kinds of Chinese herbal medicines recorded in 2000 edition *Pharmacopoeia of the People's Republic of China* (hereafter referred to as *China Pharmacopoeia*), 143 kinds are multiple origins (more than two origins), accounting for 27 % of the total recorded number. And it is far more complex in practical application than that revealed by *China Pharmacopoeia*. It is found in the long-term medical practice that even the same kind of herbals from different producing areas, wild and cultivated, shows difference in quality and efficacy. Above-mentioned problems give new challenges for the identification of Chinese herbal medicines. The identification of different origin varieties of Chinese herbal

L.-X. Duan (✉) · X. Qi
Institute of Botany, The Chinese Academy of Sciences, 100093 Beijing, China
e-mail: nlizn@ibcas.ac.cn

X. Qi
e-mail: xqi@ibcas.ac.cn

M. Chen · L.-q. Huang
Institute of Chinese Material Medica, China Academy of Chinese Medical Sciences, 100700 Beijing, China
e-mail: cm315keke@163.com

L.-q. Huang
e-mail: huangluqi@263.net

X. Qi et al. (eds.), *Plant Metabolomics*, DOI 10.1007/978-94-017-9291-2_10

medicines concerns the specific efficacy and the reproducibility of the efficacy. “Variety is wrong, all bets are off.” We can see the importance of the identification of Chinese herbal medicine.

It is very difficult to morphologically distinguish between *Astragalus membranaceus* (Fisch.) Bge. *Mongholicus* (Bge.) Hsiao. (*Mongholicus* for short) and *Astragalus membranaceus* (Fisch.) Bge. (*Membranaceus* for short), and there have been different views on their taxonomic status, which is controversial. *Astragalus* was firstly recorded in *Shen Nong's Herbal Classic* (or *The Divine Husbandman's Herbal Foundation Canon*) at top grade as a commonly used herb with functions of supplementing qi and securing the exterior, disinhibiting urine, drawing toxin and expelling pus, closing sores, and engendering flesh (Ma et al. 2004; Cho and Leung 2007; Kuo et al. 2009; Cui et al. 2003). The original plants recorded by 2005 edition *China Pharmacopoeia* (The pharmacopoeia committee of the People's Republic of China 2005) are *Astragalus membranaceus* (Fisch.) Bge. and *Astragalus membranaceus* (Fisch.) Bge. *Mongholicus* var. (Bge.) Hsiao., which belongs to the *Leguminosae* family and *Astragalus* genus, while the 2010 edition *China Pharmacopoeia* changed the name of *Astragalus membranaceus* (Fisch.) Bge. *Mongholicus* var. (Bge.) Hsiao to *Astragalus membranaceus* (Fisch.) Bge. *Mongholicus* (Bge.) Hsiao, removing var. (the pharmacopoeia committee of the People's Republic of China 2010). In addition to *Astragalus* recorded by *China Pharmacopoeia*, there are also many substitutes, such as *Astragalus floridus* Benth., *Astragalus monadelphus* Hand.-Mazz, *Astragalus chrysopterus* Bge., and *Astragalus tongolensis* Ulbr., and some local habitually prescribing herbs such as *Astragalus adsurgens* Pall., and *Astragalus complanatus* Bunge. *Membranaceus* and *Mongholicus* were used as two separate species in *Astragalus* genus, *Astragalus* subgenus; the main morphological difference is that *Membranaceus* has hairs on ovary and pod; and the leaflet is usually oval- and egg-shaped and larger, while *Mongholicus* has no hairs on ovary and pod, and the leaflet is usually oval-shaped and smaller.

In 1964, Xiao Pei-Gen et al. (Xiao Pei-Gen et al. 1964) are based on the following three points: ① From a distribution point of view, they did not form the species-independent distribution region; ② many of transitional types between them were found in small Wutai mountain and other places, indicating that they may have hybrid; ③ there is no large difference in main reproductive organs such as flowers and fruits, and there is no difference found in the external shape and internal structure of root between them, then treated *Mongholicus* as a variant of *Membranaceus*. Kun-jun Bo also treated *Mongholicus* as a variant (Fu 1993), while Yi-zhi Zhao (Zhao 2006) and Er-tong Wang et al. (Wang and Liu 1999) proposed that the two *Astragalus* species' flowering time did not meet, resulting in reproductive isolation, and suggested to restore the independent species status of *Mongholicus*.

Even it is difficult to distinguish the plant of *Astragalus*, to say nothing of the root as the medicinal parts, which is more difficult to be distinguished morphologically or microscopically, and it is necessary to have the aid of modern molecular marker technology and metabolomics technologies to distinguish Chinese herbal medicines. Main traditional methods of Chinese herbal medicines identification include origin identification, microscopic identification, physiochemical identification, and biological identification. But these identification characteristics are all the genetic phenotypes of the organism, which not only is easily influenced by genetic factors, but also closely correlated with the growth and development stages of the organism, environmental conditions, human activities such as introduction and domestication and processing, with large variation and randomness, some weakness such as subjectivity, repeatability, and poor stability are inevitable, thus bringing a certain influence to the reliability of the identification results. In recent years, with the development of molecular biology techniques, DNA molecular marker technology has been widely used in the study of medicinal plant genetic diversity, systematics, taxonomy, and gradually penetrate into the field of identification of Chinese herbal medicine, and promote the development of Chinese herbal medicine identification.

As a carrier of genetic information, DNA molecule has large information content and a high degree of genetic stability within the same species and does not be influenced by external environmental factors and biological developmental stages and organ and tissue differences, and therefore, the use of DNA molecule characteristics as genetic markers to carry out Chinese herbal medicine identification is more accurate and reliable, very ideal for the identification of closely related species, easily confusing species and rare species. Amplified fragment length polymorphism (AFLP) molecular marker technology is the combination result of random amplified polymorphic DNA (RAPD) markers and restriction fragment length polymorphism (RFLP) markers, with the reliability of RFLP and the convenience of RAPD. AFLP technique not only has characteristics of other DNA molecular markers such as rich polymorphism, co-dominant expression, without environmental impact, without multiple allelic effect, but also has special advantages such as rich bands, small sample content, high sensitivity, and quick and high efficient (Vos 1995; Qi and Lindhout 1997; Qi and Stam 1998).

Plant metabolomics mainly study metabolic phenotypes or metabotypes under specific conditions as well as the correlation between these phenotype and genotypes (Chen 2006). Plant secondary metabolic process and metabolite accumulation are subject to the regulation of various itself or environmental biotic and abiotic factors. Metabolomics study not only can help to in-depth understand the interaction between plant and environment and understand plant gene functions, plant metabolic network, and metabolic regulation, but also can reveal relationships between plant phenotype and plant growth as well as development and biodiversity. GC-MS is one of widely used and mature metabolomics analytical methods. In particular, the GC-TOF/MS is capable of fast scan, has characteristics of high throughput, high resolution, high sensitivity, and good reproducibility, very ideal for high-throughput metabolic fingerprint analysis. The use of GC-TOF/MS in the analysis of plant phenotypes has successfully distinguished mutants with phenotypic difference (Fiehn et al. 2000) or mutants without significant phenotypic difference (latent phenotype, silent mutation) (Blaise et al. 2007). Because DNA molecular markers are not affected by developmental stages of organisms, herbs with different growth years cannot be identified, and the identification of wild and cultivated herbs with the same origin (genotype) is also difficult. Therefore, the single DNA molecular marker cannot resolve all the problems in the identification of Chinese herbal medicines. Metabolomics can observe main active compounds and detect the changes in intermediate metabolites (including non-active substances), not only reflects the difference in plant genotypes, but also reflects the role of environmental factors. Thus, the "double marker" method that integrating DNA molecular markers and metabolic markers is conductive to the establishment of specific and high efficient markers platform that tightly link the quality of herbs, providing new method and technology for the identification of Chinese herbal medicines and quality control.

The following takes *Astragalus,* for example, to describe the application of metabolomics in the identification of Chinese herbal medicines.

10.2 Experimental

10.2.1 Experimental Materials

Sample collection: Plant materials of *Mongholicus* and *Membranaceus* were provided by Institute of Chinese material medica, China academy of Chinese medical sciences, which were respectively collected from Jilin province, Gansu province, and Shanxi province and identified by the above-mentioned institute. The experiment distinguishes the two species of *Astralagus* from the following three levels: The first level is the comparison between different species or variant, i.e., the comparison between *Mongholicus* and *Membranaceus*; the second level is the comparison of different producing areas, i.e., the comparison between the northeast (Jilin Province) and northwest (Gansu Province) for *Membranaceus*, and the comparison between the northwest (Gansu Province) and north China (Shanxi Province) for *Mongholicus*; the third level is the comparison between the cultivated and wild, which were divided into 8 groups (Table 10.1), and each group collected 6 individual plants as replicates.

Table 10.1 Collected *Astragalus* species and classification

Sample group	Species or subspecies	Producing area	Growth condition
A (Me, NE, C)	Me	Siping, Jilin (NE)	C
B (Me, NE, C)	Me	Tonhua, Jilin (NE)	C
C (Me, NW, W)	Me	Weiyuan, Gansu (NW)	W
D (Me, NW, W)	Me	Zhangxian, Gansu (NW)	W
E (Mo, NW, C)	Mo	Longxi, Gansu (NW)	C
F (Mo, NC, C)	Mo	Hunyuan, Shanxi (NC)	C
G (Mo, NW, C)	Mo	Zhangxian, Gansu (NW)	C
H (Mo, NC, W)	Mo	Yingxian, Shanxi (NC)	W

Abbreviation *Me Membranaceus*, *Mo Mongholicus*, *C* cultivated, *W* wild, *NE* northeast, *NW* northwest, *NC* north China

10.2.2 Experimental Methods

10.2.2.1 AFLP Part

DNA extraction using the modified CTAB, purity, and concentration was determined with a UV spectrophotometer and agarose gel electrophoresis to ensure that the sample complies with the requirements of AFLP analysis. Using EcoR I and Mse I restriction endonuclease to digest at 37 °C for 3 h, T4-DNA ligase link at 4 °C overnight, took 2 μl enzyme-linked product for preamplification, preamplification product was diluted with ddH_2O 20 times for selective amplification. Preamplification EcoR I primer sequence is 5′-gACTgCgTACCAATTCA-3′ and Mse I primer sequence is 5′-gATgAgTCCTgAgTAAA-3′. Two selective nucleotides were used for selective amplification of the 3′ end of each preamplification. Amplification system is as follows: 2 μl 10 × PCR reaction buffer, 1.8 μl dNTPs (2.5 mmol/L), 0.8 μl EcoR I connector (50 ng/μl), 0.8 μl Mse I connector (50 ng/μl), 5 μl diluted preamplification product, 0.2 μl Taq enzyme, ddH_2O made up to 20 μl, and 11 primer pairs are as follows: EcoR I-ACG/Mse I-AAC, E37M32; EcoR I-AGG/Mse I-AAC, E41M32; EcoR I-ACG/Mse I-AAG, E37M33; EcoR I-ACT/Mse I-AAG, E38M33; EcoR I-AGA/Mse I-AAG, E39M33; EcoR I-AGG/Mse I-AAG, E41M33; EcoR I-AAG/Mse I-ACT, E33M38; EcoR I-ACA/Mse I-ACT, E35M38; EcoR I-ACG/Mse I-AGC, E37M40; EcoR I-ACT/Mse I-AGC, E38M40; EcoR I-AGG/Mse I-AGC, E41M40. The denatured selective amplification products were detected with 5 % denaturing polyacrylamide gel in a vertical electrophoresis system, electrophoresed for 2 h. The gel was stained with the quick silver staining, dried at room temperature, and timely recorded.

10.2.2.2 Metabolomics Analysis Part

(1) Compound extraction. Each of the 8 *Astragalus* samples with different origins listed in Table 10.1 was taken in 6 individual replicates. The extraction method is as follows: 100 mg *Astragalus* plant material was ground in liquid nitrogen, and then the internal standards nonadecane acid 10 μl (2.1 mg/ml) and ribitol 50 μl (0.2 mg/ml) were added to the mixed 1.5 ml extraction solvent (methanol:chloroform:water, 5:2:2). Ultrasonic extraction the powder was for 30 min/time, twice. The extraction was centrifuged in 11,000 r/min for 10 min, and the supernatant was separated to a new tube. Adding 0.6 ml deionized water and 0.3 ml chloroform to the new tube, vortexing for 5 s, the new tube then was centrifuged again for 5 min. The extraction solvent separated two layers, the upper layer was aqueous phase, and the lower layer was the nonpolar aliphatic phase.

(2) Sample derivatization. Freeze-dry the aqueous phase with a freeze-drying machine, blowing dry the chloroform phase with nitrogen-blowing instrument. The sample was derivatized in two steps: The first step is to add 40 μl 20 mg/ml methoxy amino hydrochloride to the dried sample, seal, and at 37 °C

constant temperature warm bath for 2 h; the second step is to add 70 μl derivatization reagent N-methyl-N-(trimethylsilyl) trifluoro-acetamide derivatization reagent (N-Methyl-N-(trimethylsilyl) trifluoroacetamide (MSTFA), seal, and at 37 °C warm bath for 30 min.

(3) GC-TOF/MS detection. The LECO Corporation GC-TOF/MS, GC is Agilent 6890 gas chromatography with autosampler. GC chromatographic column is DB-5MS, and carrier gas was helium with a flow rate of 1.5 ml/min. The organic phase gas chromatography temperature program is 80 °C, 4 min, 5 °C/min to 330 °C, maintained for 5 min, the aqueous phase gas chromatography temperature program is 80 °C, 4 min, 5 °C/min to 180 °C, maintained for 2 min, then 5 °C/min to 220 °C, followed by 15 °C/min to 330 °C, maintained for 5 min. Transfer Line temperature and ion source temperature were set to 280 °C and 250 °C respectively. Solvent delay detection was set to 5 min. Mass-to-charge detection range is 80–500. Spectra acquisition rate is 20 spectra/s. The detector voltage is 1,700 V. EI electron bombardment ability is −70 eV.

10.2.3 The Data Processing

Each amplified band of the AFLP corresponds to a gene DNA molecule site. The presence of polymorphic amplification band indicates the variation of some species in this site. Each amplification band is treated as a trait. A polymorphic band is treated as a molecular marker, which is given a value "1," a band without polymorphism is given a value "0." Because the molecular weight of *Astragalus* AFLP amplification band is in the range of 100–500 bp, molecular weights bigger than 500 bp and smaller than 100 bp were not recorded, and only record those clear and easy to distinguish as well as stably appear in replicate experiments in the range of 100–500 bp. The hierarchical cluster analysis (HCA) was done by SPSS (SPSS 16.0, SPSS Inc.), and the similarity was calculated by squared Euclidean distance.

The pre-processing of GC-TOF/MS metabolic data includes: peak smoothing, de-noising and baseline correction using Chroma TOF software. The parameters were set to 3 s peak width, 10:1 signal-to-noise. The pre-procession result was exported into netCDF format file, then imported to the peak detection program written in Matlab 7.0, which will automatically perform baseline correction, peak detection and alignment, internal standard deduction and normalization. Finally a 3D data matrix was exported including specified peak serial number (correspond to retention time and mass-to-charge ratio), observation point (sample No.) and normalized peak intensity. The 3D matrix was imported into SIMCA-P12.0 software (Umetrics, Umeå, Sweden) to perform multi-dimensional statistical analysis. The orthogonal signal correction (OSC) was used. After center and auto-scale, the principal component analysis (PCA) was performed to observe the aggregation, discrete, and outliers of the sample. Subsequent orthogonal partial least squares discriminant analysis (OPLS-DA) was used to identify main differential variables causing this aggregation and discrete. And SPSS 16.0 software was used to perform the clustering, nonparametric tests, and area under the curve of metabolic data.

The Pearson correlation analysis between GC-TOF/MS differential metabolites and AFLP polymorphic DNA band was analyzed by SPSS 16.0 software, and the obtained correlation coefficient matrix was further generated to correlation coefficient heat map with Cluster 3.0 and Java Treeview.

10.3 Results

10.3.1 The Differentiate of Species by AFLP Molecular Makers DNA Fingerprint and GC-TOF/MS Metabolic Profile

Most of the bands in DNA fingerprint generated by the 11 primers pairs are the same. There are a total of 85 polymorphic bands in the range of 100–500 bp, in which the average polymorphic bands of each primer pair is 7.7, with most of the 13, and the least of 3. Figure 10.1a is the cluster result of molecular markers. The genetic distance of all the varieties of *Mongholicus* and *Membranaceus* is in the range of 0–25 %. The cluster result is divided into three large groups. All the *Mongholicus* samples cluster into one group (group **1**). In group **1**, the *Mongholicus* from different producing areas or growth conditions did not separate. All the *Membranaceus* samples were divided into two groups, group **2** and group **3**. The *Membranaceus* from Gansu and northeast were separate. From the collected samples, the geographical features of *Membranaceus* are more significant than that of *Mongholicus*.

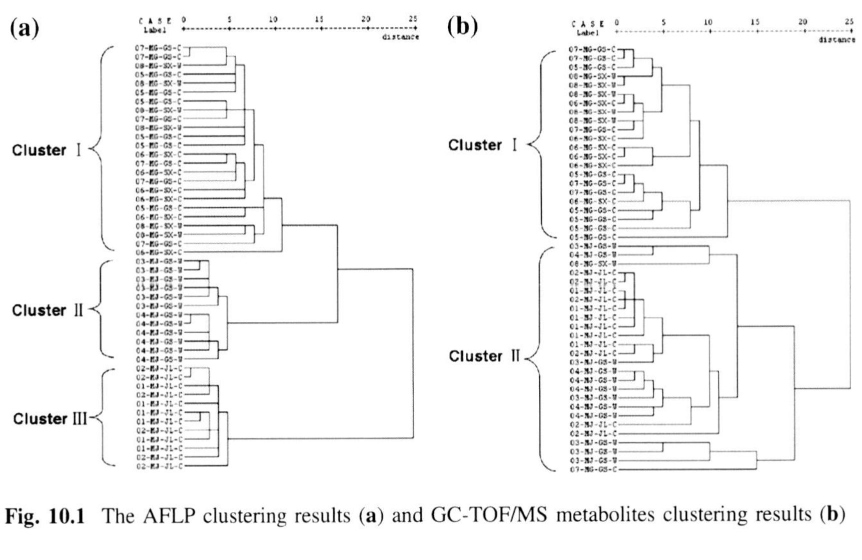

Fig. 10.1 The AFLP clustering results (**a**) and GC-TOF/MS metabolites clustering results (**b**)

As for kinship, *Mongholicus* from Gansu and Shanxi are closer with *Membranaceus* from Gansu, and more distant with *Membranaceus* from Jilin. This result may be due to the difference in geographical location, because Gansu is closer to Shanxi. And secondly, Gansu is the main producing area of China's *Astragalus*. Both the two *Astragalus* species have long history of cultivation, and the exchange of genetic information between them may occur. This result also confirmed from the side the view of Xiao Pei-Gen that *Mongholicus* is a variant of *Membranaceus*. *Membranaceus* from the northeast is relatively distant from *Astragalus* from other producing area. Literature has reported that by arbitrarily primed polymerase chain reaction (AP-PCR) method found that Astragalus from Heilongjiang is significantly different from other origins of *Astragalus* (Yip and Kwan 2006). In the AFLP differential bands, M40-E41-5 and M33-E41-2 only appear in *Membranaceus* samples, while M40-E38-1 only appears in *Mongholicus* samples. Thus, the three AFLP markers can be DNA markers distinguishing the two *Astragalus* species.

GC-TOF/MS in total analyzed 1192 peaks from the polar aqueous phase and the nonpolar chloroform phase, including 118 putative identified compounds (86 in the aqueous phase, 54 in the aliphatic phase, and 22 present both in the two phases). Figure 10.1b is the clustering result of GC-TOF/MS metabolic profile. Except for the two cross species, two large groups were significantly divided, i.e., cluster **1** and cluster **2**. The Jilin origin and Gansu origin of *Membranaceus* have the trend of clustering into subgroups.

10.3.2 The Identification of Differential Metabolites in Mongholicus *and* Membranaceus

The AFLP and GC-TOF/MS metabolites clustering results have shown to be able to clearly distinguish between the two *Astragalus* species (variants). The principal component analysis (PCA) is one of the most commonly used unsupervised pattern recognition methods. The dimension reduction of the original complex data can effectively find the most "major" information in the data and remove noise and redundancy. Figure 10.2a is the PCA result score plot of the metabolites in the two *Astragalus* species. In the score plots of the first principle component and the second principle component, there is a relative significant separation between the two *Astragalus* species, indicating that there is significant difference in metabolome of the two *Astragalus* species. The orthogonal projections to latent structures discriminant analysis (OPLS-DA) (Fig. 10.2b) is capable of differentiating the two *Astragalus* species more significantly, and the R^2X, R^2Y and Q^2 values are 0.348, 0.96, and 0.938, respectively. The following four criteria are used in the finding of differential metabolites: in the OPLS-DA model, VIP > 1.8; correlation coefficient $r > 0.8$ (Fig. 10.3); Mann–Whitney nonparametric test $P < 0.05$; area under the curve (the statistically significant test of differential metabolites) AUC > 0.8. The identification of metabolites was performed by database comparison, with similarity greater than 800 as the putative qualification. Database includes NIST 05 and Golm

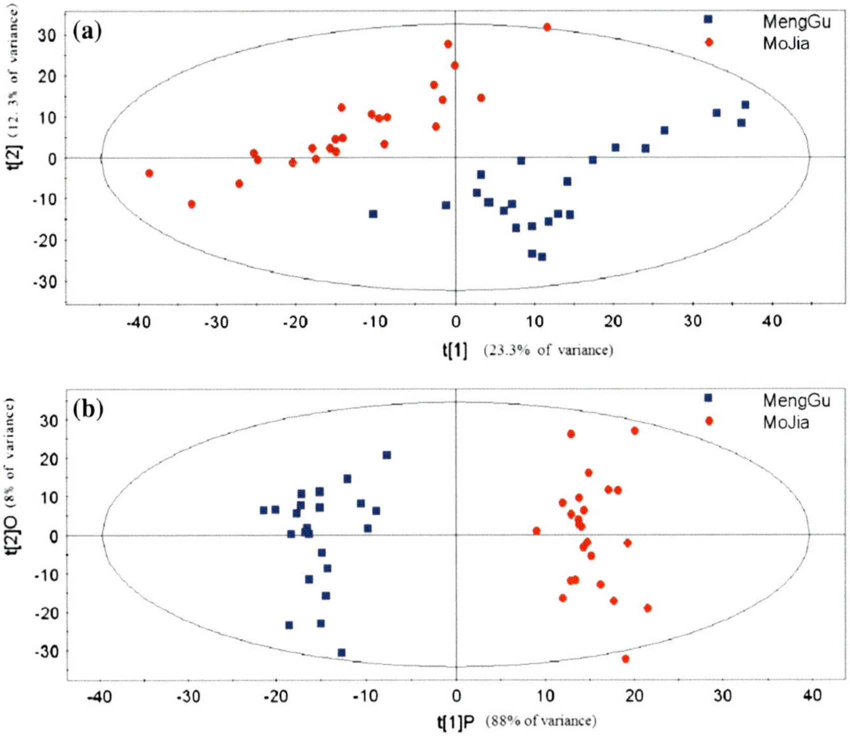

Fig. 10.2 The differentiate of *Mongholicus* and *Membranaceus* by PCA and OPLS-DA

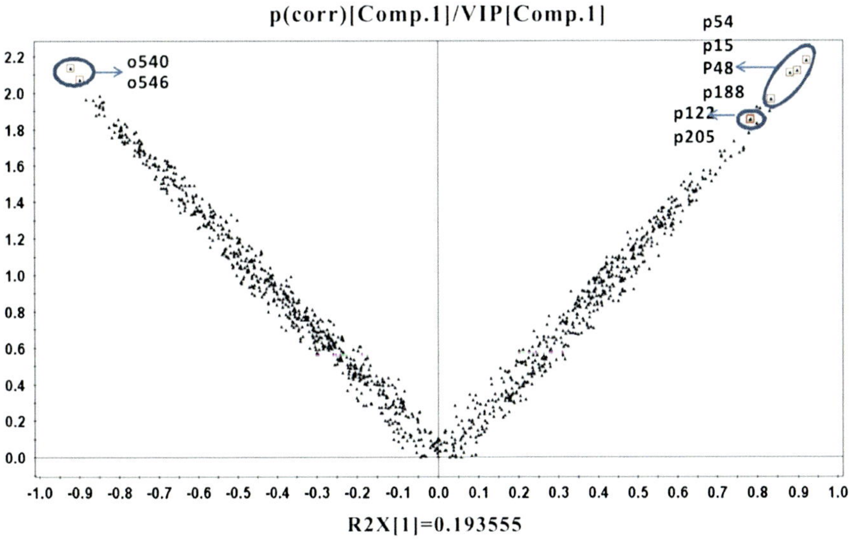

Fig. 10.3 The selection of differential metabolites

Table 10.2 The list of differential metabolites between *Mongholicus* and *Membranaceus*

No.	Retention time (min)	VIP	Compound name	Similarity	MW-U[①]	ROC[②]	Mean value (Me)	Mean value (Mo)	Me/Mo
p54	10.181	2.177	Malic acid	954	<0.001	1.000	152.819	23.971	6.375
o540	33.034	2.135	–	–	<0.001	1.000	6.765	49.950	0.135
p15	6.214	2.120	–	–	<0.001	1.000	9.103	0.955	9.529
p48	9.343	2.107	–	–	<0.001	1.000	21.718	7.886	2.754
o546	33.418	2.072	–	–	<0.001	1.000	1.259	15.401	0.082
p188	22.019	1.963	Xylose	950	<0.001	0.993	4.568	0.407	11.223
p122	16.284	1.854	–	–	<0.001	1.000	12.391	0.818	15.155
p205	23.924	1.851	Pentose phosphate	903	<0.001	0.958	12.970	2.355	5.507

① The *P* value of nonparametric test (Mann–Whitney U)
② Area under the receive operation curve

Metabolome database. There are a total of 8 metabolites as the differential metabolites distinguishing the two *Astragalus* species (Table 10.2). Identified metabolites are mainly malonic acid, pentose phosphate, and xylose, and the other 6 differential metabolites have not been identified yet, which are potential metabolic markers distinguishing the two *Astragalus* species.

The content difference of the 8 candidate differential metabolites between the two *Astragalus* species is shown in Fig. 10.4. Compared with cluster analysis, PCA or OPLS-DA is a more powerful data analysis method, which not only can reflect the difference between species, but also can find substances causing such difference between species. These substances are candidate markers rapidly distinguishing the two *Astragalus* species.

10.3.3 The Impact of Producing Area and Growth Conditions on the Metabolism of Astragalus

In order to further examine the impact of producing area and growth conditions on *Astragalus* quality, the difference in species was firstly ruled out. The roles of environmental factors were compared in the same genetic background using unsupervised PCA analysis method, showing the nature distribution characteristics of the data. Samples of the same species from different producing areas and planting were marked with different colors, and Fig. 10.5a is the PCA plot of metabolites in *Membranaceus* The black triangle symbol represents cultivars in Jilin Province, and the red dots represent the Gansu wild variety. Black triangles and red dots substantially separate on the third principal component (PC3), which is consistent with AFLP results. Due to the lack of Jilin wild *Membranaceus* and Gansu cultivated *Membranaceus*, it is difficult to determine whether the producing area or growth conditions cause the differentiate of the metabolic profile of *Membranaceus.*

Figure 10.5b is the PCA plot of metabolites in *Mongholicus*. All the samples were collected from adjacent provinces. The red dots represent Shanxi Province planted *Mongholicus*, blue diamond represents Shanxi Province wild *Mongholicus*, and black triangle represents Gansu Province cultivated *Mongholicus*. Figure 10.5b shows that the black triangles substantially separate from the red dots and blue diamonds on the fifth principal component (PC5). Red dots and blue diamonds cross, indicating that there is no significant difference between the Shanxi *Mongholicus* wild species and cultivars. Thus, the geographic location has greater impact on *Mongholicus* metabolism than growth conditions.

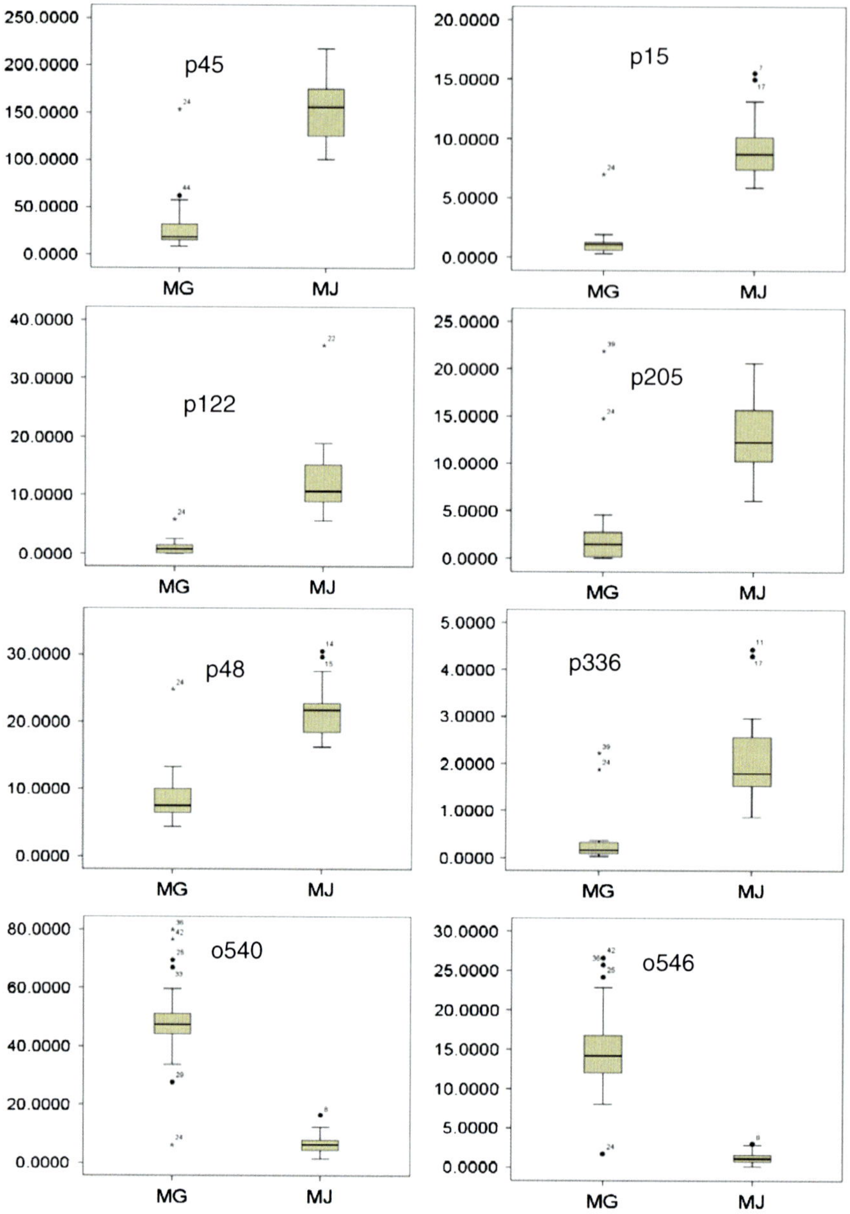

Fig. 10.4 The content difference of differential metabolites between *Mongholicus* and *Membranaceus*

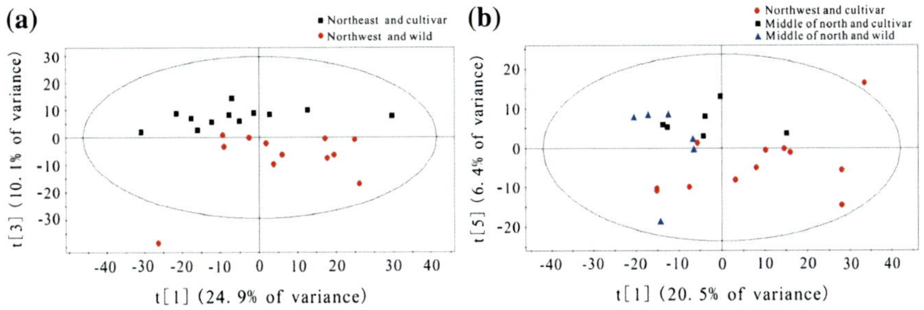

Fig. 10.5 The PCA score plot of metabolites in *Mongholicus* and *Membranaceus*

10.3.4 The Differential Analysis of the Metabolic Systems of Mongholicus *and* Membranaceus

Locate the identified metabolites in main plant metabolic pathways and compare the same producing area (Gansu) of *Mongholicus* and *Membranaceus*. Red color or blue color marks the differential metabolites with statistically significance in the ratio of metabolite content ($P < 0.01$), red represents up-regulated, blue represents down-regulated, and gray represent non-identified. Figure 10.6a is *Membranaceus* metabolites *versus Mongholicus* metabolites. It can be seen that the content of soluble sugars such as galactose and xylose in *Membranaceus* is significantly higher than that in *Mongholicus*, while the content of fatty acids and some amino acids such as aspartic acid, proline, and asparagines as well as polyamine metabolic pathway products such as ornithine and spermidine are significantly lower than that in *Mongholicus*. In contrast, the change of producing area has little impact on the main metabolic pathways of *Membranaceus* (Jilin/Gansu) (Fig. 10.6b), and the main difference lies in the three plant phytosterols and pyruvate, and homoserine. *Membranaceus* and *Mongholicus* distribute in temperature and warm temperature regions of China. *Membranaceus* is a kind of forest meadow plants, while *Mongholicus* is a kind of dry plants. Polyamine and Proline can regulate plant to adapt to different environments. The up-regulation of polyamine can improve drought tolerance of plants (Kusano et al. 2008), and the increase of soluble polysaccharide content can protect plants and proteins from cold damage and dehydration (Wanner and Junttila 1999).

10.3.5 Correlation Analysis of GC-TOF/MS Differential Metabolites and AFLP Differential Bands

The synthesis of metabolites is mainly affected by the coordination and control of genes. The correlation analysis can reveal the correlation between metabolite markers and DNAs in *Membranaceus* and *Mongholicus*. Figure 10.7 is the

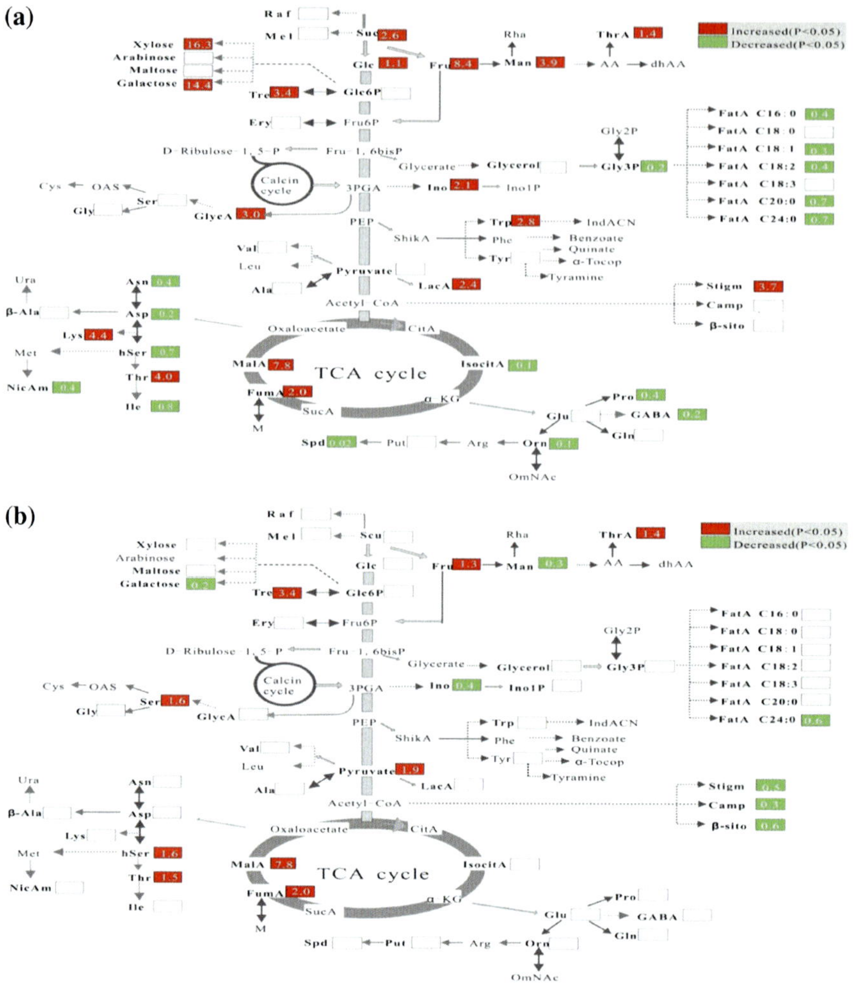

Fig. 10.6 The comparison of metabolites content in the main metabolic pathways of the two *Astragalus* species, **a** The comparison of metabolites in *Membranaceus* and *Mongholicus*, **b** the comparison of metabolites in *Menbranaceus* from different producing areas

correlation analysis between 8 metabolite markers and 85 AFLP bands. The correlation coefficients were plotted as a heat map by clustering software. Figure 10.7 shows that the differential metabolites, and AFLP bands were divided into two groups, respectively. Most of the correlation coefficients are small, only 27 (4.0 %) correlation coefficients indicate relative strong correlation ($r > 0.8$ or <-0.8). AFLP markers can be divided into two groups. The first group includes M32-E37-4 and M38-E25-3, which are positively correlated with differential metabolites o540 and o546, and negatively correlate with differential metabolites p54, p15, p188, and p48; the second group includes M40-E41-5, M40-E38-1, and M33-E41-2, which

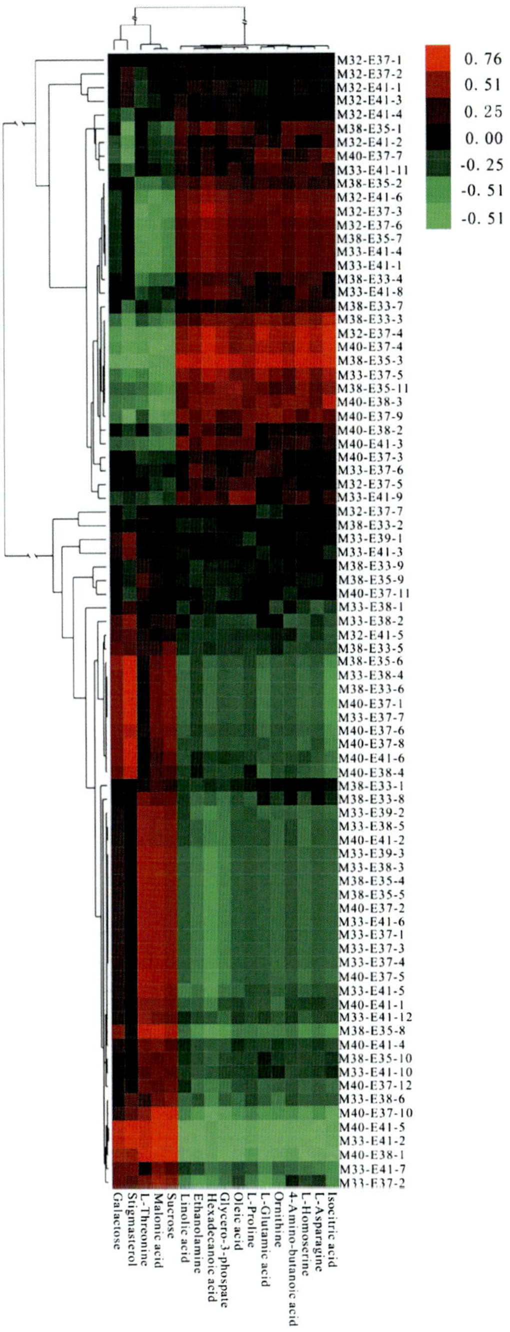

Fig. 10.7 The heat map of correlation coefficients between differential metabolites and DNA markers

are in contrast with that of the first group. The high correlation between AFLP markers and metabolite markers may involves in the complex metabolic pathways and metabolic networks in the two *Astragalus* species, which respectively accumulate specific metabolites. It has mentioned in 10.3.1 that M40-E41-5, M33-E41-2 and M38-E35-3 are DNA markers distinguishing the two *Astragalus* species. Thus, metabolites p54, p48, p15, p188, o540, o546, and AFLP markers M40-E41-5, M33-E41-2, and M38-E35-3 can be used in combination as metabolites markers and DNA markers distinguishing Membranaceus and Mongholicus, providing basics for the molecular mechanism of the metabolic difference in *Astragalus* with near kinships.

10.4 Summary and Discussion

The identification of Chinese herbal medicine is a traditional and important discipline, including the cross-linking of many disciplines such as taxonomy and pharmacology.

Along with the development of science and technology, modern molecular identification techniques and metabolomics technology gradually penetrated into the identification of Chinese herbal medicines. *Astragalus* is an important traditional Chinese herbal medicine and a kind of commonly used herbal medicines for many prescriptions. *China Pharmacopoeia* provisions two *Astragalus* species as medicinal sources, and their classification has been debated. Traditional microscopic identification and morphological identification cannot solve this problem. The study used the combination of AFLP technique and metabolomics technology to distinguish between different sources of *Astragalus* herbs, both exploring on the specific issue and providing new ways and means for the identification of Chinese herbal medicines. The results showed that both AFLP and metabolomics technologies can be well able to distinguish between *Mongholicus* and *Membranaceus.* Metabolomics and AFLP technology respectively found 8 metabolite markers (including three-identified known metabolites) and DNA markers distinguishing the two *Astragalus* species. In addition, metabolomics also explored the impact of environment (producing area, cultivated and wild) to the metabolites of *Astragalus.* Results showed that the impact of producing area is greater than that of cultivated and wild to *Mongholicus.* Identified metabolites were further located into main metabolic pathways, and it was found that the content of soluble sugars in *Membranaceus* is higher than that in *Mongholicus,* and the content of some fatty acids, amino acids, and polyamine metabolic pathway products in *Membranaceus* is significantly lower than that in *Mongholicus.* These metabolic products may be related with the distribution of the two *Astragalus* species. At the end of the experiments, the correlation between 85 AFLP bands and 8 metabolite markers was analyzed and found that there exists significant correlation between metabolite markers p54, p48, p15, p188, o540, o546, and AFLP markers M32-E37-4, M38-E35-3, M40-E41-5, M40-E38-1, and M33-E41-2 ($r > |0.8|$), in which three of the AFLP markers are the AFLP markers distinguishing *Membranaceus* and *Mongholicus*. Therefore, the 6 metabolite markers and 5 DNA markers highly correlated in metabolic profile and DNA fingerprint can be used as candidate metabolite markers and DNA markers distinguishing *Membranaceus* and *Mongholicus*, providing basics for the molecular mechanism of the metabolic difference in the two *Astragalus* species with near kinships.

As a DNA fingerprinting method, AFLP not only shows metabolism-related genes, but also can include non-metabolism-related genes. In data reading, generally polymorphic amplification bands are selected. However, AFLP cannot be used to distinguish the impact between different producing areas and environments on the quality of Chinese herbal medicines. Metabolomics is to detect plant's metabolic profile and can well reflect the impact of environmental and genetic factors on the quality of Chinese herbal medicines. As a high-throughput analytical method, metabolomics technique can quickly analyze a large of samples and find differential substances. The two methods can complement each other.

As a new detection technology, metabolomics has broad application prospects, but also has some limitations. Metabolomics is designed to detect all the metabolites in organisms or tissues, but current analysis techniques cannot reach the level of comprehensive detection. *Astragalus* is an important Chinese herbal medicine, and the main pharmacological compounds now considered are flavonoids, polysaccharides, saponins, and other substances. There are still some deviations in this study using GC-TOF/MS for the detection of these substances. Therefore, the combination of LC-MS to analyze and detect more metabolites and combined with pharmacological compounds is an important direction for the identification of Chinese herbal medicines.

References

Blaise BJ, Giacomotto J, Elena B, Dumas ME, Toulhoat P, Segalat L, Emsley L. Metabotyping of *Caenorhabditis* elegans reveals latent phenotypes. Proc Natl Acad Sci USA. 2007;104: 19808–19812.

Cho WCS, Leung KN. In vitro and in vivo immunomodulating and immunorestorative effects of *Astragalus* membranaceus. J Ethnopharmacol. 2007;113:132–3.

Chen XY. The study of plant secondary metabolism. World Sci-Tech R & D. 2006;28:1–4.

Cui R, He JC, Wang B, Zhang F, Chen GY, Yin S, Shen H. Suppressive effect of *Astragalus* membranaceus Bunge on chemical hepatocarcinogenesis in rats. Cance Chemother Pharmacol. 2003;51:75–80.

Fiehn O, Kopka J, Dörmann P, Altmann T, Trethewey RN, Willmitzer L. Metabolite profiling for plant functional genomics. Nat Biotechnol. 2000;18:1157–64.

Fu KJ. Flora of China, vol. 42. Beijing: Science Press; 1993. p. 131–133.

Kuo YH, Tsai WJ, Loke SH, Wu TS, Chiou WF. *Astragalus* membranaceus flavonoids (AMF) ameliorate chronic fatigue syndrome induced by food intake restriction plus forced swimming. J Ethnopharmacol. 2009;122:28–34.

Kusano T, Berberich T, Tateda C, Takahashi Y. Polyamines: essential factors for growth and survival. Planta. 2008;228:367–81.

Ma X, Tu P, Chen Y, Zhang T, Wei Y, Ito Y. Preparative isolation and purification of isoflavan and pterocarpan glycosides from *Astragalus* membranaceus Bge. var. mongholicus (Bge.) Hsiao by high-speed counter-current chromatography. J Chromatogr A. 2004;1023:311–7.

Pharmacopoeia Commission of the People's Republic of China. Pharmacopoeia of the People's Republic of China (a). Beijing: Chemical Industry Press; 2005. p. 212.

Pharmacopoeia Commission of the People's Republic of China. Pharmacopoeia of the People's Republic of China (a). Beijing: China Medical Science and Technology Press; 2010. p. 283.

Qi X, Lindhout P. Development of AFLP markers in barley. Mol Gen Genet. 1997;254:330–6.

Qi X, Stam P, Lindhout P. Use of locus-specific AFLP markers to construct a high-density molecular map in barley. Theor Appl Genet. 1998;96:376–84.

Schut JW, Qi X, Stam P. Association between relationship measures based on AFLP markers, pedigree data and morphological traits in barley. Theor Appl Genet. 1997;95:1161–8.

Vos P, Hogers R, Bleeker M, Reijans M, van de Lee T, Hornes M, Friters A, Pot J, Paleman J, Kuiper M, Zabeau M. AFLP: a new technique for DNA fingerprinting. Nucleic Acids Res. 1995;23:4407–4414.

Wang ET, Liu MY. The comparative biology of two medicinal *Astragalus*. Plant Res. 1999;16:85–91.

Wanner LA, Junttila O. Cold-induced freezing tolerance in *Arabidopsis*. Plant Physiol. 1999;120:391–9.

Xiao PG, Feng YX, Cheng JR, Lou ZC. Chinese herbal medicine *Astragalus* original plant and pharmacognosy. Pharmaceutical. 1964;11:114–9.

Yip PY, Kwan HS. Molecular identification of *Astragalus* membranaceus at the species and locality levels. J Ethnopharmacol. 2006;106:222–9.

Zhao YZ. The *Astragalus* plant taxonomy and floristic distribution. Plant Res. 2006;26:532–8.

Chapter 11
Metabolomics-Based Studies on Artemisinin Biosynthesis

Hong Wang, Hua-Hong Wang, Chen-Fei Ma, Ben-Ye Liu, Guo-Wang Xu and He-Chun Ye

11.1 Introduction

Artemisia annua L. is a traditional Chinese medicinal plant. In the 1970s, Chinese scientists first isolated colorless needle-like crystals with antimalarial activity from *A. annua* and named artemisinin. The chemical structure of artemisinin was confirmed as a sesquiterpene lactone with an endoperoxide bridge, of melting point 156–157 °C and molecular formula $C_{15}H_{22}O_5$. Later research showed that artemisinin and its derivatives had good effects on cerebral malaria, chloroquine-resistant falciparum malaria, and chloroquine-sensitive malaria. Artemisinin-based combination therapies (ACTs) were recommended by the World Health Organization to control the spread of malaria on 'Africa Malaria Day' in 2002, which led

H. Wang (✉)
University of Chinese Academy of Sciences, Beijing 100049, China
e-mail: hwang@ucas.ac.cn

H.-H. Wang · B.-Y. Liu · H.-C. Ye
Key Laboratory of Plant Molecular Physiology, Institute of Botany, Chinese Academy of Sciences, Beijing 100093, China
e-mail: phyto.wang@gmail.com

B.-Y. Liu
e-mail: b.liu@tu-bs.de

H.-C. Ye
e-mail: hcye@ibcas.ac.cn

C.-F. Ma · G.-W. Xu
National Chromatographic R&A Center, Dalian Institute of Chemical Physics, Chinese Academy of Sciences, Dalian 116023, China
e-mail: machenfei@petrochina.com.cn

G.-W. Xu
e-mail: xugw@dicp.ac.cn

X. Qi et al. (eds.), *Plant Metabolomics*, DOI 10.1007/978-94-017-9291-2_11

to a significantly increasing demand for artemisinin. However, the low content of artemisinin in *A. annua* (0.01–0.8 %) makes artemisinin-based antimalarial drugs expensive, while malaria endemic areas are mainly concentrated in some economically underdeveloped areas in Africa, Asia, and Latin America. So the price of the drug has become the key obstacle to popularizing ACTs. Although chemical synthesis of artemisinin is technically possible, this involves complicated steps and low yields make it of high cost and of no commercial production value. So far, extracting artemisinin from *A. annua* plants is the only viable commercial source of artemisinin-based drugs. Scientists have tried biotechnologies, such as cell culture, tissue culture, hairy root culture, and other biological techniques to obtain artemisinin. However, because biosynthesis of artemisinin mainly occurs in glandular secretory trichomes (GSTs) localized on leaves, stems, and flowers, all attempts have failed to achieve the expected effect. At the beginning of this century, bioengineering of yeast cells to produce artemisinic acid made great progress, but industrial production is still not available (Ro et al. 2006). In addition, transforming artemisinic acid to artemisinin is also a big challenge. Modifying *A. annua* plants by genetic engineering to obtain strains that have high yields of artemisinin is considered the most promising way to improve artemisinin production. Recently, with the development of genomics, the genes of the key enzymes involved in artemisinin biosynthesis have been cloned and identified, which greatly advanced study of the artemisinin biosynthetic pathway. However, the increase of artemisinin content was much lower than expected in transgenic plants. Therefore, elucidating the biosynthetic pathway and regulatory mechanism of artemisinin is a key issue to increase artemisinin content via metabolic engineering.

The biosynthesis of artemisinin belongs to the sesquiterpene branch of the plant isoprenoid pathway. Recent studies have shown at least two pathways for plant isoprenoid biosynthesis: the mevalonate and the pyruvate/glyceraldehyde-3-phosphate pathways. It is generally believed that isopentenyl diphosphate (IPP) and dimethylallyl pyrophosphate (DMAPP) used in artemisinin biosynthesis are synthesized through the mevalonate pathway. Three molecules of acetyl-CoA units generate 3-hydroxy-3-methylglutaryl-CoA (HMG-CoA) by intermolecular condensation under the catalysis of acetyl-CoA thiolase and HMG-CoA synthase. Subsequently, HMG-CoA generates mevalonic acid (MVA) in the presence of 3-hydroxy-3-methylglutary-CoA reductase (HMGR). After phosphorylation, decarboxylation, and dehydration, MVA converts into IPP, which becomes DMAPP by isomerization. Next, the two 'activated' isoprene units form geranyl diphosphate (GPP) by head-to-tail condensation according to electrophilic reaction mechanism catalyzed by farnesyl diphosphate synthase (FPS). The product GPP reacts with another IPP unit and generates fanesyl diphosphate (FPP). Recent studies indicate that the artemisinin biosynthetic precursor IPP is from both the traditional MVA pathway and the newly discovered pyruvate/glyceraldehyde-3-phosphate pathway. GPP is likely to be synthesized in plastids and transported to cytoplasm, forming FPP with another IPP unit (Schramek et al. 2010).

The biosynthetic pathway from FPP to artemisinin has not yet been fully elucidated. In recent years, with the cloning and identification of the key genes involved in artemisinin biosynthesis, its preliminary steps have been basically clarified. In 1999, Bouwmeester et al. first isolated the sesquiterpene intermediate amorpha-4,11-diene, whose structure is very similar to artemisinic acid—the artemisinin biosynthetic precursor—and speculated that it might be involved in artemisinin biosynthesis (Bouwmeester et al. 1999). Then Mercke et al. (2000), Chang et al. (2000) and Wallaart et al. (2001) cloned the amorpha-4,11-diene synthase (ADS) genes from *A. annua* successively. Subsequent catalytic mechanism studies indicated that amorpha-4,11-diene synthase used FPP as its substrate to form amorpha-4,11-diene (Picaud et al. 2006). Bertea et al. (2005) analyzed the terpenoids in leaves and GSTs of *A. annua* and found that amorpha-4,11-diene was hydroxylated to artemisinic alcohol and then oxidized to yield artemisinic aldehyde. After that, the C11–C13 double bond was reduced to yield dihydroartemisinic aldehyde and then oxidized to dihydroartemisinic acid. They proposed the following artemisinin biosynthetic pathway: amorpha-4,11-diene → artemisinic alcohol → artemisinic aldehyde → dihydroartemisinic aldehyde → dihydroartemisinic acid → artemisinin (Bertea et al. 2005). Teoh et al. reported that cytochrome P450 monooxygenase (CYP71AV1) and its reductase catalyzed the following conversion: amorpha-4,11-diene → artemisinic alcohol → artemisinic aldehyde → artemisinin and also provided direct evidence for the above hypothetical artemisinin biosynthetic pathway (Teoh et al. 2006). The C11–C13 double bond of artemisinic aldehyde could be reduced by double bond reductase (DBR2) to generate dihydroartemisinic aldehyde (Zhang et al. 2008). Then, the product was oxidized to dihydroartemisinic acid under the effect of aldehyde dehydrogenase ALDH1 (Teoh et al. 2009). In the biosynthetic pathway from FPP to artemisinin (Fig. 11.1), the steps from FPP to artemisinic acid and dihydroartemisinic acid are well understood, but the other steps remain unclear.

To sum up, the biosynthetic pathway of artemisinin and its regulation mechanism are still not clear, and metabolomics studies could clarify them. Metabolomics is a recently emerged discipline and an important part of functional genomics—it studies all metabolites and their relationships in a biological system (Sumner et al. 2003). Metabolomics involves the rapid and high-throughput characterization of small molecule metabolites found in an organism and has great potential in revealing genes' functions and analyzing the interactions of different metabolic pathways.

To further elucidate the biosynthetic pathway of artemisinin and its regulatory mechanisms, we jointly studied *A. annua* terpenoid metabolite profiles of different gene types, different growth periods, different transgenic plants, and treatment with methyl jasmonate (MeJA) using gas chromatography-mass spectrometry (GC-MS) and GC × GC-time-of-flight mass spectrometry (GC × GC-TOFMS). The detected terpenoid metabolite profiles data were analyzed by multivariate and bioinformatics analyses to determine those compounds closely related to artemisinin biosynthesis and to clarify the biosynthetic pathway of artemisinin and its regulatory mechanism. The results could provide a scientific basis for improving artemisinin yield by metabolic engineering to meet market demand.

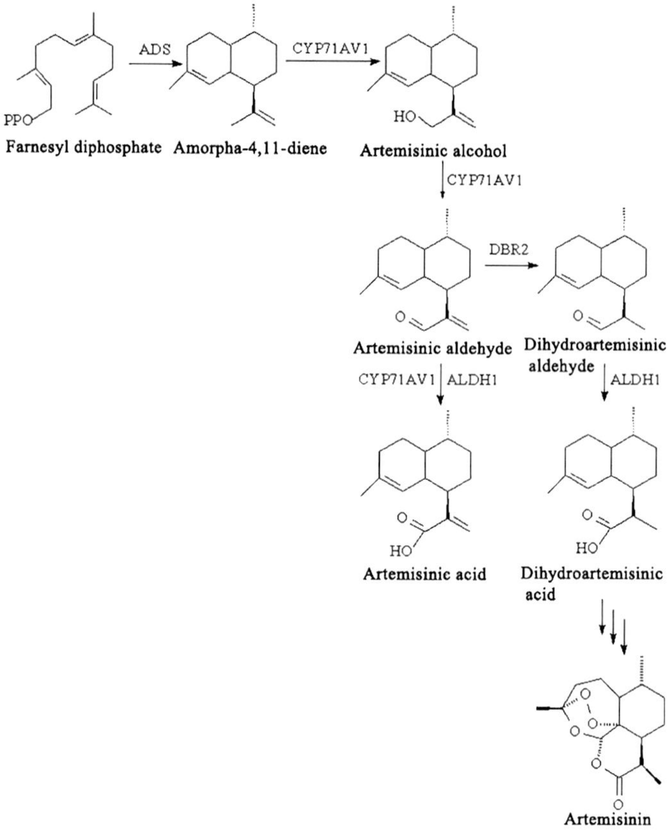

Fig. 11.1 Proposed biosynthetic pathway of artemisinin (Reprinted from Teoh et al. (2009). Copyright © 2008 Canadian Science Publishing or its licensors. Reproduced with permission)

11.2 Experiments

11.2.1 Plant Materials

Seeds of *A. annua* L. from Sichuan Province of China were sterilized by 0.1 % $HgCl_2$ and then rooted on Murashige and Skoog (MS) medium supplemented with 30 g/L sucrose (contained 0.7 % agar). After growing on MS medium for 4 weeks, the stems with axillary buds of the seedlings were cut for rapid propagation.

The *A. annua* strains used in this study include the following: ① SP18, a high-artemisinin-yielding strain (genotype) from space-flight breeding; ② 001, a high-artemisinin-yielding strain (genotype), used as an acceptor plant for genetic transformation; ③ positive control of the transgenic plants, containing the β-glucuronidase (*GUS*) reporter gene; ④ lines overexpressing the FPS gene (F lines); and ⑤ lines overexpressing the amorpha-4,11-diene synthase gene (A lines) and a line of RNA interference (RNAi) for the amorpha-4,11-diene synthase gene (Ami line). All transgenic *A. annua* plants were obtained by *Agrobacterium*-mediated genetic transformation.

11.2.2 Cultivation and Transplanting of Plant Materials

The in vitro cultured plantlets were rooted on MS medium supplemented with 30 g/L sucrose (0.7 % agar) for 15 d. Then, the sealing film was opened for 2–3 d to harden the seedlings. The rooted seedlings were transplanted into peat soil and cultivated in an experimental greenhouse using fluorescent lamps (with a light intensity of 5,000 lx) to keep a photoperiod of 16/8 h (day/night) for long day and 8/16 h for short day, with temperatures of 27 °C (day)/25 °C (night).

11.2.3 Extraction and Analysis of A. annua Terpenoids

11.2.3.1 Extraction of *A. annua* Terpenoids

The leaves and young branches of *A. annua* were frozen and milled in liquid nitrogen then dried at −50 °C for 8 h by a freeze-drying system. 50 mg dried powder was put into a 10-mL glass flask, and then 4 mL of hexane and 100 μL of trans-farnesol (77.61 μg/mL) were added to the flask. Samples were sonicated for 40 min and then cooled to room temperature. The extraction solution was centrifuged to obtain a clear supernatant, which was then dried by nitrogen and redissolved in 200 μL of dichloromethane.

11.2.3.2 GC/FID and GC/MS Analyses of *A. annua* Terpenoids

Gas chromatography/flame ionization detector (GC/FID) analysis: a DB-5 MS column (30 m × 0.25 mm × 0.25 μm) was used for separation, with helium as the carrier gas at constant flow (1.8 mL/min), injection temperature 290 °C, injection volume 1 μL, and split ratio 10:1.

GC/MS analysis: a DB-5 MS column (30 m × 0.25 mm × 0.25 μm) was used for separation, with helium as the carrier gas at constant flow (1.8 mL/min), injection temperature 290 °C, injection volume 1 μL, and split ratio 10:1. The MS scan parameters were as follows: mass range of m/z 35–400, scan interval of 0.5 s, scan speed of 1,000 amu/s, electron impact ionization of 70 eV, ion source temperature of 200 °C, and transfer interface temperature of 300 °C.

The temperature program of GC/FID and GC/MS was 100 °C for 1 min, then at 5 °C/min to 150 °C, at 2 °C/min to 200 °C, at 15 °C/min to 300 °C, and held at that temperature for 15 min.

11.2.3.3 GC × GC-TOFMS Analysis of *A. annua* Terpenoids

The system of polar/moderate polar column used the first column of DB-WAX (J&W Scientific, Folsom, CA, USA) 30 m × 250 μm × 0.25 μm and the second column of DB-1701 (J&W Scientific) 2 m × 100 μm × 0.10 μm. Helium was used as the carrier gas at a constant flow of 1.5 mL/min, injection volume 1 μL, and split ratio 10:1.

Mass spectrometry: The GC inlet and transfer line were set at 250 °C. The ion source was set to 220 °C and the modulation period of 5 s. Mass spectra from m/z 33 to 500 were collected at 50 spectra/s. The detector voltage was 1,650 V, and electron energy was 70 eV.

The temperature program: The column 1 oven was held at 100 °C for 1 min, and then ramped at 3.5 °C/min to 240 °C, held for 25 min. The column 2 oven was held at 110 °C, and the temperature program was the same as that for the column 1 oven.

Data handling: Both Chemstation data (100 Hz) and TOFMS data (50 Hz) were exported in ASCII format (*.csv files). The *.csv files were converted to *.bin files and then translated into the *.hdf files by Transform software to generate a contour plot. The components were quantified by Zoex software (Zoex Corp, Lincoln, NE, USA) and then the gained data were dealt with LECO TOFMS workstation Pegasus III.

11.3 Data Analysis and Discussion

11.3.1 Metabolic Profiling Investigation of A. annua at Different Stages of Development

Artemisinin biosynthesis and accumulation in *A. annua* are regulated by many factors, so artemisinin content changes dynamically. Ideally, an optimal period should be selected to harvest *A. annua* plants to get the highest artemisinin yield. So far, there is still no consensus on the stage of the highest content and yield of artemisinin during *A. annua* development.

In order to comprehensively understand artemisinin biosynthesis and accumulation in *A. annua*, to determine the optimal time to sample for metabolic fingerprinting analysis, and to provide some theoretical guidance for the optimum harvest period of *A. annua* plants, a metabolomics approach was applied to analyze terpenoids at different developmental stages of two genotypes SP18 and 001. More specifically, the materials of the following stages were collected and used for metabolomics analysis: vegetative growth stage (VGS) at 30 d (T1), 35 d (T2) and 45 d (T3); the pre-flower budding stage (PFBS) at 55 d (T4, switched to a photoperiod of 8/16 h) and 65 d (T5); and the flower budding stage (FBS) at 75 d (T6) and 85 d (T7).

Metabolic profiles from GC/MS were analyzed to investigate the terpenoids of *A. annua* at different development stages, and 92 peaks were identified. An internal standard method was used to calculate the concentration of each compound, and the concentration in every assay was arrayed by retention time. The data were subjected to multivariate analysis with partial least squares discriminant analysis (PLS/DA). The samples showed obvious differences among different growth stages. The score plots of PLS/DA showed that the 001 and SP18 samples at stages VGS and FBS were distributed far from each other, but samples at the same stage were tightly clustered, clearly demonstrating differences in the terpenoid contents at the two stages (Figs. 11.2 and 11.3). PFBS is a transitional stage from VGS to FBS, as shown in Figs. 11.2 and 11.3, the score of the first sample from PFBS was similar to that of VGS, but the score of the second sample from PFBS was similar to that of FBS. The variation importance in the projection (VIP) of compounds by PLS-DA reflects the components' importance in classification, those compounds with VIP > 1 were

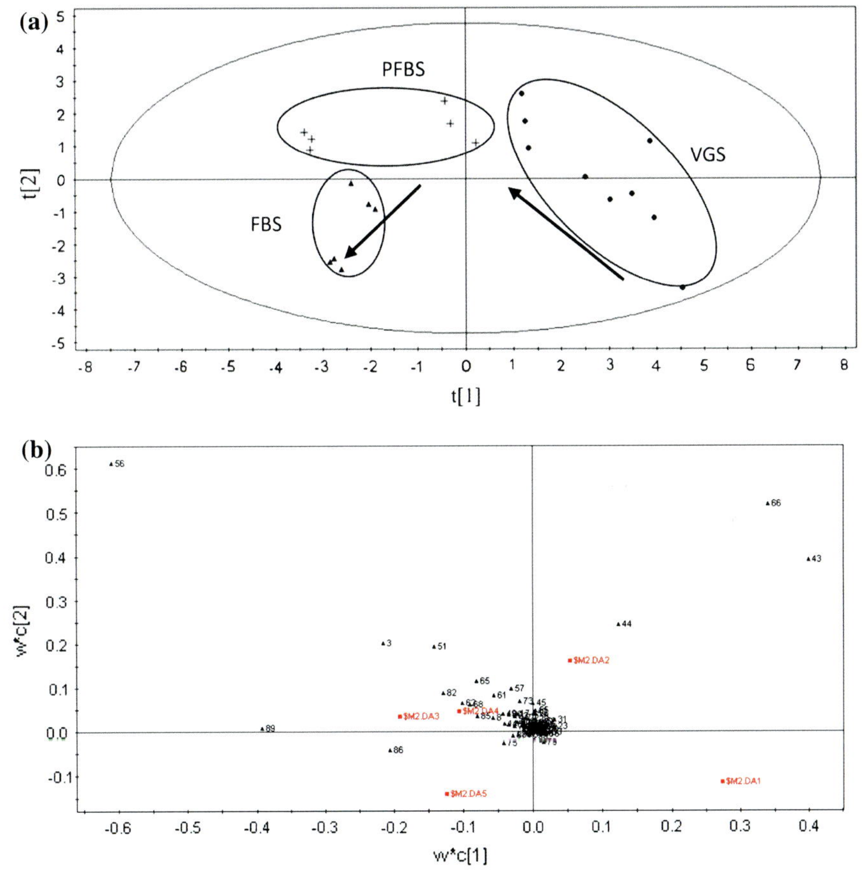

Fig. 11.2 The score (**a**) and loading (**b**) plot of PLS-DA model based on the terpenoid profiles of different growth stages of *A. annua* line 001 *VGS* ● vegetative growth stage; *PFBS* + pre-flowering budding stage; *FBS* ▲ flowering budding stage

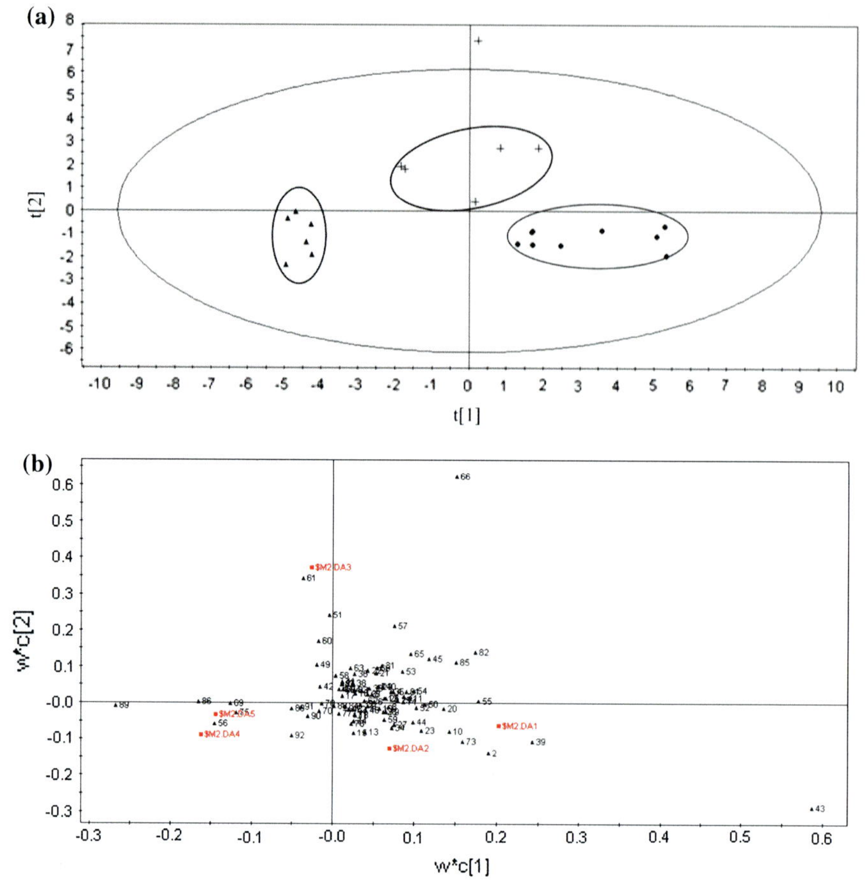

Fig. 11.3 The score (**a**) and loading (**b**) plot of PLS-DA model based on the terpenoid profiles of different growth stages of *A. annua* SP18 *VGS* ● vegetative growth stage; *PFBS* + pre-flowering budding stage; *FBS* ▲ flowering budding stage

considered to have significant changes at different stages and were used as the marker compounds in PLS-DA. Most of the marker compounds were identified as terpenoids, especially sesquiterpenoids and triterpenoids by comparison of the standard and the National Institute of Standards and Technology (NIST) database. Artemisinin and its related precursors, artemisinic acid, dihydroartemisinic acid, and arteannuin B, were all marker compounds. The contents of these marker compounds fluctuated with plant growth stages, and artemisinic acid and dihydroartemisinic acid reached the highest levels at VGS, decreased rapidly at PFBS and reached the lowest levels at FBS. Artemisinin and arteannuin B reached relatively high levels at early VGS, the highest levels at late VGS and PFBS, and then decreased slightly at FBS. In addition, the dynamics of the marker compounds with obvious changes in PLS-DA showed that the contents of these compounds increased with plant growth at VGS, but decreased at FBS. The results also showed the accumulation of metabolites reached the highest levels at late VGS or PFBS, as arrows showed in Figs. 11.2a and 11.3a.

Artemisinin synthesis and accumulation varied at different growth stages. The terpenoid metabolite profiles showed that synthesis and accumulation of terpenoids changed dynamically from VGS to FBS in thc two *A. annua* genotypes. The content of terpenoids compounds reached the highest levels at VGS, decreased at PFBS; however, the content of artemisinin reached the highest level at PFBS, and decreased at FBS, which was consistent with the report of Woerdenbag et al. (Woerdenbag et al. 1994). Interestingly, the content changes of artemisinic acid and dihydroartemisinic acid exhibited similar dynamics, and their contents decreased rapidly at PFBS. However, the dynamics of artemisinin was similar to that of arteannuin B—their contents changed only slightly during all developmental stages, suggesting that artemisinic acid and dihydroartemisinic acid might be intermediates in the artemisinin pathway, but artemisinin and arteannuin B were the final products. Lommen et al. reported that a significant conversion of dihydroartemisinic acid into artemisinin was induced during the post-harvest period, with an obvious effect on artemisinin content of leaves (Lommen et al. 2007). Thus, we speculate that the increasing content of artemisinin in PFBS was related to the decreasing content of dihydroartemisinic acid. Our results also suggest that PFBS was the optimal stage to harvest *A. annua* plants, since both plant biomass and artemisinin content were at their highest.

The biosynthesis dynamics of artemisinin were in accordance with *A. annua* terpenoid profiles at different developmental stages. This suggested that artemisinin and the terpenoids had similar regulatory mechanisms and makes the regulation of artemisinin biosynthesis complicated. Since the biosynthesis of sesquiterpenes, triterpenoids, and artemisinin all begin from FPP and might compete for the same precursor, the approach of increasing artemisinin content by increasing substrate FPP might not be effective due to the subdivision of FPP. This phenomenon has been confirmed by other studies. Han et al. overexpressed farnesyl diphosphate gene (*fps*) in transgenic *A. annua* by genetic engineering, but the artemisinin content did not increase dramatically (Han et al. 2006). However, the confirmation of the related terpenoids also provides new targets for regulating artemisinin biosynthesis, for example, by blocking the biosynthesis of these compounds to increase artemisinin accumulation.

11.3.2 Comparative Analysis of Terpene Profiles in Different A. annua Genotypes

A. annua strains of different origins have different contents of artemisinin and its precursors. Delabays et al. reported that, excluding the influence of local environment on the contents of artemisinin and its precursors, different genotypes should be the main source of different artemisinin contents (Delabays et al. 2001). Strains SP18 and 001 both have high artemisinin yields and are owned by the Laboratory of Plant

Secondary Metabolism and Metabolic Engineering of the Institute of Botany of the Chinese Academy of Sciences. These two strains have obvious differences in morphology, and the artemisinin content of SP18 is higher than 001; however, other differences in metabolites are not clear. Hence, we applied metabolomic methods to analyze the differences in terpenoids of these two genotypes.

11.3.2.1 Secondary Metabolic Profiling of SP18 and 001

In order to analyze the differences of terpenoid profiles between SP18 and 001, a total of 92 compounds were analyzed by GC/MS. Gas chromatograms are shown in Fig. 11.4. The internal standard was used to estimate the content of compounds, and their profile data were analyzed by PLS-DA. The samples of SP18 and 001 clearly formed two distinct clusters in the score plot, which indicated significant differences in metabolites of the two genotypes (Fig. 11.5). According to the PLS-DA results,

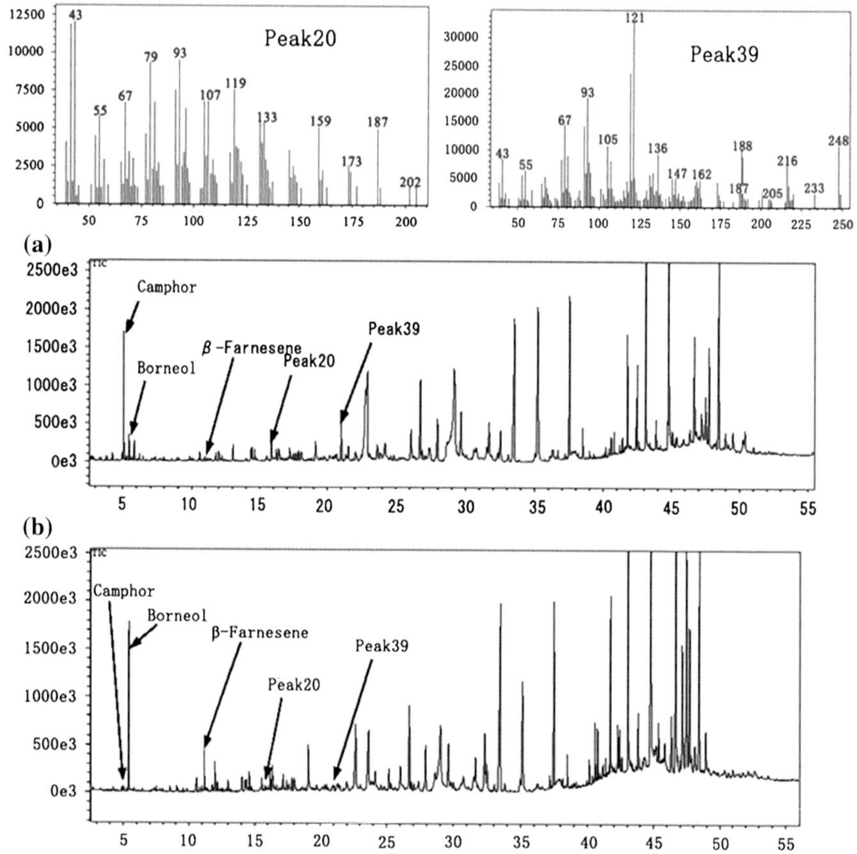

Fig. 11.4 GC/MS chromatograph of terpenoids in *A. annua* **a** SP18, **b** 001

22 compounds were selected as marker compounds based on their VIP values (VIP > 1). Among them, 12 were sesquiterpenoids, three were monoterpenes, four were triterpenoids, and the other three were unknown compounds. Artemisinin and its related precursors, dihydroartemisinic acid, artemisinic acid, and arteannuin B, were all selected as marker compounds, indicating that these compounds had different contents in the two genotypes.

Our analysis of terpenoid profiles showed that—except for borneol, β-farnesene, artemisinic acid, and arteannuin B—the contents of all the other marker compounds in SP18 were higher than those in 001, meaning that SP18 had a higher total content than 001. In addition, five compounds had contents that significantly differed in the two genotypes (Fig. 11.4), some had a very high content in one genotype but a very low content in the other genotype, and such compounds were considered as characterized compounds. The characterized compounds of 001 were borneol and β-farnesene, but those of SP 18 were camphor and two unidentified sesquiterpenoids.

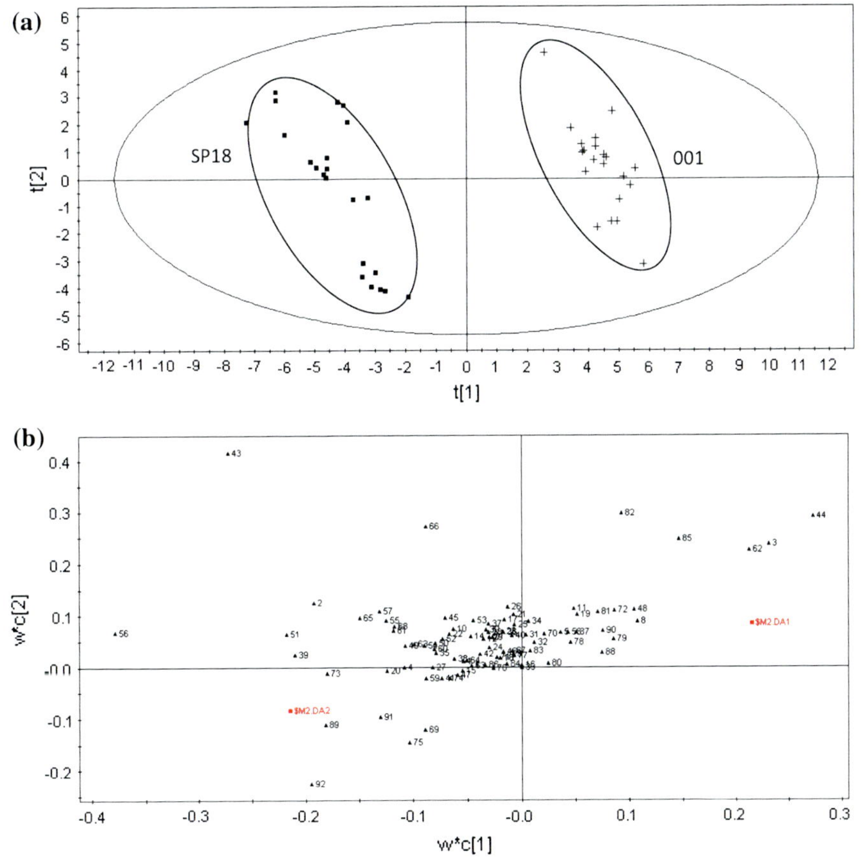

Fig. 11.5 The score (**a**) and loading (**b**) plot of PLS-DA model based on the terpenoid profiles of *A. annua* SP18 and 001

Borneol is a monoterpenoid with a relative content of 0.83 mg/g in 001, but only 0.13 mg/g in SP18; corresponding relative contents of the sesquiterpene β-farnesene were 0.27 and 0.12 mg/g. Camphor was accumulated in SP18 to a relatively high level of approximately 0.98 mg/g, but was nearly absent in 001. The other two sesquiterpenoid peaks (Nos. 20 and 39) had relatively higher contents of 0.23 and 0.60 mg/g in SP18, respectively, but only 0.03 and 0.02 mg/g in 001.

11.3.2.2 Accumulation Patterns of Artemisinin and Its Precursors in Different Genotypes

The detected artemisinin content was higher in SP18 than in 001. There were dramatically different accumulations of artemisinin and its related biosynthetic precursors, dihydroartemisinic acid, artemisinic acid, and arteannuin B between the two *A. annua* genotypes (Fig. 11.6). More specifically, dihydroartemisinic acid

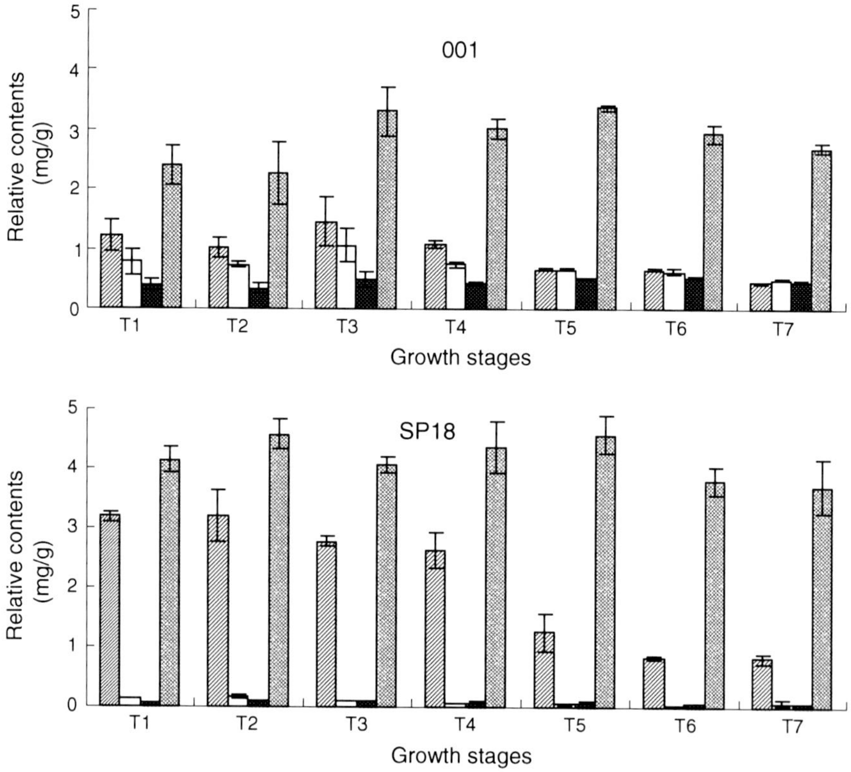

Fig. 11.6 Comparison of artemisinin and related precursors in *A. annua* SP18 and 001 Dihydroartemisinic acid ▨; Artemisinic Acid □; Arteannuin B ■; Artemisinin ⊠

accumulation was high in both genotypes, although its content was much higher in SP18 than in 001, with 3.20 and 1.45 mg/g, respectively. In contrast, the content of artemisinic acid was much higher in 001 than in SP18, with relative contents of 1.06 and 0.16 mg/g, respectively; similarly, the corresponding values for arteannuin B were 0.54 and 0.10 mg/g.

In 2000, Wallaart et al. studied the relationship between accumulation patterns of possible biosynthetic precursors and artemisinin concentration in different chemotypes of *A. annua*. They found that higher dihydroartemisinic acid content was accompanied by higher artemisinin content, but higher artemisinic acid content was accompanied by lower artemisinin content (Wallaart et al. 2000). They studied two different *A. annua* chemotypes of respectively high- and low-artemisinin yields. In the present study, the two *A. annua* genotypes had high yields of artemisinin, and hence, both had higher dihydroartemisinic acid content. However, the concentrations of artemisinic acid and arteannuin B in 001 were five times those in SP18, and their concentrations were very low in SP18. These results suggested that the content of artemisinin was more related to dihydroartemisinic acid. Recently, Bertea et al. and Brown and Sy (Bertea et al. 2005; Brown and Sy 2007) reported that artemisinin was most likely converted from dihydroartemisinic acid, and such a result is consistent with our findings of metabolic fingerprinting comparison of the two *A. annua* genotypes.

The different accumulation patterns of artemisinin and some of its possible biosynthetic precursors in different *A. annua* genotypes might imply different biosynthetic pathways. The following artemisinin biosynthetic pathway in *A. annua* was proposed (Roth and Acton 1989; Nair and Basile 1993; Sangwan et al. 1993): artemisinic acid → arteannuin B → artemisinin. Since the SP18 strain contained trace amounts of artemisinic acid and arteannuin B, but higher accumulation of dihydroartemisinic acid, our results supported the existence of the above pathway. In contrast, since higher contents of artemisinic acid and arteannuin B accumulated in 001 strain, it could be that the pathway from artemisinic acid or arteannuin B to artemisinin is possibly a rate-limiting step in 001. Brown and Sy's recent in vivo feeding experiment with artemisinic acid labeled with both ^{13}C and ^{2}H excluded artemisinic acid as a precursor of artemisinin biosynthesis in *A. annua*, but showed dihydroartemisinic acid was the precursor because the double bond in C11–C13 was difficult to deoxygenate to form a saturated single bond (Brown and Sy 2007). Considering that SP18 contained a higher level of artemisinin, and trace amounts of artemisinic acid and arteannuin B, we speculate that the pathway from amorpha-4,11-diene to artemisinic acid was partially blocked, then the carbon flow to dihydroartemisinic acid increased, and so the artemisinin content also increased. In contrast, in 001, higher amounts of carbon might flow to artemisinic acid and arteannuin B, and consequently, the contents of dihydroartemisinic acid and artemisinin were lower than those in SP18. Dihydroartemisinic acid has been reported as a direct precursor of artemisinin biosynthesis (Wallaart et al. 1999; Brown and Sy 2007). In the present study, high concentrations of dihydroartemisinic acid in both *A. annua* genotypes (Fig. 11.6) suggest that the conversion of dihydroartemisinic acid to artemisinin is a rate-limiting step, and hence, any strategies of increasing the conversion could further increase artemisinin content. Since different *A. annua* genotypes have different sesquiterpene biosynthetic pathways, different strategies should be adopted accordingly in different genotypes. For instance, blocking the carbon flow to artemisinic acid, thereby increasing the accumulation of dihydroartemisinic acid, might increase artemisinin content in 001 strain.

Since *A. annua* contains many terpenoids and some of its characteristic terpenoids are revealed by metabolomics, the study of the relationship between the biosyntheses of characteristic terpenoids and artemisinin could be another regulatory target for artemisinin biosynthesis. For example, 001 contained a relatively high level of borneol but almost no camphor. In contrast, SP18 contained a relatively high level of camphor but a relatively low level of borneol. Since camphor is an oxide of borneol, from the viewpoint of biotransformation, the above results suggest that the conversion from borneol to camphor was low in 001, but occurred easily in SP18. From the viewpoint of gene expression regulation, this might be due to the different expressions of the genes involved in the pathway in different genotypes. It remains unclear whether these genes have similar regulatory mechanisms with genes involved in the artemisinin biosynthetic pathway—studies on these issues could provide new insight into interpretation of the isoprenoid pathway.

11.3.3 Comparison of Terpenoid Metabolic Profiling Between Different Transgenic A. annua Lines

The artemisinin biosynthetic pathway belongs to a branch of plant terpenoid metabolism. The pathway from FPP to dihydroartemisinic acid is well known, i.e., FPP → amorpha-4,11-diene → artemisinic alcohol → artemisinic aldehyde → dihydroartemisinic aldehyde → dihydroartemisinic acid, and the involved genes have been cloned and validated (Covello et al. 2007; Zhang et al. 2008; Teoh et al. 2009). However, the biosynthetic pathway from dihydroartemisinic acid to artemisinin is still unclear. Cloning relevant genes in the artemisinin biosynthetic pathway is a possible way to increase artemisinin content. In 1999, through overexpression of a cotton FPS cDNA, we obtained transgenic lines with onefold to twofold increases in artemisinin content compared to controls, with the highest artemisinin content of about 1.0 % by dry weight. We also obtained transgenic lines of *A. annua* overexpressing FPS (F4, F6 and F18) and amorphpha-4,11-diene synthase (A4, A6, A8 and A9), and a RNAi line of amorphpha-4,11-diene synthase (Ami). Our previous results showed that some positive influences on artemisinin content in *A. annua* could be achieved by gene overexpression, but we did not analyze the changes in terpenoid metabolic profiling among different *A. annua* transgenic lines. In the present study, a metabolomics approach was used to study the metabolic profiling of different *A. annua* transgenic lines, their change dynamics as well as the relationship between artemisinin biosynthesis and other terpenoid biosyntheses to shed new light on the biosynthetic pathway and regulation of artemisinin.

11.3.3.1 Changes of Terpenoid Metabolic Profiling in Transgenic *A. annua* Lines by Overexpressing the Farnesyl Diphosphate Synthase Gene (F Lines)

FPP, thc common substrate for sesquiterpenoid biosynthesis, is catalyzed by FPS. In the present study, we analyzed three lines overexpressing the farnesyl diphosphate synthase gene (*FPS*): F4, F6, and F18. By deleting the impurity peaks of the GC × GC-TOFMS, we obtained an ingredient list of samples through extracting terpenoid information. Matching the list with the peaks of the control 001 and GUS lines gave 208 variables after calibration with the internal standard. PLS-DA multivariate statistical analysis on the relative concentrations showed significant differences in metabolites between the transgenic lines and controls and different distributions in distinct regions on the score plot (Fig. 11.7).

The distance between transgenic F lines and the control 001 was much larger than that between the transgenic F lines and the control GUS (Fig. 11.7), suggesting

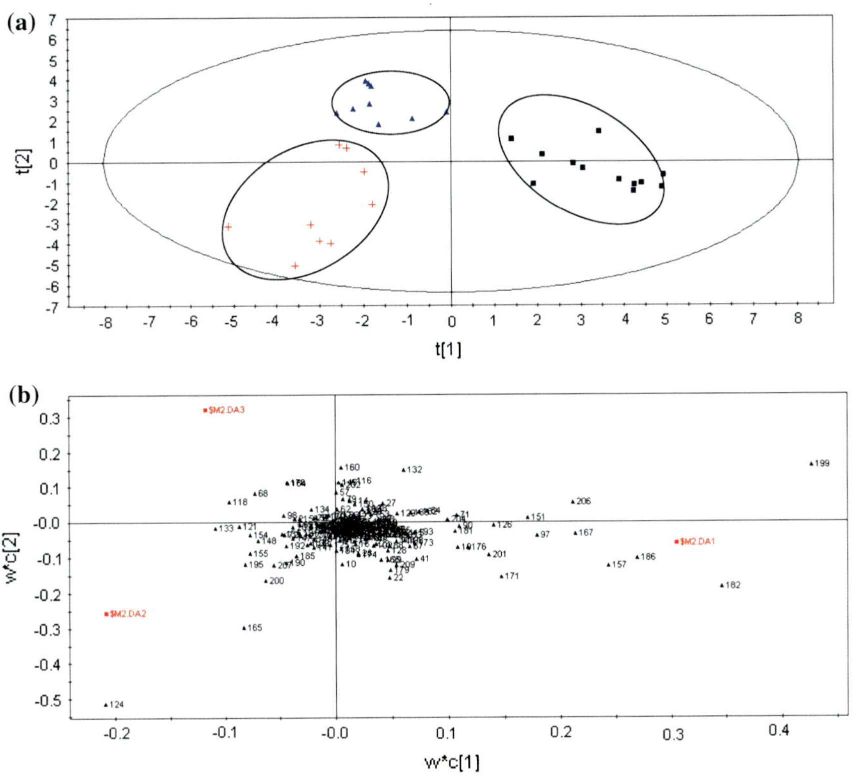

Fig. 11.7 The score (**a**) and loading (**b**) plot of PLS-DA model based on the terpenoid profiles of transgenic line F and control lines 001, *GUS* ■ 001; ▲ *GUS*; + F

Table 11.1 Marker compounds of PLS-DA between transgenic line F and controls

Peak no.	Rt		Compound name	Molecular formula	VIP
	1st	2nd			
199	3,175	2.66	Artemisinin (degraded product B)[a]	$C_{15}H_{22}O_5$	4.94
182	2,855	1.32	Artemisinic acid[a]	$C_{15}H_{22}O_2$	4.16
186	2,910	2.28	Deoxyqinghaosu[b]	$C_{15}H_{22}O_4$	3.11
165	2,600	1.3	n-Hexadecanoic acid[b]	$C_{16}H_{32}O_2$	2.92
206	3,310	2.72	Artannuin B[b]	$C_{15}H_{22}O_3$	2.39
171	2,670	1.46	Dihydroartemisinic acid[a]	$C_{15}H_{24}O_2$	2.15
97	1,930	1.76	Caryophyllene oxide[b]	$C_{15}H_{24}O$	1.99
22	875	2.76	Germacrene D[b]	$C_{15}H_{24}$	1.54
132	2,295	1.74	–	$C_{15}H_{24}O$	1.53
176	2,740	1.58	–	$C_{15}H_{24}$	1.44
179	2,800	1.52	–	$C_{15}H_{24}$	1.37
19	855	1.5	Borneol[b]	$C_{10}H_{18}O$	1.35
90	1,860	2.28	–	$C_{15}H_{22}O$	1.21
133	2,305	1.24	–	$C_{10}H_{18}O_2$	1.20
181	2,820	1.66	–	$C_{15}H_{24}O$	1.19
118	2,170	2.2	–	$C_{10}H_{16}O$	1.18
155	2,470	1.82	–	$C_{15}H_{24}O$	1.18
68	1,605	2.56	–	$C_{15}H_{22}O_2$	1.1
10	710	2.78	Caryophyllene[b]	$C_{15}H_{24}$	1.1

Note Rt retention time, VIP variable importance in the projection
[a] metabolite identical to authentic compound
[b] metabolite identical to NIST library
– mean unidentified

that the difference between F lines and control 001 was larger than that between F lines and GUS. In addition, a larger VIP value of PLS-DA suggests a large diversity of different samples. Table 11.1 lists 19 compounds with significant changes between transgenic line F and control (i.e., with VIP > 1), and they were considered as marker compounds. Preliminary analysis of qualitative MS showed that most of these compounds were sesquiterpenoids, including four sesquiterpenes and 11 sesquiterpene oxide derivatives. Since artemisinin and related compounds, such as artemisinic acid, dihydroartemisinic acid, and arteannuin B, were all marker compounds, this suggests that the network of terpenoid metabolism was affected by genetic transformation, and terpenoid metabolic profiling changed significantly. The concentrations of most of the marker compounds were down-regulated in the transgenic F lines compared with control 001, and the total concentration of the marker compounds was also down-regulated (Table 11.2), this indicates that transgenes could cause some stresses on plants and inhibitions of terpenoid metabolism. However, there were also some marker compounds in transgenic F lines with higher concentrations than those of controls, such as peak No. 165

Table 11.2 Changes of markers in F transgenic lines

Peak no.	Rt		Compounds	Content (mg/g)		
	1st	2nd		F	GUS	001
199	3,175	2.66	Artemisinin (degraded product B)[a]	0.864	0.855	1.244
182	2,855	1.32	Artemisinic acid[a]	1.156	0.988	1.505
186	2,910	2.28	Deoxyqinghaosu[b]	0.711	0.671	0.941
165	2,600	1.30	*n*-Hexadecanoic acid[b]	0.299	0.188	0.165
206	3,310	2.72	Artannuin B[b]	0.591	0.616	0.732
171	2,670	1.46	Dihydroartemisinic acid[a]	0.658	0.597	0.805
97	1,930	1.76	Caryophyllene oxide[b]	0.112	0.122	0.142
22	875	2.76	Germacrene D[b]	0.222	0.168	0.270
132	2,295	1.74	$C_{15}H_{24}O$	0.257	0.327	0.331
176	2,740	1.58	$C_{15}H_{24}$	0.017	0.000	0.057
179	2,800	1.52	$C_{15}H_{24}$	0.200	0.219	0.258
19	855	1.50	Borneol[b]	0.789	0.776	1.060
90	1,860	2.28	$C_{15}H_{22}O$	0.032	0.030	0.062
133	2,305	1.24	$C_{10}H_{18}O_2$	0.059	0.049	0.020
181	2,820	1.66	$C_{15}H_{24}O$	0.031	0.024	0.064
118	2,170	2.20	$C_{10}H_{16}O$	0.281	0.278	0.296
155	2,470	1.82	$C_{15}H_{24}O$	0.085	0.080	0.094
68	1,605	2.56	$C_{15}H_{22}O_2$	0.024	0.015	0.025
10	710	2.78	Caryophyllene[b]	0.066	0.038	0.065
Total content of VIP				6.630	6.174	8.449

Note Rt retention time
[a] identified with authentic standards
[b] identified with NIST library; repeats $n \geq 8$

(*n*-hexadecanoic acid) and peak No. 133 (monoterpene oxide derivative) with a concentration of twice that of controls. From the viewpoint of compound transformations, these two marker compounds have no direct relationship with farnesyl diphosphate in the terpenoid biosynthetic pathway. Since compared with control GUS lines, the content of terpenoid compounds in transgenic F lines increased, this suggests that the increased FPP concentration induced by overexpression of the farnesyl diphosphate gene simultaneously caused increases in some compounds in the terpenoid pathway. Note that not all were promoted—the concentrations of arteannuin B and peak No. 132 (sesquiterpene) were down-regulated.

Terpenoid metabolic profiling varied significantly in different transgenic F lines (F4, F6 and F18). The highest concentrations of artemisinin and arteannuin B were in F6 and the lowest in F18. In contrast, the highest concentrations of dihydroartemisinic acid were in F18 and the lowest in F6, but the concentrations of artemisinic acid showed no apparent changes among the three lines. The concentrations of most of the marker compounds were slightly higher in F6 than in lines F4 and F18; however, the concentrations of *n*-hexadecanoic acid, caryophyllene, one of the sesquiterpene oxide derivatives and one of the monoterpene oxide derivatives were relatively much lower. In order to study the effects of overexpressing FPS on artemisinin biosynthesis, we compared the marker compounds between transgenic line F6 and transgenic control line GUS. Table 11.3 lists the marker compounds with VIP > 1 in the PLS-DA model. The concentrations of artemisinin, dihydroartemisinic acid, a monoterpene borneol and some sesquiterpenoids were up-regulated in line F6; however, a few compounds such as diterpene phytol were down-regulated in F6.

As FPP is the common substrate for both sesquiterpenoid and triterpenoid biosynthetic pathways, the overexpression of FPS would have a great impact on terpenoid metabolism. The purpose of overexpressing FPP synthase is to increase the amount of substrate and so promote biosynthesis of the final product sesquiterpenoids. Han et al. found that although artemisinin content of transgenic *FPS* lines did increase compared with controls, the increase was not significant (Han et al. 2006). Our analysis of terpene metabolic profiling also showed that many sesquiterpene compounds in transgenic F lines had changed (Table 11.2). Compared with transgenic GUS lines, the concentrations of artemisinin, artemisinic acid, dihydroartemisinic acid, and even some other sesquiterpenoids in F lines did increase to some extent. However, the increases were not significant, suggesting that overexpression of FPS probably merely increased the concentration of the sesquiterpenoid substrate FPP, but since FPP is the common substrate for the biosynthesis of numerous sesquiterpenoids, the increase of the final product was rather limited. In addition, it was shown that not all sesquiterpenoid contents increased compared with controls; some were even down-regulated. The contents of some monoterpenes and diterpenes also changed, probably due to random insertion of the exogenous genes. These results suggest that transgenic technology might generate purposeful regulations, but also some non-intentional random effects.

Table 11.3 Comparison of marker compounds between transgenic line F6 and control line GUS

Rt		Compound name	Molecular formula	VIP	Content (mg/g)	
1st	2nd				F6	GUS
855	1.5	Borneol[b]	$C_{10}H_{18}O$	4.42	0.887	0.621
2,295	1.74	–	$C_{15}H_{24}O$	3.09	0.227	0.315
2,225	2.04	Phytol[b]	$C_{20}H_{40}O$	2.65	0.490	0.615
3,175	2.66	Artemisinin (degraded product B)[a]	$C_{15}H_{22}O_4$	2.51	1.113	0.790
2,670	1.46	Dihydroartemisinic acid[a]	$C_{15}H_{24}O_2$	2.34	0.550	0.426
2,470	1.82	–	$C_{15}H_{24}O$	2.16	0.242	0.173
2,740	2.38	–	$C_{15}H_{20}O_2$	2.15	0.080	0.027
2,595	1.84	5-Acetoxytridecane[b]	$C_{15}H_{30}O_2$	1.67	0.163	0.103
1,655	1.92	Diepicedrene-1-oxide[b]	$C_{15}H_{24}O$	1.63	0.014	0.045

Note Rt retention time, VIP variable importance in the projection
[a] metabolite identical to authentic compound
[b] metabolite identical to NIST library
– mean unidentified, repeats $n \geq 8$

11.3.3.2 Terpenoid Metabolic Profiling Changes with Overexpression and RNAi Inhibition of Amorpha-4,11-diene Synthase Gene Lines (A, Ami)

As the first enzymc located on the branching pathway to artemisinin biosynthesis, amorpha-4,11-diene synthase is considered a key enzyme. We compared terpenoid metabolic profiling of amorpha-4,11-diene synthase overexpressing lines (A) and suppressing lines (Ami) with control 001. By eliminating GC impurity peaks, then collecting terpenoid peaks, 199 matched constituents between transgenic lines and control 001 were obtained. These were then calibrated by the internal standard and analyzed with orthogonal signal correction–partial least square (OSC-PLS). There were significant differences in terpenoid metabolic profiling between transgenic lines and controls, and they were distributed in different clusters in the score plot (Fig. 11.8). The transgenic lines were located much closer, while the control was far from the transgenic lines (Fig. 11.8), indicating that the transferred genes had some

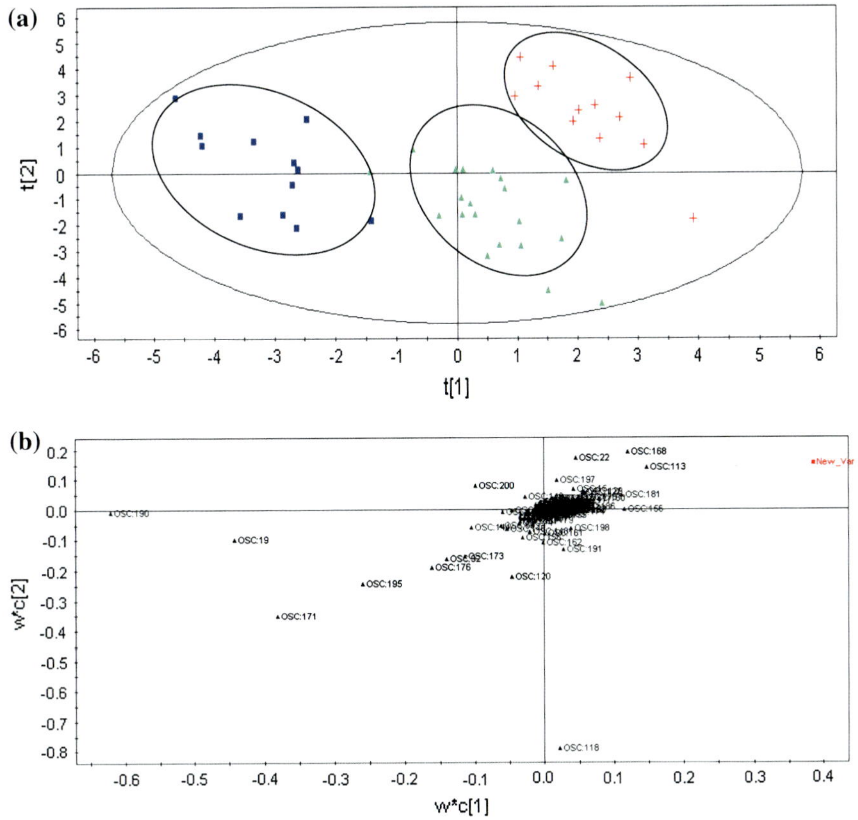

Fig. 11.8 The score (**a**) and loading (**b**) plot of OSC-PLS model based on the terpenoid profiles of transgenic lines A, Ami and control 001 ■ 001; ▲ A; + A mi

Table 11.4 Marker compounds of OSC-PLS between A, Ami transgenic lines, and its control 001

Peak no.	Rt		Compound name	Molecular formula	VIP
	1st	2nd			
190	3,175	2.66	Artemisinin (degraded product B)[a]	$C_{15}H_{22}O_5$	8.23
19	855	1.5	Borneol	$C_{10}H_{18}O$	5.79
171	2,855	1.32	Artemisinic acid[a]	$C_{15}H_{22}O_2$	5.09
118	2,225	2.04	Phytol[b]	$C_{20}H_{40}O$	4.21
176	2,910	2.28	Deoxyqinghaosu[b]	$C_{15}H_{22}O_4$	2.22
113	2,180	1.88	–	$C_{10}H_{16}O$	1.97
92	1,930	1.76	Caryophyllene oxide[b]	$C_{15}H_{24}O$	1.91
168	2,800	1.52	–	$C_{15}H_{26}O$	1.74
162	2,670	1.46	Dihydroartemisinic acid[a]	$C_{15}H_{24}O_2$	1.60
156	2,600	1.3	n-Hexadecanoic acid[b]	$C_{16}H_{32}O_2$	1.53
22	875	2.76	Germacrene D[b]	$C_{15}H_{24}$	1.00

Note Rt retention time, VIP variable importance in the projection
[a] metabolite identical to authentic compound
[b] metabolite identical to NIST library
–mean unidentified

effects on terpenoid metabolic profiling. Table 11.4 lists 10 terpenoids with VIP > 1 in OSC-PLS analysis of A and Ami transgenic lines and control 001, and uses as marker compounds to differentiate transgenic lines and controls. Artemisinin, artemisinic acid, dihydroartemisinic acid, and deoxy-artemisinin were all chosen as marker compounds. Table 11.5 lists the changes of marker compounds in transgenic A and Ami lines. The total amount of marker compounds in transgenic A lines was

Table 11.5 Changes of marker compounds in A and Ami transgenic lines

Rt		Compound name	Content (mg/g)			
1st	2nd		A	Ami	GUS	001
3,175	2.66	Artemisinin (degraded product B)[a]	1.134	0.953	0.855	1.244
855	1.5	Borneol	1.020	0.838	0.776	1.013
2,855	1.32	Artemisinic acid[a]	1.532	1.192	0.988	1.505
2,225	2.04	Phytol[b]	1.317	0.839	0.597	0.822
2,910	2.28	Deoxyqinghaosu[b]	0.872	0.808	0.671	0.941
2,180	1.88	$C_{10}H_{16}O$	0.326	0.317	0.278	0.296
1,930	1.76	Caryophyllene oxide[b]	0.138	0.135	0.122	0.141
2,800	1.52	$C_{15}H_{26}O$	0.258	0.261	0.219	0.258
2,670	1.46	Dihydroartemisinic acid[a]	0.961	0.804	0.597	0.805
2,600	1.3	n-Hexadecanoic acid[b]	0.248	0.260	0.188	0.165
875	2.76	Germacrene D[b]	0.140	0.271	0.168	0.270
Total content of VIP			6.859	5.840	4.737	6.447

Note [a] identified with authentic standards
[b] identified with NIST library; repeats $n \geq 8$

highest and was 45 % higher than the transgenic GUS line, 17 % higher than the transgenic Ami line, and slightly higher than the control 001. This suggests that overexpression of ADS could increase biosynthesis and accumulation of some compounds, in particular, phytol, dihydroartemisinic acid and *n*-hexadecanoic acid increased markedly. However, the concentration of sesquiterpene germacrene D in transgenic A lines decreased by 48 % compared to control 001, and 22 % compared to the transgenic GUS line, while its concentration in the transgenic Ami line increased to about 50 % more than that of transgenic GUS lines. Among all transgenic A lines, the highest artemisinin content was in A9 and the lowest in A6. In contrast, the highest dihydroartemisinic acid content was in A6 and the lowest in A9. For artemisinic acid, the highest content was in A8 and the lowest in A9. In order to investigate the effects of overexpressing ADS on artemisinin biosynthesis in this work, the transgenic A9 line was chosen to compare with transgenic GUS line. The concentrations of artemisinin, dihydroartemisinic acid, and deoxy-artemisinin in A9 all increased significantly compared to the GUS line; the contents of diterpenoid phytol and monoterpenoid borneol also increased substantially in A9, and the contents of only a few sesquiterpenoids were down-regulated compared with the transgenic GUS line (Table 11.6).

Although ADS is the first branching key enzyme in the artemisinin biosynthetic pathway, due to the low abundance of amorpha-4,11-diene in *A. annua*, it is hard to detect amorpha-4,11-diene in both ADS overexpressing and RNAi inhibiting transgenic lines. Thus, the only way to study the regulating effects of the transferred genes is to detect the downstream products of the pathway. Our metabolic profiling showed that levels of artemisinic acid, artemisinin, dihydroartemisinic acid, and arteannuin B in ADS overexpressing lines were all elevated to some extent,

Table 11.6 Comparison between transgenic line A9 and GUS

Rt	Compound name	Molecular formula	VIP	Content (mg/g)		
1st	2nd				A9	GUS
2,225	2.04	Phytol[b]	$C_{20}H_{40}O$	5.56	1.550	0.636
3,175	2.66	Artemisinin (degraded product B)[a]	$C_{15}H_{22}O_5$	3.65	1.291	0.786
2,670	1.46	Dihydroartemisinic acid[a]	$C_{15}H_{24}O_2$	2.53	0.608	0.463
855	1.5	Borneol	$C_{10}H_{18}O$	1.78	1.071	0.627
2,910	2.28	Deoxyqinghaosu[b]	$C_{15}H_{22}O_4$	1.51	0.841	0.631
2,855	1.32	Artemisinic acid[a]	$C_{15}H_{22}O_2$	1.35	1.453	0.988
2,800	1.52	–	$C_{15}H_{26}O$	1.31	0.217	0.199
2,465	1.8	–	$C_{15}H_{24}O$	1.05	0.048	0.069

Note Rt retention time, VIP variable importance in the projection
[a] metabolite identical to authentic compound
[b] metabolite identical to NIST library
– mean unidentified, repeats $n \geq 8$

indicating that overexpression of ADS promoted the accumulation of downstream products. Moreover, overexpression of ADS in transgenic A lines probably led to more carbon flux to the artemisinin biosynthetic pathway to down-regulate the accumulation of some sesquiterpenoids, such as caryophyllene and germacrene D. All the above results confirmed the effects of genetic modification. Transgenic RNAi lines, which are supposed to down-regulate the downstream metabolites of amorpha-4,11-diene in the artemisinin biosynthetic pathway, did not show any conspicuous decrease compared with the transgenic GUS line, but even increased to some extent. This suggests that the suppression effect of RNAi was not significant, or it might be because amorpha-4,11-diene synthase is not a rate-limiting step in the biosynthetic pathway of artemisinin, as speculated by Li (Li 2007).

Our analysis of the metabolic profiling among different transgenic lines showed that most of the marker compounds were sesquiterpenoids which appeared in all transgenic lines—and artemisinin and related compounds were almost always the marker compounds. The underlying reason could be that on the one hand, the relatively higher concentrations of sesquiterpenoids in *A. annua* made them easier to detect in data processing; on the other hand, the biosynthesis of these compounds in *A. annua* could be purposely regulated. Marker compounds changed significantly after transgenic modification and displayed varied degrees of changes among different transgenic lines. For example, the marker compounds of F lines were significantly increased compared to the A lines, which seemed to be related to the transferred gene. FPP, catalyzed by FPS, is a co-substrate for both sesquiterpenoids and triterpenoids, so its changes could cause a series of changes to downstream compounds. Sesquiterpenes, being the products converted directly by FPP, changed more markedly accordingly in transgenic F lines and might be why sesquiterpenoids showed a high frequency of appearance in the metabolic profiling. ADS, being a branching enzyme in the artemisinin biosynthetic pathway, had little effect on the biosynthesis of other terpenoids although ADS was important to the changes of artemisinin-related compounds. Hence, fewer marker compounds were obtained in metabolic profiling analysis of transgenic A lines. In addition, the metabolic profiling differed even between the different transgenic lines of the same batch, for example, F6 in transgenic F lines and A6 in transgenic A lines, where artemisinin and related metabolites showed some obvious changes. This result is consistent with the results of both Han and Li (Han 2005; Li 2007), where it was reported that the artemisinin content reached its highest in transgenic lines F6 and A9, respectively. We concluded that the above different dynamics among transgenic lines might be due to different insertion sites, expressions of transgene, as well as regulation of the whole metabolism, which resulted in not only positive effects but also random effects, and even side effects. Hence, it is indispensable to conduct screening on a large number of positive lines in order to get the desired lines when metabolite regulation by transgenic technology is employed.

11.3.3.3 Correlation Analysis of Artemisinin Biosynthesis-related Compounds in Transgenic Plants

In transgenic plants, changes of artemisinin-related compounds did not greatly accord with each other. In particular, in some lines of high artemisinin content, the content of artemisinin-related compounds was not high, and hence, we conducted correlation analysis between artemisinin and these artemisinin-related compounds. The results showed a negative correlation between artemisinin and dihydroartemisinic acid, while artemisinin and artemisinic acid had a low correlation, and dihydroartemisinic acid was highly correlated with artemisinic acid, but little correlated with artemisinin B.

Additionally, the marker compounds with significant content changes in transgenic lines included diterpenoid phytol and monoterpenoid borneol. However, these two compounds did not have the same trend in different transgenic lines, indicating that the regulation of sesquiterpenoids might affect the biosynthesis of monoterpenoids and diterpenoids although the underlying mechanism remains unclear. Our correlation results between artemisinin and related compounds showed that artemisinic acid and dihydroartemisinic acid had a positive correlation, indicating that they had the same trend in transgenic plants, and no transformation occurred between them—this is consistent with the in vivo labeling result of Brown and Sy (Brown and Sy 2007). Recently, Zhang et al. (2008) cloned and identified a dihydroartemisinic aldehyde reductase from *A. annua*, which confirmed once again that dihydroartemisinic acid was obtained from dihydroartemisinic aldehyde through artemisinic aldehyde, and excluded the possibility of the transformation from artemisinic acid to dihydroartemisinic acid in *A. annua*. As no correlation seemed to exist between arteannuin B and dihydroartemisinic acid, this implies that dihydroartemisinic acid contributed little to the biosynthesis of arteannuin B. While some correlation existed between arteannuin B and artemisinic acid, it indicates possible mutual transformation between these two compounds. Concerning the negative correlation between artemisinin and dihydroartemisinic acid, the underlying reason could be that being the precursor of artemisinin, dihydroartemisinic acid was oxidized to artemisinin; hence, the same trend of shift occurred between these two compounds in *A. annua*. The low correlation between artemisinin and artemisinic acid might imply that there was no relationship between these two compounds. Hence, we conclude that artemisinin was biosynthesized through the dihydroartemisinic acid branch pathway, while artemisinic acid was through arteannuin B (Fig. 11.9).

11.3.4 Effects of MeJA on Artemisinin Biosynthesis

MeJA, one of the signaling molecules of the jasmonate family, has multiple physiological functions. It can be used as a cellular messenger participating in plant metabolism processing and activating the expression of specific defense-related genes to produce plant defense response products. Thus, exogenous MeJA is used as a good inducer when secondary metabolites are produced by plants or plant cell

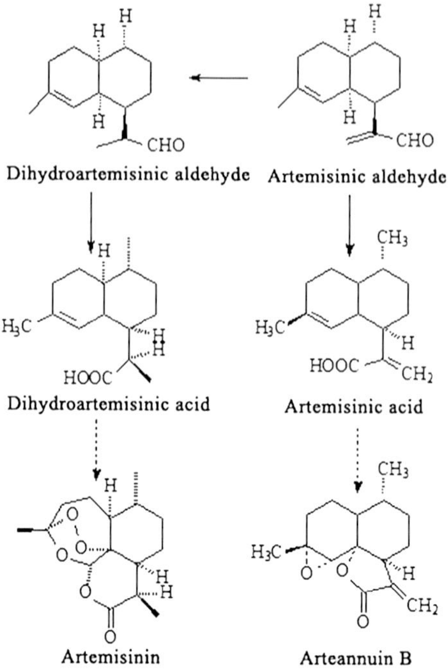

Fig. 11.9 Potential biosynthetic pathway of artemisinin and related compounds

cultures in the literature. Many studies have shown that MeJA can induce biosynthesis of terpenes (Hampel et al. 2005; Singh et al. 1998; Martin et al. 2003; Dicke et al. 1999; Kim et al. 2007). In order to promote the production of artemisinin, we investigated the effect of exogenous MeJA on the biosynthesis of artemisinin and several other terpenoid compounds.

11.3.4.1 Effects of Exogenous MeJA on Artemisinin Biosynthesis

In order to investigate the effects of different concentrations of MeJA on artemisinin biosynthesis, 100, 200, 300, 400, and 500 μmol/L MeJA were respectively sprayed on the leaf surfaces of *A. annua* plants. After treatment for 4 days, the leaves of plants were collected and the artemisinin contents were determined. All the artemisinin contents were increased to some degree under the different concentrations of MeJA treatment—notably, artemisinin content increased significantly ($p < 0.05$) after 300 μmol/L MeJA treatment. Therefore, 300 μmol/L MeJA was chosen as the treatment concentration.

To determine the appropriate timing for collecting *A. annua* leaves, after 300 μmol/L MeJA was sprayed on leaf surfaces, the leaves were collected at 0, 4, 8, 14, and 20 d after treatment and the artemisinin contents were measured respectively. For each time treatment, artemisinin accumulation was significantly induced by MeJA, and the 8-d treatment induced the maximum increase of 38 %, up to 11.9 mg/g (dry weight).

11.3.4.2 Effects of MeJA on Biosynthesis of Terpenoid Compounds

The *A. annua* plants at 8 d after 300 μmol/L MeJA treatment were chosen for metabolite profile analysis by GC and identified by GC-MS. In total, 95 peaks were analyzed, and their peak area ratios to internal standard trans-farnesol were subjected to multivariate analysis with OSC-PLS discriminant analysis using SIMCA-P software. The MeJA-treated samples and controls formed two distinct clusters in the score plot of OSC-PLS (Fig. 11.10), indicating significant changes in the secondary metabolites following MeJA treatment. Nine compounds with VIP > 1 (Table 11.7) were selected as the marker compounds to explore the different changes in terpenoids before and after MeJA treatment. These compounds included six sesquiterpenoids and three triterpenoids based on their mass spectra. The content of all marker compounds increased to some degree after MeJA treatment, among them, squalene increased by 67 % (peak No. 78), and a potential sesquiterpenoid (peak No. 51) increased by 60 %, another potential sesquiterpenoid (peak No. 44) increased by 38 % and dihydroartemisinic acid increased by 29 %.

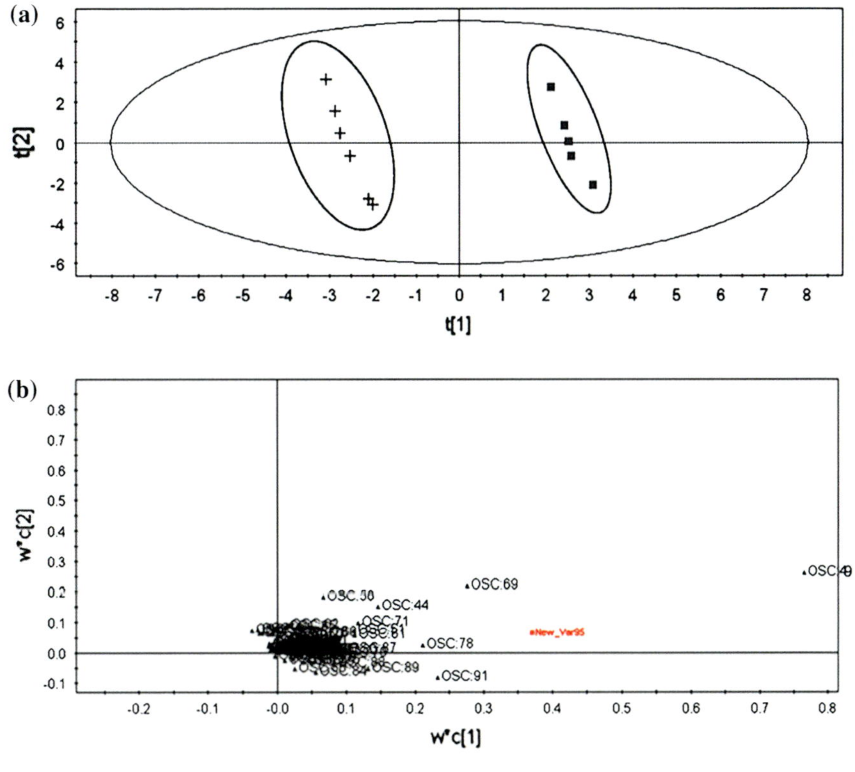

Fig. 11.10 Score (**a**) and loading (**b**) plot of OSC-PLS analysis between MJ treatment and controls + MJ; ■ CK

Table 11.7 Marker compounds in OSC-PLS between MJ treatment and control

Peak no.	Rt (min)	Compounds	Family	VIP
49	22.727	Dihydroartemisinic acid[a]	Sesquiterpenoids	7.36
69	33.435	Product B of artemisinin[a]	Sesquiterpenoids	2.67
91	48.427	$C_{30}H_{48}O$	Triterpenoids	2.32
78	42.46	Squalene[b]	Triterpenoids	2.24
44	20.985	$C_{15}H_{24}O$	Sesquiterpenoids	1.73
89	47.693	$C_{30}H_{48}O$	Triterpenoids	1.26
71	37.468	$C_{15}H_{26}O$	Sesquiterpenoids	1.13
61	29.152	$C_{14}H_{22}O_2$	Sesquiterpenoids	1.08
51	24.135	$C_{15}H_{24}O$	Sesquiterpenoids	1.07

Note Rt retention time, VIP variable importance in the projection
[a] metabolite identical to authentic compound
[b] metabolite identical to NIST library

Under environmental stresses such as drought, frost, UV irradiation, and invasion by pests and diseases, plants generate reactive oxygen species (ROS), which are harmful to plants. For this reason, plants use their own protection mechanisms to eliminate ROS, such as by inducing antioxidase to eliminate ROS to prevent further injury to plant cell membranes, or by inducing production of secondary metabolites. MeJA, one of the signaling molecules of the hormone family, can not only regulate plant growth and development, but can also cause stress effects on plants and induce the generation of ROS. Our results showed that compared with controls, the content of artemisinin significantly increased with MeJA treatment. Moreover, MeJA treatment also promoted the synthesis of other terpenoids, especially sesquiterpenes and triterpenoids. It was reported that exogenous MeJA could increase squalene synthase and squalene oxidase activity (Suzuki et al. 2002; Hayashi et al. 2003). Our results showed that squalene content was increased by 67 % under MeJA treatment, which was consistent with the above results (Suzuki et al. 2002; Hayashi et al. 2003). The biosynthesis of many sesquiterpene compounds is regulated by MeJA, and we found six sesquiterpene compounds significantly increased their contents following MeJA treatment. Exogenous MeJA-treated plants showed stress effects and resulted in accumulation of ROS. In *A. annua*, with a large amount of ROS accumulation, dihydroartemisinic acid could then be converted to dihydroartemisinic acid peroxides, and be further converted to artemisinin by auto-oxidation. Our results showed that the content of dihydroartemisinic acid increased by 29 %, while the content of artemisinin increased by 38 %—it is possible that *A. annua* plants accelerated the conversion from dihydroartemisinic acid to artemisinin to reduce the harmful effects of ROS. In summary, spraying exogenous MeJA not only stimulated the biosynthesis of artemisinin, but also promoted the biosynthesis of terpene compounds, especially the biosynthetic precursors of artemisinin. Therefore, it could be used as a means of improving artemisinin production.

11.4 Conclusion

Some concluding remarks are as follows:

① The changes in terpenoids of *A. annua* at different growth stages were analyzed by GC-MS. The number of terpenoids and the content of each type all increased with time during the vegetative growth phase, reached the highest level at the late vegetative growth phase and pre-budding stage, and then decreased rapidly with time after the reproductive growth phase. As a result, the sampling time for metabolic profiling analysis should be at the pre-flower budding stage, and this provides some guidance for the optimum harvest period of *A. annua* plants.
② The comparison of terpenoid metabolic profiles of different *A. annua* genotypes showed that the artemisinin content was positively correlated with that of dihydroartemisinic acid in lines with high artemisinin yields. Thus, the higher was the dihydroartemisinic acid content of a strain, the higher was the artemisinin content.
③ The metabolomics analyses showed that transgenic technology caused a variety of biological effects to affect the biosynthesis and accumulation of compounds in plant metabolic networks. Some of the changes were directly related to the expression of the transferred gene, and some were not. Thereby, to obtain transgenic lines with expected effects, a large number of transgenic plants should be screened.
④ Exogenous MeJA induced the biosynthesis and accumulation of terpenoids, especially sesquiterpenoids and triterpenoids. Exogenous MeJA treatment could be a useful tool to increase artemisinin production in *A. annua* plants.
⑤ Metabolomics analysis is an effective approach to understand *A. annua* terpenoid metabolic networks. Our metabolic profiling analysis showed that artemisinin biosynthesis was affected by a variety of regulatory mechanisms. Exogenous gene transformation and exogenous MeJA treatment not only affected the biosynthesis and accumulation of artemisinin and its related precursors, but they also influenced the biosynthesis of other terpenoids.

References

Bertea CM, Freije JR, van der Woude H, Verstappen FW, Perk L, Marquez V, De Kraker JW, Posthumus MA, Jansen BJ, de Groot A, Franssen MC, Bouwmeester HJ. Identification of intermediates and enzymes involved in the early steps of artemisinin biosynthesis in *Artemisia annua*. Planta Med. 2005;71:40–7.

Bouwmeester HJ, Wallaart TE, Janssen MH, van Loo B, Jansen BJ, Posthumus MA, Schmidt CO, De Kraker JW, König WA, Franssen MC. Amorpha-4,11-diene synthase catalyses the first probable step in artemisinin biosynthesis. Phytochemistry. 1999;52:843–54.

Brown GD, Sy LK. In vivo transformations of artemisinic acid in *Artemisia annua* plants. Tetrahedron. 2007;63:9548–66.

Chang YJ, Song SH, Park SH, Kim SU. Amorpha-4,11-diene synthase of *Artemisia annua*: cDNA isolation and bacterial expression of a terpene synthase involved in artemisinin biosynthesis. Arch. Biochem. Biophys. 2000;383:178–84.

Covello PS, Teoh KH, Polichuk DR, Reed DW, Nowak G. Functional genomics and the biosynthesis of artemisinin. Phytochemistry. 2007;68:1864–71.

Delabays N, Simonnet X, Gaudin M. The genetics of artemisinin content in *Artemisia annua* L. and the breeding of high yielding cultivars. Curr. Med. Chem. 2001;8:1795–801.

Dicke M, Gols R, Ludeking D, Posthumus MA. Jasmonic acid and herbivory differentially induce carnivore-attracting plant volatiles in lima bean plants. J. Chem. Ecol. 1999;25:1907–22.

Hampel D, Mosandl A, Wust M. Induction of de novo volatile terpene biosynthesis via cytosolic and plastidial pathways by methyl jasmonate in foliage of *Vitis vinifera* L. J. Agric. Food Chem. 2005;53:2652–7.

Han JL. High efficiency of genetic transformation of *Artemisia annua* L. and molecular regulation of artemisinin biosynthesis: Doctoral dissertation, Institute of Botany, Chinese Academy of Sciences, Beijing, 2005.

Han JL, Liu BY, Ye HC, Wang H, Li ZQ, Li GF. Effects of overexpression of the endogenous farnesyl diphosphate synthase on the artemisinin content in *Artemisia annua* L. J. Integr. Plant Biol. 2006;48:482–7.

Hayashi H, Huang P, Inoue K. Up-regulation of soyasaponin biosynthesis by methyl jasmonate in cultured cells of *Glycyrrhiza glabra*. Plant Cell Physiol. 2003;44:404–11.

Kim HJ, Fonseca JM, Choi JH, Kubota C. Effect of methyl jasmonate on phenolic compounds and carotenoids of romaine lettuce (*Lactuca sativa* L.). J. Agric. Food Chem. 2007;55:10366–72.

Li ZQ. Functional analysis and transformation of genes involved in artemisinin biosynthesis. Doctoral dissertation, Institute of Botany, Chinese Academy of Sciences, Beijing, 2007.

Lommen WJ, Elzinga S, Verstappen FW, Bouwmeester HJ. Artemisinin and sesquiterpene precursors in dead and green leaves of *Artemisia annua* L. crops. Planta Med. 2007;10:1133–9.

Martin DM, Gershenzon J, Bohlmann J. Induction of volatile terpene biosynthesis and diurnal emission by methyl jasmonate in foliage of Norway spruce. Plant Physiol. 2003;132:1586–99.

Mercke P, Bengtsson M, Bouwmeester HJ, Posthumus MA, Brodelius PE. Molecular cloning, expression, and characterization of amorpha-4,11-diene synthase, a key enzyme of artemisinin biosynthesis in *Artemisia annua* L. Arch. Biochem. Biophys. 2000;381:173–80.

Nair MS, Basile DV. Bioconversion of arteannuin B to artemisinin. J. Nat. Prod. 1993;56:1559–66.

Picaud S, Mercke P, He XF, Sterner O, Brodelius M, Cane DE, Brodelius PE. Amorpha-4,11-diene synthase: mechanism and stereochemistry of the enzymatic cyclization of farnesyl diphosphate. Arch. Biochem. Biophys. 2006;448:150–5.

Ro DK, Paradise EM, Ouellet M, Fisher KJ, Newman KL, Ndungu JM, Ho KA, Eachus RA, Ham TS, Kirby J, Chang MC, Withers ST, Shiba Y, Sarpong R, Keasling JD. Production of the antimalarial drug precursor artemisinic acid in engineered yeast. Nature. 2006;440:940–3.

Roth RJ, Acton N. A simple conversion of artemisinic acid into artemisinin. J. Nat. Prod. 1989;52:1183–5.

Sangwan RS, Agarwal K, Luthra R, Thakur RS, Singh SM. Biotransformation of arteannuic acid into arteannuin B and artemisinin in *Artemisia annua*. Phytochemistry. 1993;34:1301–2.

Schramek N, Wang HH, Römisch-Margl W, Keil B, Radykewicz T, Winzenhörlein B, Beerhues L, Bacher A, Rohdich F, Gershenzon J, Liu BY, Eisenreich W. Artemisinin biosynthesis in growing plants of *Artemisia annua*. A $^{13}CO_2$ study. Phytochemistry. 2010;71:179–87.

Singh G, Gavrieli J, Oakey JS, Curtis WR. Interaction of methyl jasmonate, wounding and fungal elicitation during sesquiterpene induction in *Hyoscyamus muticus* in root cultures. Plant Cell Rep. 1998;17:391–5.

Sumner LW, Mendes P, Dixon RA. Plant metabolomics: large-scale phytochemistry in the functional genomics era. Phytochemistry. 2003;62:817–36.

Suzuki H, Achnine L, Xu R, Matsuda SP, Dixon RA. A genomics approach to the early stages of triterpene saponin biosynthesis in *Medicago truncatula*. Plant J. 2002;32:1033–48.

Teoh KH, Polichuk DR, Reed DW, Nowak G, Covello PS. *Artemisia annua* L. (Asteraceae) trichome-specific cDNAs reveal CYP71AV1, a cytochrome P450 with a key role in the biosynthesis of the antimalarial sesquiterpene lactone artemisinin. FEBS Lett. 2006;580:1411–6.

Teoh KH, Polichuk DR, Reed DW, Covello PS. Molecular cloning of an aldehyde dehydrogenase implicated in artemisinin biosynthesis in *Artemisia annua*. Botany. 2009;87:635–42.

Wallaart TE, van Uden W, Lubberink HG, Woerdenbag HJ, Pras N, Quax WJ. Isolation and identification of dihydroartemisinic acid from *Artemisia annua* and its possible role in the biosynthesis of artemisinin. J. Nat. Prod. 1999;62:430–3.

Wallaart TE, Pras N, Beekman AC, Quax WJ. Seasonal variation of artemisinin and its biosynthetic precursors in plants of *Artemisia annua* of different geographical origin: proof for the existence of chemotypes. Planta Med. 2000;66:57–62.

Wallaart TE, Bouwmeester HJ, Hille J, Poppinga L, Maijers NC. Amorpha-4,11-diene synthase: cloning and functional expression of a key enzyme in the biosynthetic pathway of the novel antimalarial drug artemisinin. Planta. 2001;212:460–5.

Woerdenbag HJ, Pras N, Chan NG, Bang BT, Bos R, van Uden W, Van YP, Van Boi N, Batterman S, Lugt CB. Artemisinin, related sesquiterpenes, and essential oil in *Artemisia annua* during a vegetation period in Vietnam. Planta Med. 1994;60:272–5.

Zhang Y, Teoh KH, Reed DW, Maes L, Goossens A, Olson DJ, Ross AR, Covello PS. The molecular cloning of artemisinic aldehyde Delta 11(13) reductase and its role in glandular trichome-dependent biosynthesis of artemisinin in *Artemisia annua*. J Biol Chem. 2008;283:21501–8.

Chapter 12
NMR-Based Metabolomic Methods and Applications

Chao-Ni Xiao and Yulan Wang

12.1 Introduction

^{1}H NMR spectroscopy is a technique that can measure the biochemical composition of samples and provide "metabolic fingerprint." The content difference of thousands of metabolites can be monitored, and the dynamic biochemical profile characteristics of metabolites can be observed by comparing the NMR spectra of different samples. Such a metabolomics (metabonomics or metabolic fingerprinting) method has been successfully applied in drug toxicology (Nicholson et al. 2002), clinical chemistry (Bamforth et al. 1999), the environmental study (Griffin et al. 2000; Bundy et al. 2001), plant metabolism (Aranibar et al. 2001; Bailey et al. 2003), natural products metabolism (Bailey et al. 2002), and many other fields. It is also used in creating a variety of mathematical models and predicting the relationships between structure and activity, structure and metabolism (Ghauri et al. 1992; Kessler 1997; Hammond and Kubo 1999; Bailey 2000). These applications have confirmed that metabolomics is a reliable and high-throughput method that can directly analyze complex samples such as plant medicine without requiring complicated separation process and study the entire chemical composition of the samples with a high degree of accuracy and consistency. This chapter will describe applications of metabolomics with examples in plant metabolic composition affected by environmental and processing factors and in predicting efficacy and toxicity of drugs.

C.-N. Xiao
Northwest University, Xi'an 710127, China
e-mail: xiaocn95@hotmail.com

Y. Wang (✉)
Wuhan Institute of Physics and Mathematics, Chinese Academy of Sciences, Wuhan 430071, China
e-mail: yulan.wang@wipm.ac.cn

X. Qi et al. (eds.), *Plant Metabolomics*, DOI 10.1007/978-94-017-9291-2_12

12.2 The Study of Plant Metabolic Composition Affected by Environmental and Processing Factors with Rosemary for Example (Xiao et al. 2008, 2009)

As perennial evergreen shrub in the Lamiaceae family growing widely in the Mediterranean basin and part of Europe, rosemary (*Rosmarinus officinalis* L.) has been used as herb and folk medicine for centuries around the world. For a long period of time, synthetic antioxidants such as butylated hydroxytoluene (BHT) and butylated hydroxyanisole (BHA) have been used to inhibit lipid oxidation and to prevent off-flavor compound formation. But these antioxidants have chronic toxicological effects such as carcinogenic and mutagenic. The rosemary plant is rich in polyphenolic acids, flavonoids, terpenes, volatile oils, etc., which have strong antioxidant activity and are safe and non-toxic, usually as additives used in food, cosmetics, and pharmaceutical industries. Rosemary is the only approved natural antioxidants in Europe and the USA and has replaced synthetic antioxidants in many industries. So far, many studies have shown that rosemary has antimicrobial, anti-inflammatory, anti-viral, and anti-tumor biological activities. The antioxidant and biological activity of rosemary is mainly owing to the radical scavenging effect of a number of secondary metabolites.

12.2.1 NMR and HPLC-MS Analysis of Rosemary Extracts

Figure 12.1 shows the typical ^{1}H NMR spectra of rosemary extracts obtained from ambient temperature water, boiling water, 50 % aqueous methanol, and chloroform/methanol (v/v, 3:1). The metabolites in these spectra were assigned according to the literature (Cuvelier et al. 1994; Fan 1996; Pukalskas et al. 2005; Sobolev et al. 2005) and extensive analysis of the ^{1}H-^{1}H COSY, ^{1}H-^{1}H TOCSY, ^{1}H-^{13}C HSQC, and ^{1}H-^{13}C HMBC 2D NMR spectra. Corresponding chemical shift data are listed in Table 12.1.

To compare the metabolite composition of different extracts, ^{1}H NMR spectra were acquired for the extracts from four different solvent systems. It was apparent that the spectra of rosemary extracts from three polar solvents (Fig. 12.1a–c) were dominated by sugars such as sucrose, fructose, α-glucose, and β-glucose; hydroxycarboxylic acids such as lactate, acetate, malate, succinate, citrate, α-ketoglutarate, malonate, tartrate, fumarate, and quinate; and amino acids such as alanine, asparagine, proline, and valine. Choline was also observable. Although the signal intensity of these metabolites appeared in the aliphatic region varied to a certain degree among the three extracts, marked differences were more evident in the aromatic region. For example, the resonances from rosmarinic acid, caffeic acid, and syringate had much higher intensities for the aqueous methanol extract than the water extracts. The spectrum of the chloroform/methanol extract was drastically different from the other three extracts, in which the major components were

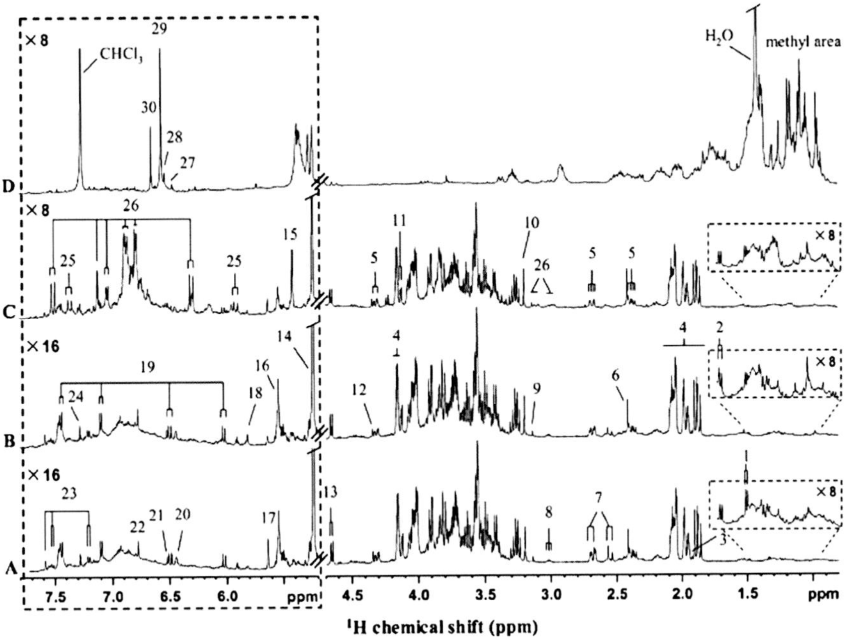

Fig. 12.1 500 MHz ^{1}H NMR spectra of rosemary extracts from different solvents, namely ambient temperature water *A*, boiling water *B*, 50 % aqueous methanol *C*, and chloroform/methanol (v/v, 3:1) *D*. The region δ7.8–5.2 (see the *dot box*) in **a** and **b** and **c** and **d** was expanded 16 and 8 times, respectively, whereas the region δ1.5–0.8 was expanded 8 times in comparison with the region δ4.7–0.8 (Reprinted from Xiao et al. (2008). Copyright © 2008 American Chemical Society. Reprinted with permission)

carnosic acid, carnosol, methylcarnosate, and rosmadial. Only small amounts of hydroxycarboxylic acid, sugars, amino acids, and polyphenolic acids were detectable in this extract, although they were abundant in the other three ones. This is not surprising because these metabolites are much more hydrophilic and not dissolved in chloroform-based solvent extraction. Even from the above qualitative analysis, nevertheless, it is clear that the metabolite profile of the plant extracts is critically dependent on the solvents employed. However, the complexity of rosemary extracts and multiple spectral data, which are necessary for statistical purposes, makes it prohibitively difficult for the spectra to be analyzed with the naked eye; multivariate data analysis is more appropriate for mining such complex data.

12.2.2 Multivariate Data Analysis of the NMR Data for Rosemary Extracts

Initially, PCA was conducted on the spectral data, and two principal components were calculated for the extracts obtained from four different solvents with a total of

Table 12.1 NMR data for rosemary metabolites (Reprinted from Xiao et al. (2008). Copyright © 2008 American Chemical Society. Reprinted with permission)

No.	Metabolite	Group	δ¹H	δ¹³C	Assigned with
1	Lactate	CH_3	1.33 (d, 6.9 Hz)	22.3	TOCSY, HSQC
		CH	4.12 (q, 6.9 Hz)	71.4	
		COOH		185.3	
2	Alanine	CH_3	1.48 (d, 7.3 Hz)	19.1	TOCSY, HMBC
		CH	3.78 (q, 7.3 Hz)	53.3	
		COOH		179.3	
3	Acetate	CH_3	1.92 (s)	26.3	HSQC, HMBC
		COOH		184.3	
4	Quinate①	1 C		78.3	TOCSY, HSQC, HMBC
		2 CH_2	1.88 (m), 2.09 (m)	43.6	HPLC-MS
		3 CH	4.03 (m)	70.0	
		4 CH	3.57 (m)	80.1	
		5 CH	4.16 (q, 3.4 Hz)	73.3	
		6 CH_2	1.98 (m), 2.06 (m)	40.3	
		COOH		184.3	
5	Malate	α-CH	4.31 (dd, 3.1, 10.2 Hz)	73.2	TOCSY, HSQC, HMBC
		β-CH	2.68 (dd, 3.1, 15.4 Hz)	45.2	
		β′-CH	2.37 (dd, 10.2,15.4 Hz)	45.2	
		COOH		183.6	
6	Succinate	CH_2	2.40 (s)	36.5	HMBC
		COOH		184.6	
7	Citrate	α, α′-CH_2	2.56 (d, 15.8 Hz)	47.9	TOCSY, HMBC
		γ, γ′-CH_2	2.70 (d, 15.8 Hz)	47.9	
		β-C		78.8	
		6 COOH		184.5	
		1,5 COOH		181.7	
8	α-ketoglutarate	β-CH_2	3.01 (t, 6.9 Hz)	39.3	TOCSY, HSQC, HMBC
		γ-CH_2	2.45 (t, 6.9 Hz)	33.0	
		C=O		183.9	
		COOH		199.8	
9	Malonate	CH_2	3.13 (s)		HMBC
		COOH		180.0	
10	Choline	N–CH_3	3.20 (s)	56.8	HSQC, HMBC
		α-CH_2		70.4	
11	Fructose	1 CH	4.11 (d, 3.7 Hz)	78.0	TOCSY, HSQC
12	Tartrate	CHOH	4.34 (s)	76.9	HSQC, HMBC
		COOH		181.2	
13	β-glucose	1 CH	4.26 (d, 8.0 Hz)	99.0	TOCSY, HSQC

(continued)

Table 12.1 (continued)

No.	Metabolite	Group	δ^1H	$\delta^{13}C$	Assigned with
14	α-glucose	1 CH	5.23 (d, 3.7 Hz)	95.1	TOCSY, HSQC
15	Sucrose	1 CH	5.40 (d, 3.9 Hz)	94.5	TOCSY, HSQC
		1′ CH	4.22 (d, 8.8 Hz)	79.0	
16	U1		5.54 (t, 2.0 Hz)	129.7	TOCSY, HSQC, HMBC
			2.11 (#)	57.4	
			2.36 (#)	27.6	
			2.44 (#)	41.3	
			2.49 (#)	30.1	
			4.07 (t, 6.6 Hz)	71.9	
17	U2		5.64 (t, 1.8 Hz)	110.0	TOCSY, HSQC, HMBC
			4.72 (#),4.34 (#)	68.3	
		COOH		172.2	
18	U3		5.8 (d)	122.5	TOCSY, HSQC
			2.13 (#), 2.86 (#)	44.5	
			2.06 (#)	25.9	
			2.57 (#)	52.8	
			2.62 (#)	60.5	
19	*Cis*-4-glucosyloxy-cinnamic acid①	=CH–COO	6.03 (d,12 Hz)	128.8	TOCSY, HSQC, HMBC
		Ar–CH=	6.50 (d,12 Hz)	132.4	HPLC-MS
		1 Ar		134.0	
		2,6 Ar–H	7.45 (d, 8.0 Hz)	119.0	
		3,5 Ar–H	7.10 (d, 8.0 Hz)	132.0	
		4-Ar–O		158.8	
		COOH		180.7	
		1′ CH	5.16 (d, 7.3 Hz)	97.1	
		2′ CH	3.87 (#)	#	
		3′ CH	3.89 (#)	#	
		4′ CH	3.98 (#)	#	
		5′ CH	3.64 (#)	#	
20	Shikimate①	1 C		138.9	TOCSY, HSQC, HMBC
		2 CH=	6.46 (m)	133.6	
		3 CH	4.42 (t)	75.2	
		4 CH	3.99 (m)	72.2	
		5 CH	3.72 (m)	66.3	
		6 CH	2.20 (m)	35.6	
		6 CH′	2.78 (Ghauri et al.)	35.6	
		COOH		178.0	
21	Fumarate	CH–CH	6.52 (s)	140.1	HSQC, HMBC
		COOH		177.5	

(continued)

Table 12.1 (continued)

No.	Metabolite	Group	$\delta^{1}H$	$\delta^{13}C$	Assigned with
22	3,4,5-trimethoxy-phenyl methanol	1 Ar		138.4	HSQC, HMBC
		2,6 Ar–H	6.78 (s)	107.1	HPLC-MS
		3,5 Ar		150.4	
		CH_2–O	#	76.9	
		OCH_3	3.88 (s)	#	
23	Vanillate①	1 Ar		120.1	TOCSY, HSQC, HMBC
		2 Ar–H	7.58 (d, 1.9 Hz)	116.1	HPLC-MS
		3 Ar		150.3	
		4 Ar–O		#	
		5 Ar–H	7.20 (d, 8.0 Hz)	118.2	
		6 Ar–H	7.54 (dd, 1.9, 8.0 Hz)	125.6	
		OCH_3	3.93 (s)	58.9	
		COOH		177.4	
24	Syringate①	1 Ar		138.4	HSQC, HMBC
		2,6 Ar–H	7.28 (s)	109.4	HPLC-MS
		3,5 Ar		154.7	
		COOH		176.9	
		OCH_3	3.91 (s)	59.7	
25	Caffeate	=CH–COOH	5.90 (d, 16.0 Hz)	118.6	TOCSY, HSQC, HMBC
		–CH=	7.37 (d, 16.0 Hz)	150.7	HPLC-MS
		COOH		171.6	
26	Rosmarinate①	1 Ar		129.7	TOCSY, HSQC, HMBC
		2 Ar–H	7.15 (d, 1.9 Hz)	118.1	HPLC-MS
		3 Ar		#	
		4 Ar		#	
		5 Ar–H	6.88 (d, 8.0 Hz)	117.5	
		6 Ar–H	7.00 (dd, 1.9, 8.0 Hz)	125.4	
		7 –CH=	7.51 (d, 15.9 Hz)	148.9	
		8=CH–COO	6.31 (d, 15.9 Hz)	116.8	
		9 COO		171.9	
		1′ Ar		133.1	
		2′ Ar–H	6.89 (#)	124.6	
		3′ Ar		#	
		4′ Ar		#	
		5′ Ar–H	6.75 (#)	#	
		6′ Ar–H	6.71 (#)	120.6	
		7′ CH	2.98 (dd, 8.7,16.7 Hz)	39.5	
		7′ CH′	3.01 (dd, 3.9,16.7 Hz)	39.5	
		8′ CH	5.01 (dd, 3.9, 8.7 Hz)	79.1	
		9′ COOH		180.2	

(continued)

Table 12.1 (continued)

No.	Metabolite	Group	δ^1H	$\delta^{13}C$	Assigned with
27	Rosmadial①	13 Ar		137.9	HMBC, HPLC-MS
		14 Ar–H	6.46 (s)	#	
		15 CH	3.19 (sep)	33.3	
		16, 17 CH_3		26.9	
		CHO		180.7	
28	Methylcarnosate①	13 Ar		142.3	HMBC, HPLC-MS
		14 Ar–H	6.53 (s)	#	
		15 CH	3.18 (sep)	26.5	
		16 CH_3, 17 CH_3		32.0	
29	Carnosic acid①	13 Ar		140.5	HSQC, HMBC
		14 Ar–H	6.54 (s)	119.5	HPLC-MS
		15 CH	3.20 (sep)	27.3	
		16 CH_3,17CH_3	1.22 (d), 1.20 (d)	22.3	
		18 CH_3	0.98 (s)	28.1	
		19 CH_3	0.78 (s)	15.7	
30	Carnosol①	1 CH, CH′	2.92 (m), 2.40 (m)	29.0	TOCSY, HSQC, HMBC
		2 CH, CH′	2.04 (m), 1.69 (m)	18.4	HPLC-MS
		3 CH, CH′	1.55 (m), 1.29 (m)	40.7	
		4 C		34.4	
		5 CH	1.73 (Ghauri et al.)	45.2	
		6 CH, CH′	2.21 (m),1.18 (m)	29.4	
		7 CH	5.41 (Ghauri et al.)	77.5	
		8 Ar		131.9	
		9 Ar		121.3	
		10 C		48.2	
		11 Ar-OH		#	
		12 Ar-OH		132.7	
		13 Ar		141.3	
		14 Ar–H	6.65 (s)	112.0	
		15 CH	3.10 (sep)	27.0	
		16 CH_3, 17 CH_3	1.18 (d),1.20 (d)	22.3	
		18 CH_3	0.91 (s)	19.4	
		19 CH_3	0.87 (s)	31.4	
31②	Valine	γ-CH_3	1.05 (d, 7.0 Hz)	20.6	TOCSY, HMBC
		γ′-CH_3	0.99 (d, 7.0 Hz)	19.6	
		β-CH	2.25 (m)	32.0	

(continued)

Table 12.1 (continued)

No.	Metabolite	Group	δ^1H	$\delta^{13}C$	Assigned with
32②	Proline	α-CH	4.14 (m)	64.9	TOCSY, HSQC, HMBC
		β-CH_2	2.34 (m)	32.3	
		γ-CH_2	2.02 (m)	26.6	
		δ-CH	3.37 (m)	49.9	
		δ'-CH	3.41 (m)	49.9	
		COOH		176.3	
33②	Asparagine	α-CH	4.01 (dd, 4.6, 7.6 Hz)	54.2	TOCSY, HSQC, HMBC
		β-CH	2.86 (dd, 7.6, 16.1 Hz)	37.7	
		β'-CH	2.96 (dd, 4.6, 16.1 Hz)	37.7	
		γ-$CONH_2$		177.6	
		COOH		176.8	

① metabolite structure and numbering in Fig. 12.2
② signals found in sun-drying extracts
U un-identified signal
undetermined signals or multiplicities
Ar aromatic ring; *s* singlet; *d* doublet; *t* triplet; *q* quartet; *dd* doublet of doublets; *sep* septet; *m* multiplet

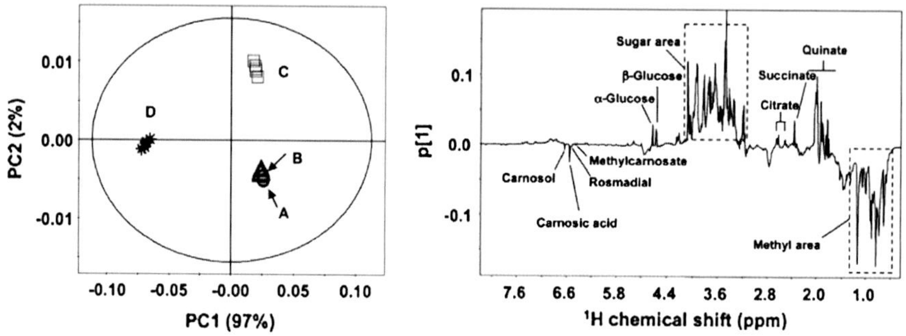

Fig. 12.2 PCA score plot (*left*) and corresponding loading plot (*right*) of rosemary extracts from the four different solvents. *A* ambient temperature water (*white circle*); *B* boiling water (*white triangle*); *C* 50 % aqueous methanol (*white square*); and *D* chloroform/methanol (v/v, 3:1) (*asterisk*) (Reprinted from Xiao et al. (2008). Copyright © 2008 American Chemical Society. Reprinted with permission)

97.0 and 1.8 % of variables being explained by PC1 and PC2, respectively. The score plot (Fig. 12.2) showed that, in the first principal component PC1, obvious differences were present for the samples obtained from the strong polar solvents (A–C) and from the weak polar solvent D. It is also interesting to note that samples from the individual solvents were clustered together closely, indicating the excellent reproducibility in the extraction procedures and NMR measurements. The corresponding loading plot showed that, compared with the extracts from the three aqueous solvents, the chloroform/methanol extract contained a higher level of carnosic acid and its derivatives together with much lower levels of sugars and hydroxycarboxylic acids.

To understand the significance of variables (i.e., metabolites) contributing to classification, the spectral data were further subjected to orthogonal projections to latent structures discriminant analysis (OPLS-DA). The coefficient plots showed the metabolites having contributions to the class difference, and the correlation coefficients (with color-coded scale) for NMR signals indicated the significance of the metabolites' contribution. Because there were no established criteria to assess the collective significances when all variables were considered in the case of multivariate data analysis, here each variable (i.e., metabolite) was assessed only individually and, respectively, using the criteria for univariate analysis. According to number of samples ($n = 5$), the coefficient of 0.81 the cutoff value that was calculated on the basis of discrimination significance at the level of $p = 0.05$. By doing so, nevertheless, our results will be conserved; thus, the statistical significance is underestimated to some extent. Figure 12.3 showed the OPLS-DA score plots and corresponding coefficient plots for the aqueous solvent extracts. Clear separations were observed for two water extracts (Fig. 12.3a), for extracts from methanol/water (50 %) and boiling water (Fig. 12.3b), and samples from methanol/water (50 %) and ambient temperature water (Fig. 12.3c), respectively. The values of R^2X and Q^2 listed on the score plots indicate that these models were of reasonable quality. The coefficient plots were color-coded with the absolute value of correlation coefficients, where a hot-colored signal (red) indicates more significant contribution to the class separation than a cold-colored one (blue); the positive and negative signs indicate the direction of the changes (i.e., positive and negative correlation) for the metabolites. The coefficients for some metabolites, indicating the importance of their contributions, are summarized in Table 12.2. Between two water extracts, most of the metabolites showed no significant difference except for citrate and alanine judged by the coefficients ($|r| > 0.81$). However, compared with the boiling water extracts, the aqueous methanol extracts contained significantly lower levels of glucose, fructose, quinate, citrate, and alanine ($r < -0.81$), but significantly higher levels of rosmarinate, caffeate, *cis*-4-glucosyloxycinnamic acid, syringate, 3,4,5-trimethoxyphenylmethanol, sucrose, shikimate, succinate, fumarate, and malonate ($r > 0.81$). The difference between the chloroform/methanol extract and the extracts from aqueous solvents was so huge that we did not perform OPLS-DA further.

The concentrations of most metabolites were also calculated using the integration areas of the selected NMR signals (least overlapping ones) relative to the internal standard TSP (for aqueous extracts) or TMS (for chloroform extracts). All data are expressed as mean (standard deviation (mean ± SD mg/g of dried rosemary material) from the five parallel samples (Table 12.2). The statistical significances between them were analyzed by one-way ANOVA ($p < 0.05$). Compared with ambient water extracts, most metabolites appeared to have higher levels in boiling water extracts except for citrate, tartrate, malonate, quinate, and lactate. However, significant difference occurred only for citrate and alanine, which is in good

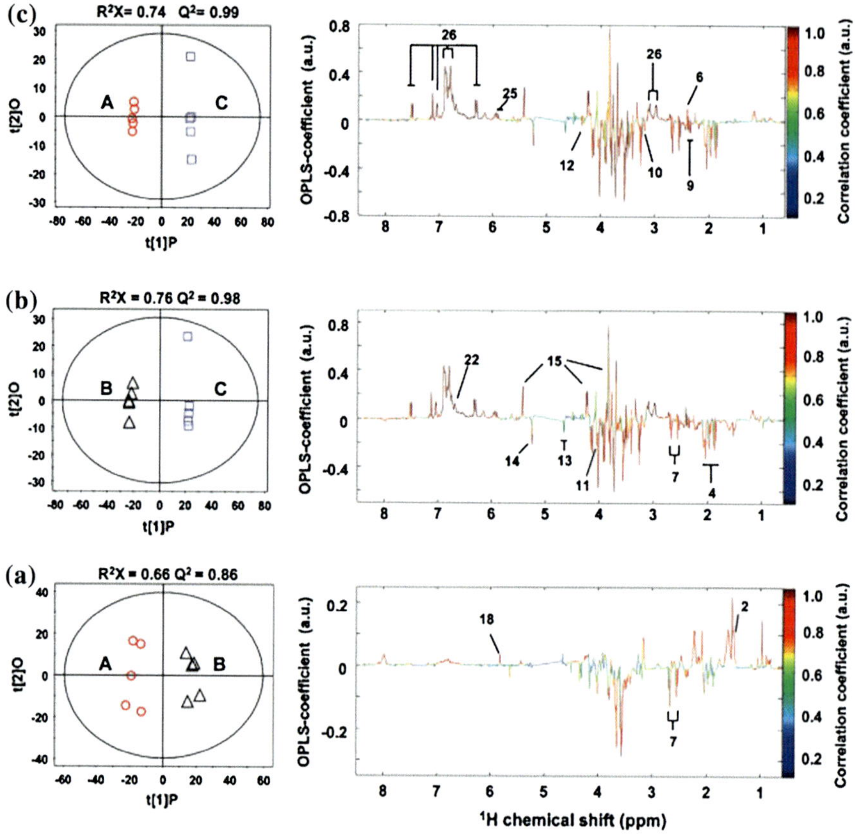

Fig. 12.3 The OPLS-DA score plot (*left*) and coefficient plot (*right*) of the NMR spectra of rosemary extracts in three solvents. The metabolites numbers see Table 12.1. *A* Ambient temperature water (*white circle*); *B* boiling water (*white triangle*); and *C* 50 % aqueous methanol (*white square*) (Reprinted from Xiao et al. (2008). Copyright © 2008 American Chemical Society. Reprinted with permission)

agreement with the OPLS-DA results. Such a minor compositional difference between two extracts probably resulted mainly from metabolite solubility. Using the same approach, the significant differences were assessed for aqueous methanol extracts in comparison with the boiling water extracts. The results were highlighted by significantly higher levels of rosmarinate, caffeate, syringate, *cis*-4-glucosyl-oxycinnamic acid, 3,4,5-trimethoxyphenylmethanol, sucrose, shikimate, succinate, fumarate, and malonate together with lower levels of glucose, fructose, quinate, and citrate, which is also consistent with the OPLS-DA results. Because aqueous methanol (1:1) was generally accepted as a solvent to purge enzymatic activity by precipitate proteins (including enzymes) and boiling water was expected to denature enzymes, the major compositional differences between extracts from boiling water and methanol/water (1:1, v/v) were probably also from extraction efficiency or

Table 12.2 OPLS-DA coefficients and metabolite content of rosemary extracts with different solvents (Reprinted from Xiao et al. (2008). Copyright © 2008 American Chemical Society. Reprinted with permission)

Metabolite	Coefficient (r)[1]			Mean ± SD[2] (mg/g)			
	B/A[3]	C/A	C/B	A	B	C	D
Rosmarinic acid	0.6	0.99	0.99	4.32 ± 0.56	4.86 ± 0.14	21.72 ± 1.91[4]	–
Caffeic acid	0.32	0.97	0.99	1.62 ± 0.10	1.78 ± 0.22	4.50 ± 0.65[4]	–
Syringate	0.58	0.91	0.9	0.75 ± 0.04	0.89 ± 0.06	1.24 ± 0.01[4]	–
Vanillic acid	0.36	0.23	0.45	0.91 ± 0.06	0.92 ± 0.01	0.98 ± 0.13	–
Cis-4-glucosyloxycinnamic acid	0.58	0.88	0.78	5.24 ± 0.41	5.56 ± 0.35	8.56 ± 0.19[4]	–
3,4,5-trimethoxyphenylmethanol	0.7	0.99	0.99	1.06 ± 0.06	1.29 ± 0.05	4.73 ± 0.21[4]	–
Carnosic acid	–[7]	0.99	–	–	–	–	40.72 ± 12.13
Carnosol	–	–	–	–	–	–	14.01 ± 2.07
Methylcarnosate	–	–	–	–	–	–	7.54 ± 2.89
Rosmadial	–	–	–	–	–	–	4.58 ± 2.14
Sucrose	0.68	0.99	0.99	3.46 ± 0.22	4.16 ± 0.17	19.74 ± 1.27[4]	–
Glucose[6]	0.29	−0.99	−0.99	45.68 ± 3.25	48.08 ± 4.72	37.04 ± 3.09[4]	–
Fructose	0.14	−0.99	−0.99	32.28 ± 0.56	33.02 ± 2.47	24.12 ± 0.02[4]	–
Quinate	−0.45	−0.96	−0.95	49.84 ± 0.76	50.22 ± 3.61	43.78 ± 2.02[4]	–
Shikimate	0.71	0.99	0.99	2.18 ± 0.14	2.32 ± 0.16	3.00 ± 0.06[4]	–
Citrate	−0.88	−0.98	−0.98	10.16 ± 0.18	8.80 ± 1.35[5]	6.46 ± 1.19[4]	–
Succinate	−0.45	0.85	0.87	2.40 ± 0.05	2.44 ± 0.16	2.72 ± 0.01[4]	–
Malate	−0.44	0.56	0.76	14.54 ± 1.04	14.32 ± 2.34	14.82 ± 0.27	–
Fumarate	0.65	0.99	0.99	0.32 ± 0.03	0.34 ± 0.03	0.68 ± 0.01[4]	–
Tartrate	0.57	−0.76	0.63	2.48 ± 0.23	2.36 ± 0.46	2.44 ± 0.09	–
Malonate	−0.64	0.97	0.97	1.09 ± 0.03	1.08 ± 0.06	1.24 ± 0.04[4]	–

(continued)

Table 12.2 (continued)

Metabolite	Coefficient (r)①			Mean ± SD② (mg/g)			
	B/A③	C/A	C/B	A	B	C	D
Lactate	−0.4	−0.61	−0.59	1.36 ± 0.21	1.16 ± 0.15	1.02 ± 0.04	–
Acetate	0.68	0.21	−0.12	0.62 ± 0.03	0.70 ± 0.12	0.66 ± 0.21	–
Alanine	0.84	−0.67	−0.88	0.70 ± 0.01	1.08 ± 0.26 ⑤	0.65 ± 0.03	–
Choline	0.35	0.79	0.82	1.40 ± 0.01	1.52 ± 0.10	1.38 ± 0.08	–

① The coefficients from OPLS-DA results; positive and negative signs indicate positive and negative correlation in the concentrations, respectively. The coefficient of 0.81 was used as the cutoff value for the significant difference evaluation ($p < 0.05$)

② The average concentration and standard deviation (mean ± SD, mg/g of dried rosemary material) were obtained from five parallel samples

③ The extracts were from ambient temperature water (A), boiling water (B), 50 % aqueous methanol (C), and chloroform/methanol (D)

④ Significant difference compared with B by one-way ANOVA ($p < 0.05$)

⑤ Significant difference compared with A by one-way ANOVA ($p < 0.05$)

⑥ The sum of α-glucose and β-glucose concentrations

⑦ —means no resonances were present in the corresponding extracts solubility differences rather than enzymatic changes. The chloroform/methanol extract contained almost exclusively carnosic acid and its derivatives (the concentration of carnosic acid was in the range of 40.72 ± 12.13 mg/g of dried rosemary material) with few metabolites observed in common with the other three extracts. Previous studies reported that methanol and DMSO solvents effectively extracted carnosic acid and carnosol as well polyphenolic acids (del Bano et al. 2003). Our study has shown that chloroform/methanol is an excellent solvent system for the selective extraction of some phenolic diterpenes, which are lipophilic antioxidants. In addition, aqueous methanol was more efficient for extracting hydrophilic antioxidants such as rosmarinate, caffeate, and syringate than water.

12.2.3 Effects of Seasonal Variation on the Rosemary Metabolite Composition

The samples collected in February, April, June, and August were extracted with 50 % aqueous methanol to investigate the seasonal effects on the rosemary metabolite composition. The ^{1}H NMR spectra showed that the extracts of rosemary harvested in different months had similar chemical constituents (data not shown), although with differences highlighted in their concentrations. A PCA model constructed with two principal components (PC1, 90 %; and PC2, 7 %) showed clear separation for the extracts obtained from four different months and tight clustering for each group (Fig. 12.4), indicating the sensitivity and powerfulness of such classification methods (Xiao et al. 2008).

Further analysis using OPLS-DA (Fig. 12.5) showed the clear monthly differentiation for the plant extracts with the values of R^2X/Q^2 indicating the model

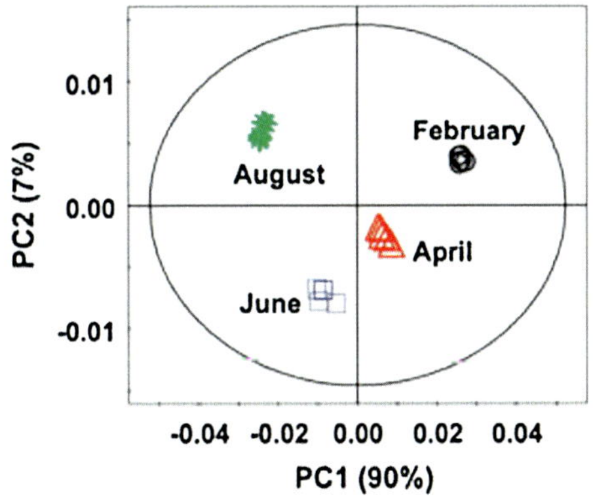

Fig. 12.4 PCA score plot for the rosemary samples harvested in different months. February: (*white circle*), April: (*white triangle*), June: (*white square*), and August: (*asterisk*). Metabolite numbers are listed in Table 12.1 (Reprinted from Xiao et al. (2008). Copyright © 2008 American Chemical Society. Reprinted with permission)

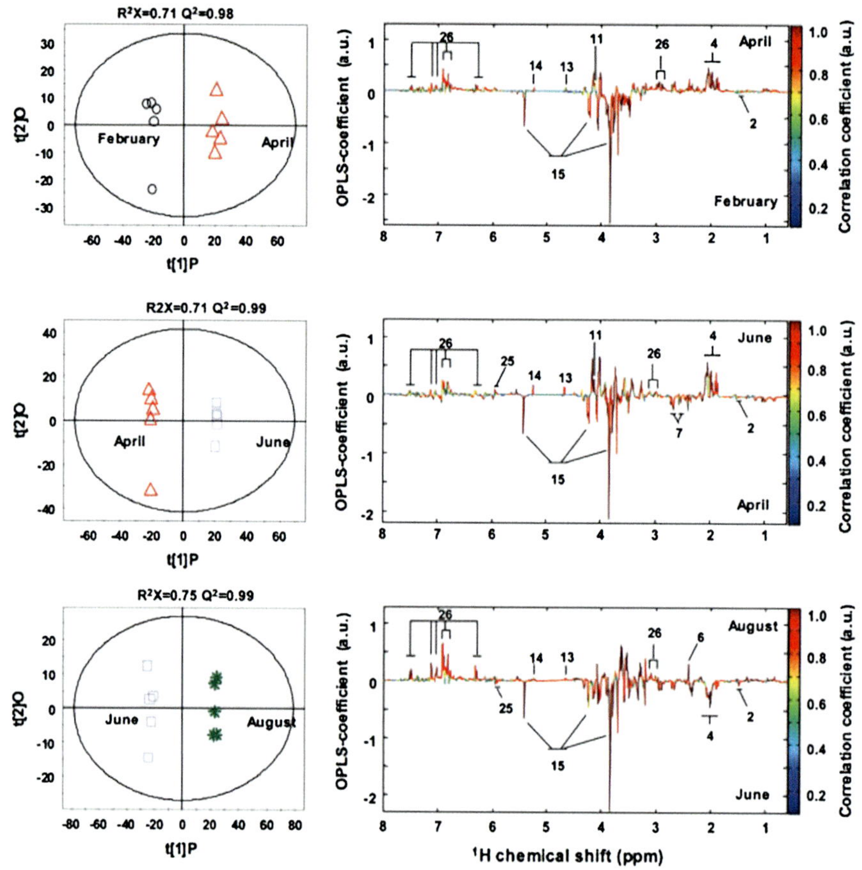

Fig. 12.5 NMR OPLS-DA score plot (*left*) and coefficient plot (*right*) for rosemary samples harvested in different months. Metabolite numbers see Table 12.1. February: (*white circle*), April: (*white triangle*), June: (*white square*), and August: (*asterisk*) (Reprinted from Xiao et al. (2008). Copyright © 2008 American Chemical Society. Reprinted with permission)

validity and the metabolites having significant contributions to the monthly classification ($|r| > 0.81$). The concentrations of some metabolites were also calculated (see Table 12.3, mean ± SD, mg/g of dried rosemary material). Compared with the February samples, the extracts from April showed significantly higher levels of romarinate, syringate, glucose, fructose, succinate, and quinate and lower levels of sucrose. In addition, extracts obtained in June showed statistically higher concentrations of rosmarinate, caffeate, glucose, fructose, and quinate and lower concentrations of syringate, sucrose, citrate, succinate, and alanine compared with the samples from April. Furthermore, the samples collected in August had statistically higher quantities of rosmarinate, syringate, and succinate and lower amounts of sucrose, fructose, quinate, and alanine than the samples harvested in June. The

Table 12.3 OPLS-DA coefficients and metabolite content of rosemary extracts from harvesting months (Reprinted from Xiao et al. (2008). Copyright © 2008 American Chemical Society. Reprinted with permission)

Metabolite	Coefficient (r)①			Mean ± SD② (mg/g)			
	Apr/Feb	June/Apr	Aug/June	Feb	Apr	June	Aug
Rosmarinic acid	0.95	0.81	0.98	4.84 ± 0.28	7.48 ± 0.81③	8.64 ± 0.35④	13.64 ± 0.58⑤
Caffeic acid	0.54	0.94	−0.65	1.06 ± 0.04	1.28 ± 0.19	2.08 ± 0.06④	1.76 ± 0.01
Syingic acid	0.91	−0.83	0.91	0.42 ± 0.01	0.76 ± 0.15	0.51 ± 0.02④	0.68 ± 0.01⑤
Sucrose	−0.98	−0.97	−0.98	41.06 ± 0.19	28.1 ± 1.77③	15.56 ± 0.35④	2.60 ± 0.93⑤
Glucose⑥	0.93	0.96	0.77	2.96 ± 0.32	4.32 ± 1.07③	7.06 ± 0 .53④	7.94 ± 1.12
Fructose	0.96	0.98	−0.95	5.34 ± 0.23	7.24 ± 0.82③	9.76 ± 0.26④	6.58 ± 0.71⑤
Citrate	0.79	−0.95	0.77	3.84 ± 0.02	4.22 ± 0.13	2.94 ± 0.01④	3.36 ± 0.25
Succinate	0.98	−0.99	0.99	0.68 ± 0.02	0.88 ± 0.06③	0.66 ± 0.03④	0.90 ± 0.07⑤
Quinate	0.96	0.98	−0.99	8.84 ± 0.19	12.3 ± 0.60③	15.42 ± 0.7④	13.98 ± 0.90⑤
Alanine	−0.51	−0.84	−0.99	0.58 ± 0.01	0.56 ± 0.01	0.48 ± 0.01④	0.34 ± 0.03⑤

① The coefficients from OPLS-DA results; positive and negative signs indicate positive and negative correlation in the concentrations, respectively. The coefficient of 0.81 was used as the cutoff value for the significant difference evaluation ($p < 0.05$)
② The average concentration and standard deviation (mean ± SD, mg/g of dried rosemary material) were obtained from five parallel samples
③ The significant difference by one-way ANOVA ($p < 0.05$): April *versus* February
④ The significant difference by one-way ANOVA ($p < 0.05$): June *versus* April
⑤ The significant difference by one-way ANOVA ($p < 0.05$): August *versus* June
⑥ The sum of α-glucose and β-glucose concentrations
⑤ The significant difference by one-way ANOVA ($p < 0.05$): August *versus* June
⑥ The sum of α-glucose and β-glucose concentrations

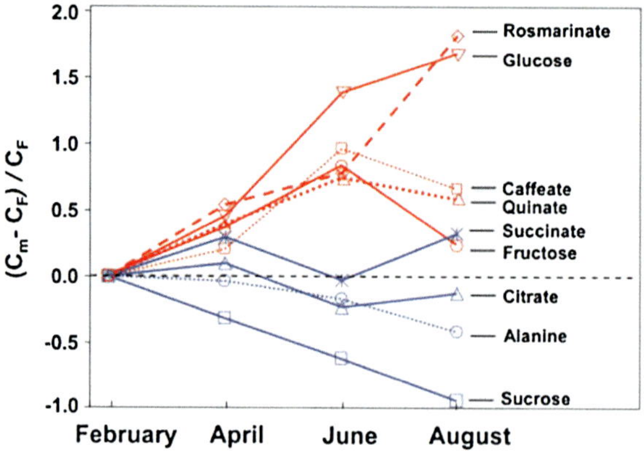

Fig. 12.6 Metabolite concentration changes relative to that in February for samples collected from different time points (Xiao et al. 2008). c_m represents the metabolite content in samples collected in each month; c_F represents the metabolite content in samples collected in February (Reprinted from Xiao et al. (2008). Copyright © 2008 American Chemical Society. Reprinted with permission)

metabolite concentrations calculated from the integral of the metabolite resonances were also subjected to ANOVA ($p < 0.05$); the results were broadly consistent with those of OPLS-DA.

The seasonal concentration variations of some rosemary metabolites are shown in Fig. 12.6 relative to that in February. Compared with in February, the concentration of succinate rose in April and August, whereas the concentration of citrate showed a slight increase in April but some decrease in June and August. Overall, the levels of sucrose and alanine showed a steady decrease from February to August, whereas the concentrations were increased for rosmarinate, caffeate, quinate, glucose, and fructose. Specifically, from February to August, the mean concentration of rosmarinate showed an almost twofold increase (from 4.84 ± 0.28 to 13.64 ± 0.58 mg/g of dried rosemary material), which broadly agreed with the previous observations about the seasonal variations, probably owing to the weather conditions (Del Bano et al. 2003; Luis and Johnson 2005). In contrast, the level of sucrose decreased more than an order of magnitude (from 41.06 ± 0.19 to 2.60 ± 0.93 mg/g of dried rosemary material) in the same period. The opposite trend for the levels of rosmarinate and sucrose was observed previously in the cultured cell systems, which suggested that sucrose affected the biosynthesis of rosmarinate (del Bano et al. 2003). Such inverse correlation was also evident here between sucrose level and the levels of glucose and fructose, probably owing to sucrose breakdown as observed in suspension cultures of Coleus cells (del Bano et al. 2003). It has been reported that the biosynthesis of quinate can result from D-glucose (Draths et al. 1992); thus, the same trend of changes for the quinate and glucose levels was understandable. These results further suggested that the biosynthesis and content of secondary metabolites were influenced by primary metabolites.

12.2.4 *Effects of Drying Processes on Rosemary Metabolite Composition*

To investigate the dependence of rosemary metabolite composition on the drying processes, the aqueous methanol extracts from freeze-dried and sun-dried rosemary samples were analyzed with ^{1}H NMR followed by multivariate data analysis. The NMR spectra of two different extracts showed some clear differences in their signal intensities. The score plot from OPLS-DA (Fig. 12.7) shows that the freeze-dried samples were clearly discriminated from the sun-dried ones in terms of their metabolite composition, and the correlation coefficients and quantitative information about the relevant metabolites are listed in Table 12.4. Compared with those from the sun-dried samples, the extracts from freeze-dried ones contained significantly lower amounts of rosmarinate, sucrose, alanine, asparagine, valine, and proline ($r < -0.81$) together with significantly higher levels of glucose, fructose, malate, choline, succinate, quinate, and lactate ($r > 0.81$). No significant differences were observed for some metabolites such as caffeate, syringate, 3,4,5-trimethoxyphenylmethanol, tartrate, citrate, and acetate. The levels of these metabolites were also subjected to ANOVA, and similar conclusions were obtained.

These metabolite compositional differences were probably attributable to the different effects of two drying processes on the cellular enzyme activities. During the freeze-drying process, enzymes cause little metabolite changes due to low temperature and lack of water availability. In contrast, during the sun-drying process, gradual water depletion will cause drought stress to the plant cells, inducing metabolic changes to various degrees. In fact, elevations of proline and sucrose were also reported when they were subjected to drought stress (Pinheiro et al. 2001; Valliyodan and Nguyen 2006). This study indicates that NMR-based metabonomic analysis is probably an effective way to investigate such stress-induced biochemical processes for plants. This also strongly suggests that the effects of drying processes have to be taken into consideration when plant metabolite compositions are studied in terms of phytomedicines and food.

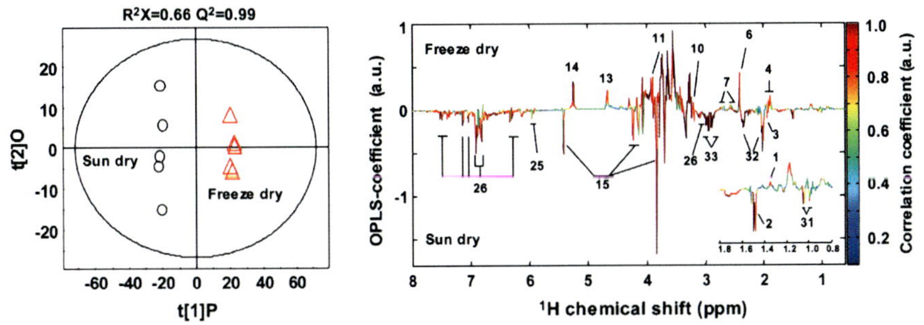

Fig. 12.7 NMR OPLS-DA score plot (*left*) and coefficient plot (*right*) for rosemary samples from different drying methods. Metabolite numbers see Table 12.1. Sun-dried: (*circle*), freeze-dried: (*white triangle*) (Reprinted from Xiao et al. (2008). Copyright © 2008 American Chemical Society. Reprinted with permission)

Table 12.4 OPLS-DA coefficients and metabolite content of the rosemary extracts from different drying processes (Reprinted from Xiao et al. (2008). Copyright © 2008 American Chemical Society. Reprinted with permission)

Metabolite	Coefficient (r)①	Mean ± SD② (mg/g)	
	Freeze /Sun dry	Freeze dry	Sun dry
Rosmarinic acid	−0.96	6.32 ± 0.10	8.45 ± 0.54③
Caffeic acid	−0.23	1.94 ± 0.29	2.08 ± 0.08
Syringic acid	−0.47	0.46 ± 0.01	0.44 ± 0.04
Cis-4-glucosyloxycinnamic acid	0.81	3.69 ± 0.08	4.32 ± 0.0③
3,4,5-trimethoxyphenylmethanol	−0.51	2.04 ± 0.09	1.75 ± 0.03
Sucrose	−0.96	7.40 ± 0.37	15.36 ± 2.00③
Glucose④	0.99	15.1 ± 2.38	7.00 ± 0.44③
Fructose	0.94	10.42 ± 0.01	9.64 ± 0.42③
Tartrate	0.23	1.06 ± 0.05	0.88 ± 0.06③
Malate	0.86	5.76 ± 0.10	4.52 ± 0.18③
Choline	0.97	0.60 ± 0.03	0.52 ± 0.16③
Manolate	−0.85	0.54 ± 0.01	0.56 ± 0.01
Citrate	0.58	3.82 ± 0.58	2.94 ± 0.01
Succinate	0.99	1.18 ± 0.01	0.66 ± 0.03③
Acetate	−0.77	0.16 ± 0.08	0.20 ± 0.01
Quinate	0.81	18.9 ± 0.89	15.42 ± 0.7③
Alanine	−0.99	0.30 ± 0.02	0.48 ± 0.01③
Latate	0.86	0.42 ± 0.01	0.34 ± 0.01③
Asparagine	−0.99	2.46 ± 0.03	5.46 ± 0.01③
Valine	−0.96	0.32 ± 0.01	0.38 ± 0.01③
Proline	−0.99	0.96 ± 0.05	4.69 ± 0.19③

① The coefficients from OPLS-DA results; positive and negative signs indicate positive and negative correlation in the concentrations, respectively. The coefficient of 0.81 was used as the cutoff value for the significant difference evaluation

② The average concentration and standard deviation (mean ± SD, mg/g of dried rosemary material) were obtained from five parallel samples

③ Significant difference compared with freeze-dried samples ($p < 0.05$)

④ The sum of α-glucose and β-glucose concentrations

12.3 *Artemisia annua* L. as an Example to Study and Predict the Efficacy and Toxicity (Bailey et al. 2004)

Artemisia annua L. (also known as sweet wormwood or by its Chinese name *qing hao*) and the isolated constituent artemisinin have well-documented anti-plasmodial activity. It has been suggested that the efficacy of the *A. annua* plant itself derives from a synergistic effect and that it is a combination of constituents in the plant that confer the total anti-plasmodial activity. Several polymethoxyflavones have been found to have activity in combination with artemisinin, and it has been reported that flavonoids enhance the anti-plasmodial activity of artemisinin. Further, the clinical efficacy of *A. annua* extracts as a treatment for malaria has been demonstrated, with 92 % of malaria patients in a study showing disappearance of parasitaemia within 4 days.

However, where plant extracts themselves are to be used, it is necessary to determine the reproducibility and information regarding the content of such extracts. It has been shown, for example, that the levels of artemisinin and related compounds fluctuate due to both seasonal and geographical variation.

12.3.1 Experimental Profile

Nineteen *Artemisia annua* L. accessions sourced from different locations were obtained by Oxford Natural Products Plc. (Oxford, UK). After extraction, the sample is divided into three parts: ① for detecting the anti-plasmodial activity of the extract, which is represented by half inhibitory concentration (IC_{50}) (Desjardins et al. 1979; Ponnudurai et al. 1981; O'Neill M et al. 1985). The higher the IC_{50} value, the worse the activity; ② for detecting the cell toxicity, which is represented by $ToxED_{50}$. The higher the $ToxED_{50}$, the lower the toxicity; ③ one-dimensional NMR spectrum was collected for each sample, and multivariate data analysis was conducted to analyze the results obtained.

12.3.2 The Metabolic Fingerprint of Artemisia Annua Predicts Its Anti-Plasmodial Activity

The biochemical content of *Artemisia annua* L. along with many other natural products, shows variation depending on geography or season. It is thus important than any technique used for the analysis of such samples is able to first and foremost discern the differences between samples of differing origin. This then enables some form of quality control to be performed to ensure that quality and content of the extracts are maintained. Three representative ^{1}H NMR spectra are shown in Fig. 12.8. These spectra (representing IC_{50} value of different extracts) show that while the artemisinin resonances themselves are readily apparent, particularly in the more potent extract 7, there are many other regions of the spectrum where large differences occur between the different extracts (for example, signal intensity in the region between δ6.0 and 8.0). The PCA plot obtained from analysis of the ^{1}H NMR spectra of all the extracts is shown in Fig. 12.9. Each point on the plot represents one ^{1}H NMR spectrum of an extract, with points of the same number indicating replicate samples of the same origin. It can be seen from this figure that each of the samples within a particular group of replicate samples readily cluster together

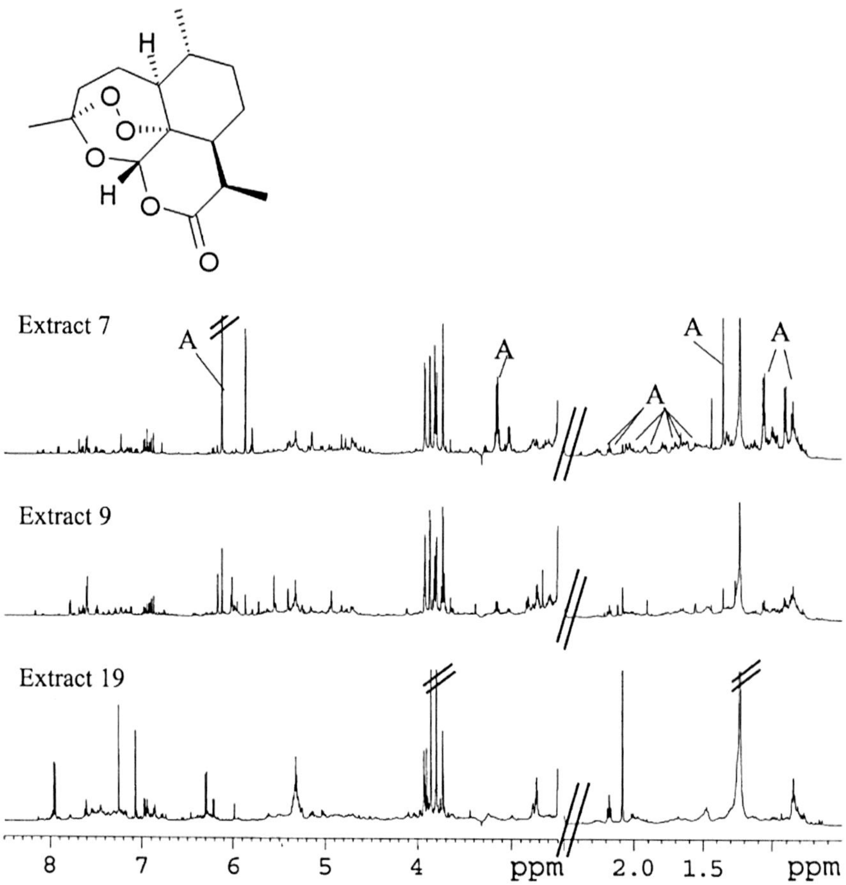

Fig. 12.8 Representative ^{1}H NMR spectra for the three IC_{50} classes, $IC_{50} < 0.1$ μg/ml (extract 7), 0.1 μg/ml $< IC_{50} <$ 1 μg/ml (extract 9), and $IC_{50} > 1$ μg/ml (extract 19). Region δ2.5–8.5 is expanded vertically by a factor of 6 to allow observation of lower level aromatic resonances. Resonances attributable to artemisinin are indicated with an "A" (Reprinted from Bailey et al. (2004). Copyright © 2004, with permission from Elsevier)

indicating the reproducibility of the extraction process and the analytical methodology. Groups of samples of differing origin can be discriminated from one another based on inherent differences in their biochemical profiles. Using this approach, therefore, it is possible to monitor the overall makeup of a sample, compare the whole extract with that of previous samples, samples of different origins, or seasonal differences, and use this information to implement quality control of natural products.

It is possible to re-label the data points to reflect the anti-plasmodial activities of each of the samples. A coding was performed whereby the data were split into three classes based on their IC_{50} value, with the cutoff values being 0.1 and 1 μg/ml. The

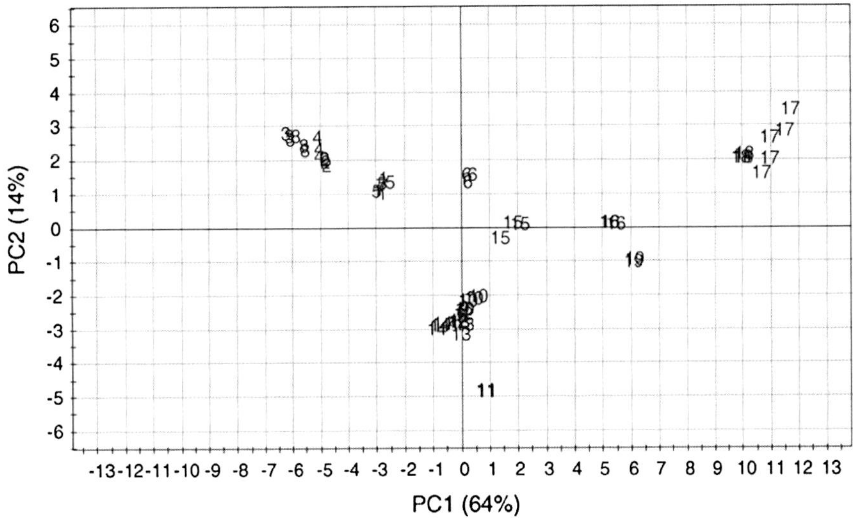

Fig. 12.9 PCA score plot for *A. annua* plant extracts. Samples are separated according to original extract number (Reprinted from Bailey et al. (2004). Copyright © 2004, with permission from Elsevier)

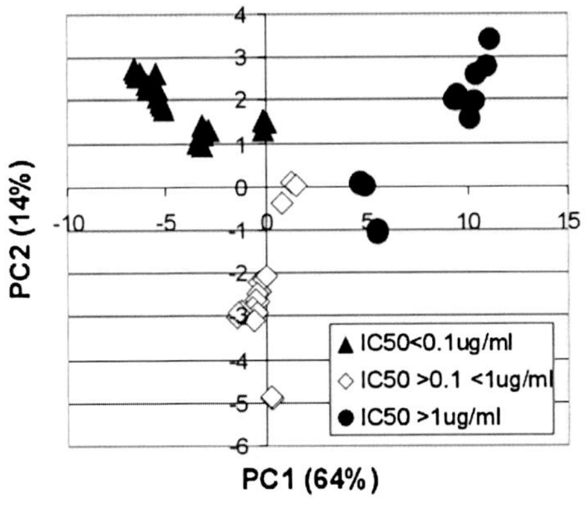

Fig. 12.10 PCA score plot for *A. annua* plant extracts. Samples are separated into three groups, $IC_{50} < 0.1$ μg/ml (*black filled triangle*), 0.1 μg/ml $< IC_{50} <$ 1 μg/ml (*white diamond*), and $IC_{50} > 1$ μg/ml (*black filled circle*) (Reprinted from Bailey et al. (2004). Copyright © 2004, with permission from Elsevier)

resulting PCA score plot can be seen in Fig. 12.10. It is apparent that this model (containing 78 % variance in the first two PCs) is able to discriminate between the three classes used. This suggests that ^{1}H NMR spectra contain sufficient information relating to the physico-chemical properties of the extract to be able to predict the potential magnitude of anti-plasmodial activity found in a plant extract. By analyzing the PCA loading plot, it was possible to determine the variables (spectral regions) that are responsible for this separation and are thus the regions that have greatest variation between the extracts. The loading plot for this model indicates that the spectral regions containing artemisinin are chiefly responsible for this separation. In order to ascertain the degree of influence of artemisinin on the model, regions containing artemisinin resonances were removed from the dataset and the analysis repeated. A very similar separation is achieved, and the corresponding loading plot contains variables largely associated with the resonances around δ3.7–4.0. The identity of the molecules for which these resonances are attributable was not determined. While these additional resonances and the ability to achieve separation using them are of interest and further demonstrate the presence of synergy at work, the chief aims of the work are to consider the extracts as a whole, and so all further work considers the entire spectral region.

While PCA clearly demonstrates the potential of this technique, a more robust approach to obtaining predictive data is to employ supervised methods. These methods involve providing the model with the values for the variable to be predicted (i.e., IC_{50} value) for part of the dataset (the training set), with the model then being optimized based on those values. Because the algorithm in effect uses the answers to create the model, it is then necessary to validate this model using the remaining unused samples (the test set). Those samples with an IC_{50} value >1 μg/ml were excluded from this analysis, for two reasons. This class is the smallest of the three, and having larger values means that the model is likely to be skewed in order to take them into account. In addition, the higher values mean that these samples are not of interest anyway, as they essentially have no activity. Using the remaining two classes as above ($IC_{50} < 0.1$ μg/ml and $IC_{50} > 0.1$ μg/ml, respectively), it is possible to construct a "dummy" y-matrix whereby the two classes are represented by a 1 or a 0. Partial least squares discriminant analysis (PLS-DA) can then be performed on the data to construct a new model using this additional data. By excluding data at random from this model, predictions as to the likely class membership of the excluded data points can be made in order to validate the model. As the two classes are designated either a 1 ($IC_{50} < 0.1$ μg/ml) or a 0 ($IC_{50} > 0.1$ μg/ml), any data point with a predicted value of greater than 0.5 is considered to be a member of class 1, and any point with a value of less than 0.5 is considered to belong to class 0. The model shown in Fig. 12.11 was constructed using 36 of the samples (the training set), while 10 were used to validate the model (test set). It can be seen from the figure that all the data points from both training and test sets are correctly predicted in what is overall a robust model with $R^2 = 0.90$ and $Q^2 = 0.89$, where R^2 is the variance and Q^2 is the cross-validated variance or predictive ability of the model. In general, a value of $Q^2 > 0.5$ is generally considered to be good (Eriksson et al. 1999). The analysis was repeated five times (with samples split into different training and test sets each time), with all samples correctly classified in all models, with >91 % of samples predicted with >99 % confidence. R^2 values ranged from

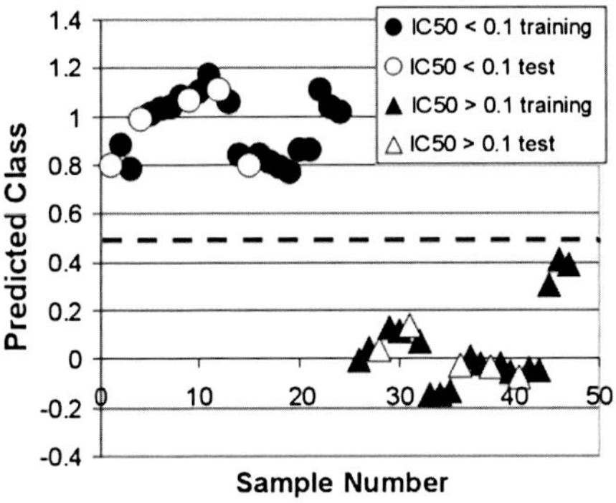

Fig. 12.11 PLS-DA Y-predicted scatter plot for classes $IC_{50} > 0.1$ μg/ml (class 0, triangles) and $IC_{50} < 0.1$ μg/ml (class 1, *circles*). Training set represented by closed shapes, (*black filled triangle*) and (*black filled circle*), test set represented by open shapes, (*white triangle*) and (*white circle*) (Reprinted from Bailey et al. (2004). Copyright © 2004, with permission from Elsevier)

0.90 to 0.92 and Q^2 values from 0.88 to 0.90. Two components were used in all models. While it may appear that only the same amount of information is available from both the PCA and PLS-DA models, the fact that more robust validation is available for the supervised PLS-DA method means that the results have greater credibility. Thus, using this PLS-DA model, it would be possible to predict those samples that are likely to have anti-plasmodial activities of <0.1 μg/ml.

The PCA and PLS-DA models discussed above give a good indication as to the likely magnitude of anti-plasmodial activity. This approach however is based on an artificially imposed classification, i.e., $IC_{50} < 0.1$ μg/ml or > 0.1 μg/ml, which while giving a clear indication of potential anti-plasmodial activity, classes may or may not be significant. This can be taken one step further, however, with the prediction of the actual IC_{50} value for each of the extracts. Instead of the dummy y-matrix constructed for the PLS-DA analysis, it is possible to construct a model using the IC_{50} values obtained for each of the extracts from a biological assay. The result can then be used to predict values for test data. Three components were used for all models, with >87 % of samples predicted with >99 % confidence. R^2 values ranged from 0.83 to 0.93 and Q^2 values from 0.62 to 0.91. The model construction process was repeated in order to exclude every extract from the training set once (with additional samples being removed on a random basis). The overall predictions for each extract are summarized in Table 12.5. It can be seen that in general, the predicted value is reasonably close to the actual value (the correlation between average predicted versus actual values is 0.90), and this clearly demonstrates the potential of such an approach. If the predicted IC_{50} values were used to classify the extracts into two classes, IC_{50} greater or less than 0.1 μg/ml, then all extracts would be placed in the same class as if the actual data were used. This illustrates the fact that this method of analysis can be used as a first step filtering technique in order to identify key extracts worth pursuing using other approaches

Table 12.5 Summary of predicted IC_{50} and $ToxED_{50}$ values for nineteen *A. annua* extracts (Reprinted from Bailey et al. (2004). Copyright © 2004, with permission from Elsevier)

Extract	IC_{50}value (μg/ml)	IC_{50} predicted	$ToxED_{50}$ value (μg/ml)	$ToxED_{50}$ predicted	Relative artemisinin level①
1	0.01	0.05 ± 0.04	27	42 ± 6	0.47
2	0.01	0.01 ± 0.02	9	20 ± 3	0.66
3	0.020	0.027 ± 0.00	8	22 ± 6	1
4	0.02	0.00 ± 0.03	19	17 ± 6	0.69
5	0.020	0.065 ± 0.005	65	26 ± 11	0.44
6	0.02	0.06 ± 0.02	50	33 ± 2	0.34
7	0.0296	0.0262 ± 0.0003	171	–	0.45
8	0.04	0.03 ± 0.02	11	12 ± 2	0.91
9	0.169	0.209 ± 0.008	69	54 ± 4	0.18
10	0.13	0.17 ± 0.01	26	28 ± 5	0.14
11	0.29	0.16 ± 0.04	41	75 ± 4	0.15
12	0.30	0.20 ± 0.03	66	62 ± 2	0.21
13	0.31	0.22 ± 0.04	8	33 ± 6	0.18
14	0.32	0.22 ± 0.02	76	72 ± 3	0.20
15	0.47	0.23 ± 0.04	50	60 ± 5	0.12
16	8.55	0.21 ± 0.03	46	30 ± 1	0.04
17	24.67	0.20 ± 0.03	72	40 ± 12	0.02
18	4.2	0.10 ± 0.01	20	48 ± 14	0.02
19	3.9	0.243 ± 0.002	–	–	0.04

① Obtained from the ^{1}H NMR peak intensity for the artemisinin peak at ca. 6.1 ppm. Values expressed relative to the highest peak that of extract 3

12.3.3 The Metabolic Fingerprint of Artemisia Annua Predicts Its Toxicity

In drug research and development, the drug toxicity and activity are equally important. In addition to the IC_{50} measurements, therefore, each extract was analyzed by measuring the cytotoxicity ($ToxED_{50}$) of the extract. It is possible to predict values for $ToxED_{50}$ using the same NMR spectra. An example PLS-DA plot is shown in Fig. 12.12, with the samples separated into those with a $ToxED_{50}$ value of > or < 65 μg/ml. The analysis was repeated five times. Three, four, or five components were used in all models. Results show that >84 % of samples predicted with >99 % confidence. R^2 values ranged from 0.78 to 0.94 and Q^2 values from 0.69 to 0.84. Although the statistics reveal that the models are not as good as the ones created for the IC_{50} values, these models are still able to give a good indication as to the likely toxicity of a particular plant extract.

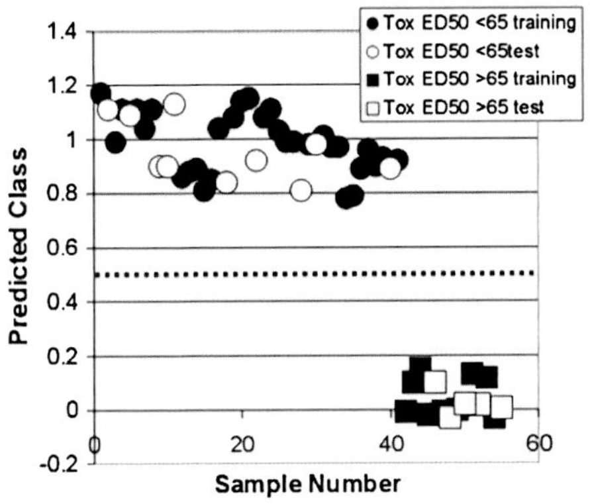

Fig. 12.12 PLS-DA Y-predicted scatter plot for classes $ToxED_{50} > 65$ μg/ml (class 0, *squares*) and $ToxED_{50} < 65$ μg/ml (class 1, *circles*). Training set represented by closed shapes, (*black filled square*) and (*black filled circle*), test set represented by open shapes, (*white square*) and (*white circle*) (Reprinted from Bailey et al. (2004). Copyright © 2004, with permission from Elsevier)

In order to obtain predicted $ToxED_{50}$ values for the extracts, PLS was again performed using $ToxED_{50}$ experimental data to construct models for the y-matrix. Four, five, or six components were used for all PLS models, with >78 % of samples predicted with >99 % confidence. R^2 values ranged from 0.89 to 0.98 and Q^2 values from 0.73 to 0.94. The average predicted $ToxED_{50}$ values for each extract are summarized in Table 12.5. It can be seen that as with the IC_{50} predictions, the predicted values are, overall, in close agreement with the actual values, although the average predicted values versus actual values correlation are only 0.60, compared with the 0.90 value obtained from the IC_{50} models. The important point, however, is that using the same NMR spectra in modeling both IC_{50} and $ToxED_{50}$ values, it has been possible to predict reasonable values for two different parameters which would normally require two separate assays to be run to obtain the same information. Using this information, it may be that criteria could be set up to identify the most promising extracts for further study.

12.4 Conclusion

It is not difficult to see from the above two cases that NMR-based metabolomics analysis method can be used to comprehensively study the metabolite composition of plant influenced by extraction methods, sample collection seasons, and drying methods. Metabolomics NMR analysis is one of the most effective methods in efficient detection of metabolites derived from various environments and by post-collection treatment factors, which has a broad application prospect in the quality control of herbal medicine based on molecular composition. In the mean time, NMR-based metabolomics methods can predict the activity and toxicity of plant extracts, which is a very promising tool in drug development.

References

Aranibar N, Singh BK, Stockton GW, Ott KH. Automated mode-of-action detection by metabolic profiling. Biochem Biophys Res Commun 2001;286:150–5.

Bailey NJ, Oven M, Holmes E, Nicholson JK, Zenk MH. Metabolomic analysis of the consequences of cadmium exposure in Silene cucubalus cell cultures via ^{1}H NMR spectroscopy and chemometrics. Phytochemistry. 2003;62:851–8.

Bailey NJ, Sampson J, Hylands PJ, Nicholson JK, Holmes E. Multi-component metabolic classification of commercial feverfew preparations via high-field ^{1}H-NMR spectroscopy and chemometrics. Planta Med. 2002;68:734–8.

Bailey NJC. Imperial College of Science Technology and Medicine. Ph.D. thesis, 2000.

Bailey NJC, Wang YL, Sampson J, Davis W, Whitcombe I, Hylands PJ, Croft SL, Holmes E. Prediction of anti-plasmodial activity of *Artemisia annua* extracts:application of ^{1}H NMR spectroscopy and chemometrics. J Pharm Biomed Anal. 2004;35:117–26.

Bamforth FJ, Dorian V, Vallance H, Wishart DS. Diagnosis of inborn errors of metabolism using ^{1}H NMR spectroscopic analysis of urine. J Inherit Metab Dis. 1999;22:297–301.

Bundy JG, Osborn D, Weeks JM, Lindon JC, Nicholson JK. An NMR-based metabonomic approach to the investigation of coelomic fluid biochemistry in earthworms under toxic stress. FEBS Lett. 2001;500:31–5.

Cuvelier ME, Berset C, Richard H. Antioxidant constituents in sage (salvia-officinalis). J Agric Food Chem. 1994;42:665–9.

del Bano MJ, Lorente J, Castillo J, Benavente-Garcia O, del Rio JA, Ortuno A, Quirin KW, Gerard D. Phenolic diterpenes, flavones, and rosmarinic acid distribution during the development of leaves, flowers, stems, and roots of Rosmarinus officinalis. Antioxidant activity. J Agric Food Chem. 2003;51:4247–53.

Desjardins RE, Canfield CJ, Haynes JD, Chulay JD. Quantitative assessment of antimalarial activity in vitro by a semiautomated microdilution technique. Antimicrob Agents Chemother. 1979;16:710–8.

Draths KM, Ward TL, Frost JW. Biocatalysis and 19th-century organic-chemistry-conversion of D-glucose into quinoid organics. J Am Chem Soc. 1992;114:9725–6.

Eriksson L, Johansson E, Kettaneh-Wold N, Wold S. Introduction to multi-and megavariate data analysis using projection methods (PCA & PLS). Umeå, Umetrics AB; 1999.

Fan WMT. Metabolite profiling by one- and two-dimensional NMR analysis of complex mixtures. Prog Nucl Magn Reson Spectrosc. 1996;28:161–219.

Ghauri FY, Blackledge CA, Glen RC, Sweatman BC, Lindon JC, Beddell CR, Wilson ID, Nicholson JK. Quantitative structure-metabolism relationships for substituted benzoic acids in the rat. Computational chemistry, NMR spectroscopy and pattern recognition studies. Biochem Pharmacol. 1992;44:1935–46.

Griffin JL, Walker LA, Garrod S, Holmes E, Shore RF, Nicholson JK. NMR spectroscopy based metabonomic studies on the comparative biochemistry of the kidney and urine of the bank vole (*Clethrionomys glareolus*), wood mouse (*Apodemus sylvaticus*), white toothed shrew (Crocidura suaveolens) and the laboratory rat. Comp Biochem Physiol B Biochem Mol Biol. 2000;127:357–67.

Hammond DG, Kubo I. Structure-activity relationship of alkanols as mosquito larvicides with novel findings regarding their mode of action. Bioorg Med Chem. 1999;7:271–8.

Kessler H. Structure-activity relationships by NMR: A new procedure for drug discovery by a combinatorial-rational approach. Angew Chem Int. 1997;36:829–31 (Edition English).

Luis JC, Johnson CB. Seasonal variations of rosmarinic and carnosic acids in rosemary extracts. Analysis of their in vitro antiradical activity. Span J Agric Res. 2005;3:106–12.

Nicholson JK, Connelly J, Lindon JC, Holmes E. Metabonomics: a platform for studying drug toxicity and gene function. Nat Rev Drug Discov. 2002;1:153–61.

O'Neill MJ, Bray DH, Boardman P, Phillipson JD, Warhurst DC, Plants as sources of antimalarial drugs part. 1. In vitro test method for the evaluation of crude extracts from plants. Planta Med. 1985;51:394–398.

Pinheiro C, Chaves MM, Ricardo CP. Alterations in carbon and nitrogen metabolism induced by water deficit in the stems and leaves of *Lupinus albus* L. J Exp Bot. 2001;52:1063–70.

Ponnudurai T, Leeuwenberg AD, Meuwissen JH. Chloroquine sensitivity of isolates of Plasmodium falciparum adapted to in vitro culture. Trop Geogr Med. 1981;33:50–4.

Pukalskas A, van Beek TA, de Waard P. Development of a triple hyphenated HPLC-radical scavenging detection-DAD-SPE-NMR system for the rapid identification of antioxidants in complex plant extracts. J Chromatogr A. 2005;1074:81–8.

Sobolev AP, Brosio E, Gianferri R, Segre AL. Metabolic profile of lettuce leaves by high-field NMR spectra. Magn Reson Chem. 2005;43:625–38.

Valliyodan B, Nguyen HT. Understanding regulatory networks and engineering for enhanced drought tolerance in plants. Curr Opin Plant Biol. 2006;9:189–95.

Xiao CN, Dai H, Liu HB, Wang YL, Tang HR. Revealing the metabonomic variation of rosemary extracts using ^{1}H NMR spectroscopy and multivariate data analysis. J Agric Food Chem. 2008;56:10142–53.

Xiao CN, Liu HB, Dai H, Tseng LH, Wang YL, Tang HR. Rapid and efficient identification of metabolites using HPLC-DAD-SPE-CryoNMR-MS method. Chin J Magn Reson. 2009;26:1–16.

Chapter 13
Metabolomics Research of Quantitative Disease Resistance Against Barley Leaf Rust

Li-Juan Wang and Xiaoquan Qi

Abstract Plant quantitative resistance is believed to be more durable in protection of crops. But until now, the reports on the mechanism of quantitative resistance remain limited. In the present study, *Rphq2* is a major quantitative trait loci (QTL) conferring quantitative resistance to leaf rust isolate 1.2.1 in barley-resistant cultivar "Vada." To further understanding the metabolic mechanisms of resistance conferred by *Rphq2*, metabolic profiling was performed on barley susceptible cultivar L94 and the near-isogenic line (NIL) L94-*Rphq2* with (infected) or without (mock) leaf rust infection using GC-TOF/MS. The PLS-DA analysis of the metabolites showed that the mock and infected materials started to clearly separate at 12 h after inoculation, which suggest that the metabolite profiles of inoculation group changes after the induction of pathogens. The levels of glucose, xylose, many amino acids, *and fatty acids in all infected leaves were higher than those of in the mock.* After infection, the differences between L94-*Rphq2* and L94 mainly include xylose, mannose, hexadecanoic acid, octadecanoic acid, and 9,12-octadecadienoic acid. Our results suggest that leaf rust deploys a common metabolic reprogramming strategy in L94 and L94-*Rphq2*, and fatty acids may be associated with the resistance conferred by *Rphq2*.

13.1 Introduction

Plant disease resistance refers to the mutual adaptation and selection between plant and pathogen during long-term coevolution and the formation of an inheritable character for preventing, suspending, or blocking pathogen invasion and expansion, so as to reduce the incidence and extent of the loss. Production practices have manifested that the breeding and exploiting of disease resistance cultivars is the most economic and efficient approach to the prevention and control of plant disease.

L.-J. Wang (✉) · X. Qi
Institute of Botany, Chinese Academy of Sciences, Beijing 100093, China
e-mail: wlj307@caf.ac.cn

X. Qi
e-mail: xqi@ibcas.ac.cn

X. Qi et al. (eds.), *Plant Metabolomics*, DOI 10.1007/978-94-017-9291-2_13

According to the genetic performance and the sensitivity to environmental conditions, plant disease resistance can be divided into qualitative disease resistance and quantitative disease resistance (Ou et al. 1975; Parlevliet 1979; Ribeiro do Vale et al. 2001). Qualitative disease resistance, also known as vertical disease resistance or isolate specific disease resistance, is controlled by one or a few major genes, with mechanism in keeping with the "gene for gene hypothesis" in the interaction between plant and pathogen (Flor 1971), i.e., plant disease resistance phenotype is the consequence of the interaction between resistance gene (*R* gene) product of host and avirulence gene (*Avr* gene) product of the pathogen. However, plants with such a resistance usually exert greater selection pressure to pathogens, consequently accelerate the variation of pathogenic factor within pathogen populations, and lead to the loss of the plant resistance because of the break of *R-Avr* incompatibility interaction (Kiyosawa 1982). Therefore, there is a certain degree of limitations for *R* gene's application in breeding practice. Quantitative resistance is also referred to as partial resistance, horizontal resistance, multi-gene resistance, or field resistance (Ou et al. 1975; Ribeiro do Vale et al. 2001), controlled by several minor genes, and shows typical quantitative genetic characteristics. The multi-gene genetic characteristics exert smaller selection pressure to pathogens, greatly reducing the mutation frequency of pathogens (Leonards-Schippers et al. 1994). Experience has shown that quantitative disease resistance has a broad-spectrum and good durability, thus has a good prospect of application in production practice. However, because the quantitative disease resistance is controlled by a number of minor genes, and usually the effect of a single resistance quantitative trait locus (QTL) is small (less than 30 %), which brings about difficulties for research work. In-depth study of quantitative disease resistance mechanism usually requires the separation of target QTL, such as the building of near-isogenic lines (NILs), map-based cloning, and gene function analysis. Several disease resistance QTLs have been cloned in the past 2 years (Fu et al. 2009; Fukuoka et al. 2009; Krattinger et al. 2009; Manosalva et al. 2009). But most of the research works stay on the initial positioning and fine mapping of QTL. So far, the knowledge to the possible characteristic and quantitative resistance mechanism of quantitative resistance gene is still quite vague, which largely limits the application of quantitative resistance gene in breeding practice.

Previous studies have shown that metabolites play a very important role in the plant–microbe interaction. After the invasion by the pathogen, the plant's primary and secondary metabolic pathways will become more active and produce more energy and antimicrobial secondary metabolites to prevent pathogen invasion and expansion (O'Connell and Panstruga 2006; Bolton 2009). In contrast, pathogens can also interfere with the normal metabolism of plants to meet their nutritional needs (Solomon et al. 2003; Swarbrick et al. 2006; Divon and Fluhr 2007) and produce some pathogenic biochemical factors such as toxins and extracellular polysaccharides (EPSs), which has pathogenicity to plant. Metabolomics is a new branch of systems biology following the rise of genomics, transcriptomics, and proteomics. By examining the metabolites profiles and their dynamic changes before and after the external stimuli or disturbances to study the metabolic network of biological system, metabolomics provide a new way of thinking for the study of plant disease resistance. Hamzehzarghani et al. (2005) and Swarbrick et al. (2006) analyzed the nontarget metabolomic profile of resistant cultivar and susceptible cultivar of cereal crops and obtained several metabolite biomarkers associated with disease resistance. These markers would play roles as assisting breeding in production practice. Allwood et al. (2006) and Parker et al. (2009) studied the changes of metabolomic profiles of barley (*Hordeum vulgare*), rice (*Oryza sativa*), and brachypodium (*Brachypodium distachyon*) before and after the inoculation of rice blast fungus (*Magnaporthe grisea*). The results show that there are significant changes in the metabolomic profiles of barley, rice, and brachypodium after the infection by rice blast fungus. Rice blast fungus adopted a similar strategy of "interference" to the metabolomic profiles of above-mentioned three species. Hofmann et al. (2010) studied the changes of metabolomic profiles of *Arabidopsis* (*Arabidopsis thaliana*) after the parasitism of nematode (*Heterodera schachtii*) and found that 1-Kestose, Raffinose, and α-Trehalose are metabolite biomarkers for the diagnosis of *Arabidopsis* nematodosis. These preliminary studies prove that metabolomics is a very effective way for the study of plant–microbe interaction.

Barley–barley leaf rust (*Puccinia hordei*) is a good model for the study of quantitative resistance of cereal crops. Cytological observations show that the leaf rust urediospore germinate on the surface of host leaf to produce germ tube. The germ tube forms appressorium above stomata once it finds host stomata. Then, substomatal vesicle, primary infection hypha, secondary infection hypha, haustorial mother cell (HMC), and penetration peg are formed. After penetrating through host cell wall, the haustorium, a nutrient absorbing organ, are formed to absorb nutrients from the host (Niks 1986).

A cytological significant characteristic of barley resistance against leaf rust is that host cell wall will get thicker in the area where the penetration peg have contacted and form hemispherical sediment between the cell wall and the cell membrane, i.e., papillae, thus delaying and hindering the formation of haustorium (Niks 1986; Jacobs 1989a, b; O'Connell and Panstruga 2006; Marcel et al. 2007). Therefore, this type of resistance is also called pre-haustorial resistance (Heath 1974). The study uses the barley susceptible cultivar L94, the near-isogenic line (NIL) L94-*Rphq2* which contains a barley resistance QTL *Rphq2* against leaf rust, and the barley leaf rust isolate 1.2.1 as experimental materials to conduct metabolomics research based on gas chromatography time-of-flight mass spectrometry (GC-TOF/MS) in the hope of uncovering barley quantitative resistance mechanism against leaf rust from the perspective of metabolites change.

13.2 Experimental Section

13.2.1 Experimental Materials

L94 is a susceptible cultivar to barley leaf rust. Vada is a quantitative resistant cultivar against barley leaf rust, which shows prolonged median latent period (LP_{50}) and reduced infection rate in phenotype. Preliminary analysis indicates that the resistance of Vada against barley leaf rust isolate 1.2.1 is controlled by 6 QTLs, in which the *Rphq2* (which is located on the long arm of barley chromosome 2B) is the major QTL expressing in seedling stage (Qi et al. 1998). By multi-generation backcross, NILs of *Rphq2* in L94 background and Vada background, i.e., L94-*Rphq2* and Vada-*rphq2* were obtained. The fragment length of Vada and L94 in L94-*Rphq2* and Vada-*rphq2* are 4.6 and 5.2 cm, respectively (Van Berloo et al. 2001).

13.2.2 Experimental Methods

13.2.2.1 Resistance Identification

L94, Vada, L94-*Rphq2*, and Vada-*rphq2* were sown in the greenhouse. When the first leaf is fully unfolded, collected fresh barley leaf rust (*P. hordei* Otth.) from the susceptible L94, diluted 20 times with talcum powder, and then used a writing brush to spread evenly to the blade surface. The control was spread in the same way with only talcum powder. After inoculation with pathogens, 1000 × Tween water was gently sprayed to leaves to form a layer of water film on the blade surface, then kept it relative humidity in the dark for 12 h. Greenhouse conditions were set as temperature 20 ± 1 °C, 16 h light/8 h darkness. When the first spore was seen, a linear area of about 2 cm was marked in the middle part of each leaf, counted the spore number at 12 h intervals, until the number no longer increased. Subsequently, the time of 50 % of spores showing were counted, respectively in L94, Vada, L94-*Rphq2*, and Vada-*rphq2*, i.e., LP_{50}. The LP_{50} in susceptible cultivar L94 is set to 100. Then the relative median latency period (RLP_{50}) of barley leaf rust 1.2.1 to Vada, L94-*Rphq2*, and Vada-*rphq2* were obtained by that of Vada, L94-*Rphq2*, and Vada-*rphq2* divided by LP_{50} in L94. When apparent chlorotic spots appeared on the leaf, but not formed new spores, i.e., at 110 h after inoculation, 2 leaves from each material were taken to subject to Uvitex 2B fluorescent dye staining and microscopic observation, (the staining method, see Niks 1982), counted the number of infection colonies of HMC is less than 6, i.e., the percentage of early aborted (EA %) in all of the infection colonies of each leaf (Niks 1982).

13.2.2.2 Metabolite Extraction and Pretreatment

The susceptible cultivar L94 and the NIL L94-*Rphq2* in L94 background were analyzed by metabolomics. The control and the inoculated material were sampled, respectively, at 0, 12, 24, 48, 96, and 120 h after inoculation. About 130 mg of plant powder were used for the extraction of the metabolites. Three samples from each material at each time point were chosen for metabolites extraction. Samples were put into liquid nitrogen rapidly and stored at −80 °C for future use. After the 130 mg of sample was grounded in liquid nitrogen, 2 ml of extraction solution 1 [methanol: chloroform:water = 2.5:1:1 (v:v:v)] was added, which was prechilled at −20 °C, then

10 μl of nonadecanoic acid (2 mg/ml) (lipid phase internal standard) and 50 μl ribitol (0.2 mg/ml) (aqueous phase internal standard) were added, fully mixed by a shaker at 220 rpm for 30 min, centrifuged at 11,000 g for 10 min, then transferred the supernatant to a clean 10-ml centrifuge tube. Added 1 ml of solution 2 [methanol: chloroform = 1:1 (v:v)] which was prechilled at −20 °C to the pellet, fully mixed by a shaker at 220 rpm for 30 min, following the second 10 min centrifugation at 11,000 g. The second supernatant was collected and mixed with the first supernatant collected. Then, 500 μl of sterilized water were added. The mixture was centrifuged at 5000 rpm for 3 min. After the centrifugation, the upper layer is a methanol/water phase, the lower is chloroform phase. The methanol/water phase was vacuum freeze-dried, and the chloroform phase was dried using nitrogen. Dry samples were derivatized using a two-step method. The first step is that 50 μl of methoxyamino hydrochloride (20 mg/ml) dissolved with pyridine was added to the resulting phase above, sealed, vortexed, and short centrifuged, and following a 90-min warm bath at 30 °C. Then 80 μl of silylation reagent N-methyl-N-(trimethylsilyl)-trifluoroacetamide (MSTFA) was added and warm bath at 37 °C for 30 min.

13.2.2.3 GC-TOF/MS and Statistical Analysis

The LECO Corporation GC-TOF/MS was used. The GC was fitted with an Agilent 6,890 column with an auto-sampler. The helium carrier gas flow was adjusted to 1.5 ml/min. The oven temperature program for the chloroform phase is 100 °C (2 min)–200 °C (1 min) at 8 °C/min, followed by 200–260 °C (1 min) at 3 °C/min, finally from 260 to 320 °C (4 min) at 8 °C/min. The oven temperature program for the methanol/water phase is 100 °C (2 min)–260 °C (1 min) at 4 °C/min, followed by 260–320 °C (3 min) at 20 °C/min. The inlet and ion source temperature were set to 280 and 200 °C, respectively. The spectra delay for 5 min. Mass spectra data at the range of *m/z* 30–600 were acquired, at a scan rate of 10 spectra per second. The detection voltage was set to 1,600 V. The sample was bombarded with 70 eV EI.

GC-TOF/MS metabolomic data preprocessing: ChromaTOF software was used to perform the baseline correction and de-noising. The signal-to-noise ratio is 30:1. The preprocess results were exported into NETCDF format files, then imported into software Matlab7.0 with the peak discrimination program which will automatically perform the baseline correction, peak identification, peak alignment, internal standard deduction, normalization, etc., and ultimately get a three-dimensional matrix, which is composed of specified peak serial number (which corresponds to retention time and mass-to-charge ratio), the observation point (sample No.) and the normalized peak intensity. The three-dimensional matrix was imported into SIMCA-P11.5 software (Umetrics, Umeå, Sweden) to do multi-dimensional statistical analysis. After using of orthogonal signal collation algorithm (OSC), centralization (CENTR), standardization (AUTO-SCALE), and principal components analysis (PCA) were conducted to observe the aggregation, discretion, and outliers of samples. Supervised partial least squares-discriminant analysis (PLS-DA) was used to identify major differential variables causing this aggregation and discrete. On the

PLS-DA loading plot, discrete metabolites with the variable importance in projection (VIP) >1 were selected as potential differential metabolites, and the average peak area of these metabolites was verified on single-dimensional statistical level, i.e., one-way ANOVA $P < 0.05$. The one-way ANOVA was done using SPSS 17.0 software. The differential metabolites were identified by comparing the mass spectra with those of standard metabolomics database NIST and GOLM.

13.3 Results and Analysis

13.3.1 The Prolonged Latent Period and Reduced Infection Rate Mediated by Rphq2

At 115 h after inoculation, the first spore appeared on L94 under (20 ± 1) °C. The phenotype observation showed that the rust urediospores development process and the infection rate on L94-*Rphq2* and Vada are lower than that on L94 and Vada-*rphq2*, respectively. Five plants from each of L94, Vada, L94-*Rphq2,* and Vada-*rphq2* were selected for phenotypic observation. The LP_{50} of *Puccinia triticina* 1.2.1 in L94, L94-*Rphq2,* Vada, and Vada-*rphq2* are 151, 163, 193, and 175 h,

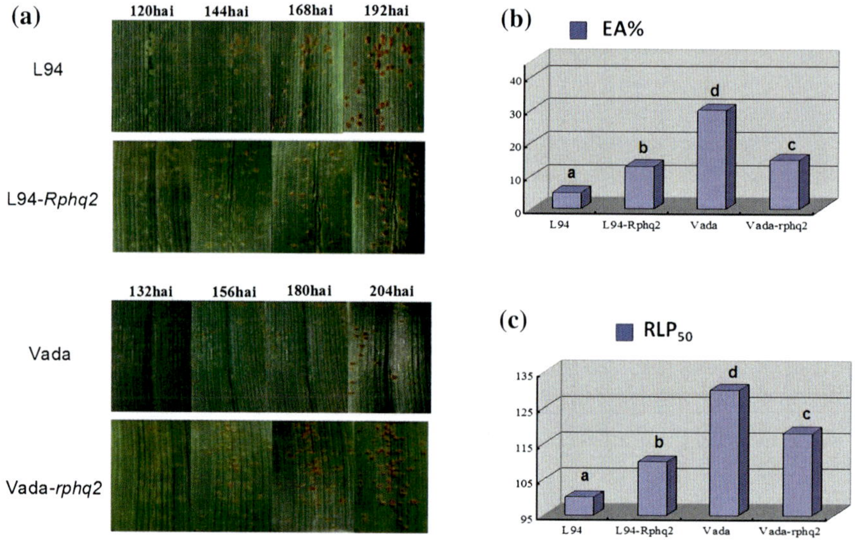

Fig. 13.1 Phenotypic and microscopic observation of L94, L94-*Rphq2*, Vada, and Vada-*rphq2* inoculated with barley *Puccinia triticina* 1.2.1. **a** The development of disease after inoculation (hours after inoculation, hai). **b** The percentage of the early aborted infection colonies (EA %) in L94, L94-*Rphq2*, Vada, and Vada-*rphq2*. The horizontal axis represents the different varieties and the vertical axis represents EA %. **c** The statistical results of RLP_{50}. The horizontal axis represents the different varieties and the vertical axis represents the relative latent periods

respectively (Fig. 13.1a). The LP_{50} of *P. triticina* 1.2.1 in L94-*Rphq2* was 12 h longer than that in L94, while the LP_{50} in Vada-*rphq2* was 18 h shorter than that in Vada. The RLP_{50} of L94, Vada, L94-*Rphq2*, and Vada-*rphq2* were shown in Fig. 13.1c. 60–80 infection colonies were selected from L94, Vada, L94-*Rphq2*, and Vada-*rphq2*, respectively. The percentage of the early aborted infection colonies (EA %) in L94, L94-*Rphq2*, Vada, and Vada-*rphq2* are 4, 11, 28, and 12 %, respectively (Fig. 13.1b).

13.3.2 Changes of Barley Metabolomic Profiles Before and After Inoculation with Pathogens

At 0 h, 12 h, 24 h, 48 h, 96 h and 120 h after inoculation, total metabolites were extracted from the inoculated materials and the mock. The extracted metabolites are divided into two categories, i.e., methanol/water phase (polar metabolites) and chloroform phase (nonpolar metabolites). After analysis, 320 nonpolar metabolites and 240 polar metabolites were obtained. Figure 13.2 shows the PLS-DA analysis of the metabolites of uninoculated and inoculated materials. There is no separation between the control group and the inoculation group at 0 h. The control group and the inoculation group started to clearly separate at 12 h after inoculation, which suggest that the metabolite profiles of inoculation group changes after the induction of pathogens. In the PLS-DA score plot, L94 and L94-*Rphq2* were in different phase suggesting the differences between the metabolites profiles of the materials. On the loading plot of the PLS-DA, discrete points with VIP > 1 were selected as potential differential metabolites. These potential differential metabolites were further confirmed using ANOVA ($P < 0.05$). The results showed that differential substances between L94-CK and L94-*Rphq2*-CK included thymol, ribofuranose, D-xylopyranose, D-glucopyranose, and unknown compound NA11 (non-identified analytes) (Table 13.1 lists compounds with significant differences). Mass spectra alignment showed that NA11 belonged to carbohydrate. Compared with the control group, several kinds of metabolite were significantly elevated in the inoculated group. These metabolites included energy-related carbohydrates, such as glucose and fructose, cell wall components, such as mannose and xylose, fatty acids, such as palmitic acid. Substances with significantly decreased contents are malic acid, sitosterol, L-arabinose, etc. There are a series of differential substances between L94 and L94-*Rphq2* after pathogen induction. Table 13.2 lists 14 substances with the greatest differences and the dynamic changes of their ratio in L94-*Rphq2* and L94 over time, wherein NA1 and NA2 are two unknown compounds.

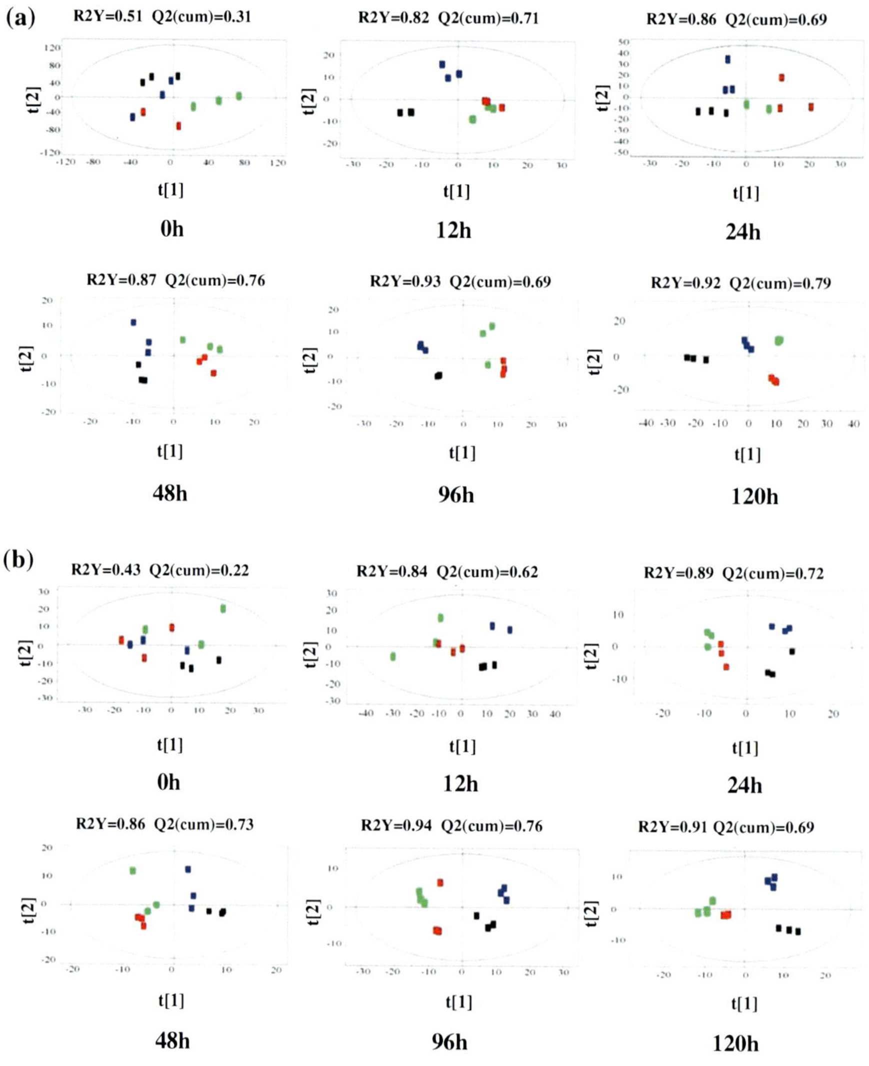

Fig. 13.2 The PLS-DA score plots of L94-CK, L94-*Rphq2*-CK, L94, and L94-*Rphq2* at different time after inoculation. **a** Polar metabolites. **b** Nonpolar metabolites. *Red* L94-CK; *green* L94-*Rphq2-CK*; *black* L94; *blue* L94-*Rphq2*; Q^2 (cum): the cumulative predictive ability of the model; R2Y is the percentage of all response variables explained by mode

13.3.2.1 The Difference of Primary Metabolic Pathways Before and After Inoculation

The identified differential metabolites were positioned into plant primary metabolic pathways, as shown in Fig. 13.3. Differential metabolites between the susceptible cultivar L94-CK and L94-*Rphq2*-CK are sucrose and glutamine. Compared with

Table 13.1 Differential metabolites between L94-CK and L94-*Rphq2*-CK

Retention time (min)	Extraction phase	Putative name	Similarity	VIP value	L 94-*Rphq2*-CK/L94-CK ratio					
					0 h	12 h	24 h	48 h	96 h	120 h
36.5120	Methanol/water	Thymol	837	2.13	1.19①	1.43	1.18①	1.1	1.25	1.45①
26.8283	Methanol/water	D-Xylopyranose	802	2.59	1.61①	1.50	1.27①	1.17	1.23①	1.51①
36.3667	Methanol/water	Sucrose	919	2.59	1.12①	1.41①	1.27①	1.11	1.13	1.44①
12.2383	Methanol/water	Glutamine	925	1.11	0.72	1.46①	0.88	1.16①	1.39①	1.44
43.8100	Methanol/water	NA11	–	1.65	0.81	0.76①	0.82①	1.15①	1.08	0.70

① The significance of one-way ANOVA, $P < 0.05$

Table 13.2 Differential metabolites between L94 and L94-*Rphq2*

Retention time (min)	Extraction phase	Putative name	Similarity	VIP value	L 94-*Rphq2*/L94 ratio					
					0 h	12 h	24 h	48 h	96 h	120 h
20.6383	Methanol/ water	Glucose	917	1.3	1.11	0.85①	0.84	1.33	1.45	3.04②
14.6717	Chloroform	Hexadecanoic acid	914	1.6	0.97	0.73①	0.65①	1.21	1.37①	0.91
10.2367	Methanol/ water	Proline	908	2.1	1.05	1.45②	0.76	0.67①	0.81①	0.45②
16.9167	Chloroform	9,12-octadecadienoic acid	906	1.04	1.03	0.76①	0.84①	1.43	1.51	0.72①
11.3600	Methanol/ water	Cysteine	883	1.98	1.08①	1.37①	1.14	0.68①	0.83	1.11
13.2450	Methanol/ water	Glutamine	877	2.05	1.01	1.36①	0.86	0.78①	0.81①	0.49②
15.7970	Methanol/ water	1H-Indole-3-acetamide	874	1.76	1.09	1.04	1.25①	1.79①	1.19①	2.11②
12.9400	Methanol/ water	Phenyalanine	864	1.65	1.06	1.29①	0.81①	0.71①	0.84①	0.48②
21.3717	Methanol/ water	Galactose	861	2.49	1.10	1.17	1.23	1.19①	0.46②	0.72①
42.5883	Methanol/ water	Xylose	814	1.99	1.11①	1.25①	1.26①	1.11	1.31①	1.37①
5.0750	Chloroform	NA1	–	2.09	1.09	5.48①	4.58①	0.90	1.11	1.02
17.1117	Chloroform	NA2	–	1.45	1.06	0.77①	0.61②	1.13	1.36①	1.54①

① The significance of one-way ANOVA, $P < 0.05$; ② $P < 0.01$

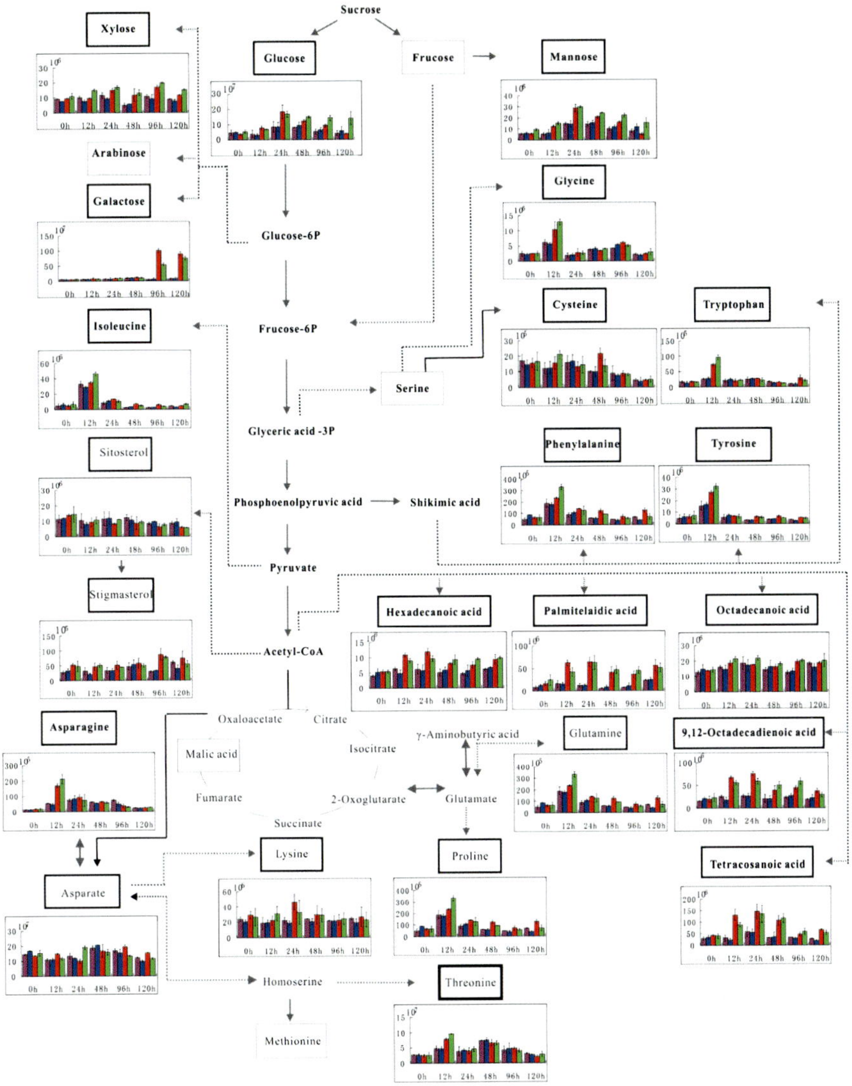

Fig. 13.3 The difference of metabolic pathways in the control group and the inoculation group at different time points after inoculation. In the histogram, the horizontal axis represents different time points after inoculation; the vertical axis represents the chromatography peak intensity of analyzed substance. Substances that are not marked in box represent they are not identified in this study. The *thin line* represents the identified substances without significant change. The *thick lines* represent the identified substances with significant change. Under substances with significant change, the bar graph was used to represent the dynamic change of peak intensities in the control group and the inoculation group at different time points after inoculation. *Purple* L94-CK; *blue* L94-*Rphq2*-CK; *red* L94; *green* L94-*Rphq2*

the control group, there are significant changes in the inoculation group in energy metabolism, amino acid synthesis, fatty acid synthesis, and secondary metabolite synthesis pathway. Among them, the greatest change is in the amino acids pathway. At 12, 24, 48, 96, and 120 h after inoculation, the amino acid content of glutamine branch and aromatic amino acid family in L94 and L94-*Rphq2* is higher than those in L94-CK and L94-*Rphq2*-CK. The content of aspartic acid branch and serine branch in inoculation group also briefly increased (12 h after inoculation). Glucose is the key substance of glycolysis pathway, an important energy producing pathway. The contents of glucose in the inoculation group are significantly higher than that in the control group. Similarly, compared with the control group, the content of xylose (which is derived from glucose) and mannose (which is derived from fructose) is also significantly elevated. Saturated fatty acids and unsaturated fatty acids synthesis pathways are significantly activated by pathogens. The level of hexadecanoic acid, palmitoleic acid, tetracosanoic acid, octadecanoic acid, and 9, 12-octadecadienoic acid is significantly increased compared with the control group. Interestingly, the level of sitosterol and stigmasterol was significantly changed after pathogen induction. Compared with the control group, sitosterol content decreased while stigmasterol content increased in the inoculation group.

After infection, the difference between L94-*Rphq2* and L94 mainly included sugars, saturated fatty acids, and unsaturated fatty acids. The content of glucose, xylose, and mannose in the inoculation group is significantly higher than that in control group, but the variation trend of the former and the latter in L94-*Rphq2* and L94 is different. At 12 and 24 h after inoculation, glucose content in L94 is significantly higher than that in L94-*Rphq2*. From 48–120 h after inoculation, glucose content in L94 decreased to lower than that in L94-*Rphq2* (Fig. 13.3), while xylose and mannose contents in L94 are always lower than that in L94-*Rphq2*. The variation trend of saturated fatty acids, such as hexadecanoic acid, octadecanoic acid, and tetracosanoic acid, and unsaturated fatty acids, such as palmitoleic acid and 9,12-octadecadienoic acid, is similar in inoculated L94 and L94-*Rphq2*, i.e., in the early period of inoculation (12 and 24 h), the content in L94 is significantly higher than that in L94-*Rphq2*, while in the later period (96 and 120 h), the content in L94 is significantly lower than that in L94-*Rphq2* (Fig. 13.3).

13.4 Discussion

The study based on gas chromatography–mass spectrometry investigated the metabolites profiles change of the susceptible cultivar L94, and NIL L94-*Rphq2* at different time points after inoculation. With the use of combination statistical methods of multi-dimensional and one-dimensional, the dynamic changes of differential substances in the control group of L94-CK and L94-*Rphq2-CK* and in the inoculation group of L94 and L94-*Rphq2* were obtained. The mechanism of quantitative disease resistance to barley leaf rust was tentatively discussed from the perspective of metabolomics.

Plant disease resistance is multi-level and multi-aspect. Even for a particular pathogen, the plant will also mobilize all factors that can be used to prevent the occurrence of infection. In this paper, differential substances between L94-CK and L94-*Rphq*-CK were identified. Thymol is a kind of monoterpene phenol, which can destroy the permeability of the pathogenic microorganism's biological membrane, causing the overflow of cell contents, further eliciting internal environmental imbalance (Pramila and Dubey 2004). D-xylopyranose is an important component of the hemicellulose in the plant cell wall. The difference in these substances indicates that there may exist differences in the cell wall components between L94 and L94-*Rphq2*. This difference may affect the penetration through into host cell wall by leaf rust penetration peg, thereby affecting the formation of haustorium in host cells.

After being invaded by pathogens, the plant will initiate a series reaction of recognizing alien and subsequent disease resistance reactions. Plant's disease resistance strategies include physical changes, such as cell wall thickening, callose accumulation, cork layer formation, and xylem vessel fibrosis and biochemical changes, such as generating active oxygen species, producing small metabolites for signal transduction such as salicylic acid (SA), jasmonic acid (JA), abscisic acid (ABA), and ethylene, as well as producing resistance-related proteins and metabolites (Jones and Dangl 2006; Van Loon and Van Strien 1999). All the above reactions need the support of large amounts of energy, thus plant energy metabolism will be significantly enhanced after being infected by pathogen.

In this study, glucose content in the inoculation group L94 and L94-*Rphq2* is significantly increased after pathogen induction, which may be associated with the acceleration of energy metabolism. Secondly, glucose in the plant is the main source of the pathogen nutrient. After absorbing glucose, pathogens first convert it into mannitol and then utilize it (Solomon et al. 2003). The increase of glucose is often observed in the early period of affinity interaction of plant–pathogen (Solomon et al. 2003; of Parker et al. 2009). The haustoria were formed at 12 and 24 h after inoculation. The pathogens start to absorb nutrients from the host once the haustoria formed. The plant will rapidly accumulate glucose and fructose to meet the demand of its own and the pathogen, so glucose content in L94 is significantly higher than that in the L94-*Rphq2*. While in the later period, the decrease of glucose in L94 may be because a part of glucose was absorbed by pathogen and converted into mannitol. Xylose is the main component of the hemicellulose in the plant cell wall. After inoculation, xylose content in L94-*Rphq2* is always significantly higher than that in L94, which may be associated with the papillae formation and the prevention of haustoria formation during the infection in L94-*Rphq2*.

The synthesis of fatty acids starts from acetyl CoA. The first product is a hexadeca-saturated fatty acid, namely hexadecanoic acid, which converts to malonyl-CoA as a two carbon unit to generate octadecanoic acid and longer fatty acids (Kachroo and Kachroo 2006). Octadecanoic acid not only can produce linolenic acid which is the precursor of JA, but also precursors of many secondary metabolites with antibacterial activity (Blerchert et al. 1995). Studies on the legumes melanose (*Diaporthe phaseolorum*) and gray mold (*Botrytis cinerea*) showed that resistant plants accumulate more octadecanoic acids and linolenic acids (Xue et al. 2006; Ongena et al. 2004) than susceptible plants. Figure 13.3 shows that fatty acids levels are all increased significantly in L94 and L94-*Rphq2* after inoculation. From 24 h after inoculation, contents of hexadecanoic acid, palmitoleic acid, octadecanoic acid, and 9, 12-octadecadienoic acid in L94-*Rphq2* are significantly higher than that in L94, indicating that the octadecanoic acid pathway in L94-*Rphq2* is more active, which may be associated with secondary metabolites accumulation and JA pathway activation.

Amino acids levels in L94 and L94-*Rphq2* instantaneously increased at 12 h after inoculation. Amino acids are important for the production of structural proteins (such as microtubule protein and microfilament protein) and pathogenesis-related (PR) proteins. Meanwhile, amino acids are also the major source of nitrogen nutrition for pathogens (Solomon et al. 2003). Within 24 h after the inoculation of potato late blight fungus (*Phytophthora infestans*), amino acids levels in disease-resistant and susceptible varieties are all significantly raised (Abu-Nada et al. 2007). Phenylalanine is an aromatic amino acid. It is the precursor of disease resistance signaling molecule SA. Meanwhile, phenylalanine also involves in the phenylpropanoid metabolic pathways to produce intermediates such as *trans*-cinnamic acid, coumalic acid, ferulic acid, and sinapic acid, which can be converted into coumarin and chlorogenic acid, or converted into CoA esters and further converted into secondary metabolites with antibacterial activities such as lignin, flavonoids, isoflavones, and alkaloids (Kaminaga et al. 2006). Glutamic acid can recirculate ammonium ions from the phenylpropanoid metabolic pathway, which guaranteed the rapid and effective operation of the phenylpropanoid metabolic pathway. High concentration of glutamic acid is a strong evidence of the activation of phenylpropanoid pathway (Buchanan et al. 2000). In this study, contents of phenylalanine and glutamate in L94 and L94-*Rphq2* in early period (12 h) after inoculation are all significantly higher than that in the control group of L94-CK and L94-*Rphq2-CK*, which indicate that the phenylpropanoid metabolic pathway is instantaneously activated in the early period of inoculation.

SA, JA, and ethylene pathways are the main defense signaling pathways in plants. Many studies proved that different defense signaling pathways interplay with each other. They either inhibit each other or promote each other (Kachroo and Kachroo 2006; Thomma et al. 2001; Flors et al. 2008). In addition, studies on the interaction of *Arabidopsis* and *Alternaria brassicicola* found that, in the early period of pathogen invasion, SA pathway was induced, but did not prevent the extension of the pathogen, then depended on the JA pathway to produce resistance. This phenomenon is known as "Shoot First-Ask Questions Later" strategy of host plant (Thomma et al. 1998). In this paper, the results showed that phenylpropanoid metabolic pathway was active in L94-*Rphq2* in the early period after inoculation (12 h), and octadecanoic acid pathway was active in the later period of inoculation (48–120 h). This phenomenon may imply that *Rphq2*-mediated quantitative resistance may be associated with SA pathway and JA pathway. However, this type of association needs further validation.

After inoculation, sitosterol content was significantly decreased while stigmasterol content was significantly increased in L94. Stigmasterol is produced from sitosterol catalyzed by cytochrome P450 CYP710A1 and subsequently by C22 desaturation and is associated with plant susceptibility. Studies on the interaction of *Arabidopsis* and *Pseudomonas syringae* found that once *Arabidopsis* recognized the pathogen-associated molecular patterns (PAMPs) of pathogen, stigmasterol synthesis was activated. The stigmasterol content is higher in susceptible plants. The resistance of a mutant of stigmasterol synthesis deletion *cyp710A1* to *P. syringae* is enhanced (Griebel and Zeier 2010). The specific role of stigmasterol in plant–pathogen interaction is not fully understood, but studies have shown that stigmasterol is not dependent on SA, JA, or ethylene pathway (Griebel and Zeier 2010). In this study, L94-*Rphq2* is quantitatively resistant to leaf rust, which also has a part of susceptible characteristic; therefore, stigmasterol content in L94-*Rphq2* is a little higher than that in L94-*Rphq2*-CK, but the increase of content is significantly lower than that in L94.

In the present study, by retrieving opening Mass database NIST and Golm, the majority of the metabolites were identified. However, a portion of resistance-related metabolites has not been identified. Further, appropriate means of separation and analysis are need to identify these unknown metabolites. In addition, there are large differences in the objects of metabolomics analysis, such as size, quantity, functional group, volatility, charge, electric mobility, polarity, and other physical and chemical parameters. Unbiased comprehensive analysis should be adopted because a single means of separation and analysis are difficult to guarantee. This needs to combine with gas chromatography–mass spectrometry (GC/MS), liquid chromatography–mass spectrometry (LC/MS), nuclear magnetic resonance (NMR), and fourier transform infrared spectroscopy mass spectrometry (FTIR/MS), etc., to obtain more complete information of metabolites.

References

Abu-Nada Y, Kushalappa EA, Marshall W, Al-Mughrabi K, Murphy A. Temporal dynamics of pathogenesis-related metabolites and their plausible pathways of induction in potato leaves following inoculation with *Phytophthora infestans*. Eur J Plant Pathol. 2007;118:375–91.

Allwood JW, Ellis DI, Heald JK, Goodacre R, Mur LAJ. Metabolomic approaches reveal that phosphatidic and phosphatidyl glycerol phospholipids are major discriminatory non-polar metabolites in responses by *Brachypodium distachyon* to challenge by *Magnaporthe grisea*. Plant J. 2006;46:351–68.

Blerchert S, Brodchelm W, Holder S, Kammerer L, Kutchan TM, Mueller MJ, Xia ZQ, Zenk MH. The octadecanoid pathway: signal molecules for the regulation of secondary pathways. Proc Natl Acad Sci USA. 1995;92:4099–105.

Bolton MD. Primary metabolism and plant defense—fuel for the fire. Mol Plant-Microbe Interact. 2009;22:487–97.

Buchanan M, Starrs L, Egelhaaf SU, Cates ME. Kinetic pathways of multiphase surfactant systems. Phys Rev E. 2000;62:6895–905.

Divon HH, Fluhr R. Nutrition acquisition strategies during fungal infection of plants. FEMS Microbiol Lett. 2007;266:65–74.

Flor H. Current status of the gene-for-gene concept. Annu Rev Phytopathol. 1971;9:275–96.

Flors V, Ton J, Doorn R. Interplay between JA, SA and ABA signalling during basal and induced resistance against *Pseudomonas syringae* and *Alternaria brassicicola*. Plant J. 2008;54:81–92.

Fu D, Uauy C, Distelfeld A, Blechl A, Epstein L, Chen X, Sela H, Fahima T, Dubcovsky J. A kinase-START gene confers temperature-dependent resistance to wheat stripe rust. Science. 2009;323:1357–60.

Fukuoka S, Saka N, Koga H, Ono K, Shimizu T, Ebana K, Hayashi N, Takahashi A, Hirochika H, Okuno K, Yano M. Loss of function of a proline-containing protein confers durable disease resistance in rice. Science. 2009;325:998–1001.

Griebel T, Zeier J. A role for ß-sitosterol to stigmasterol conversion in plant-pathogen interactions. Plant J. 2010;63:254–68.

Hamzehzarghani H, Kushalappa AC, Dion Y, Rioux S, Comeau A, Yaylayan V, Marshall WD, Mather DE. Metabolic profiling and factor analysis to discriminate quantitative resistance in wheat cultivars against fusarium head blight. Physiol Mol Plant Pathol. 2005;66:119–113.

Hofmann J, El Ashry A, Anwar S, Erban A, Kopka J, Grundler F. Metabolic profiling reveals local and systemic responses of host plants to nematode parasitism. Plant J. 2010;62:1058–71.

Heath MC. Light and electron microscope studies of the interactions of host and non-host plants with cowpea rust *Uromyces phaseoli* var. vignae. Physiol Plant Pathol. 1974;4:403–14.

Jacobs TH. The occurrence of cell wall appositions in flag leaves of spring wheats, susceptible and partially resistant to wheat leaf rust. J Phytopathol. 1989a;127:239–49.

Jacobs TH. Haustorium formation and cell wall appositions in susceptible and partially resistant wheat and barley seedlings infected with wheat leaf rust. J Phytopathol. 1989b;127:250–61.

Jones JDG, Dangl JL. The plant immune system. Nature. 2006;444:323–9.

Kachroo A, Kachroo P. Salicylic acid-, jasmonic acid- and ethylene-mediated regulation of plant defense signaling. In: Setlow J, ed. Genetic regulation of plant defense mechanisms. New York: Springer. vol. 28; 2006. pp. 55–83.

Kaminaga Y, Schnepp J, Peel G, Kish CM, Ben-Nissan G, Weiss D, Orlova I, Lavie O, Rhodes D, Wood K, Porterfield DM, Cooper AJ, Schloss JV, Pichersky E, Vainstein A, Dudareva N. Plant phenylacetaldehyde synthase is a bifunctional homotetrameric enzyme that catalyzes phenylalanine decarboxylation and oxidation. J Biol Chem. 2006;281:23357–66.

Kiyosawa S. Genetic and epidemiological modeling of breakdown of plant disease resistance. Annu Rev Phytopathol. 1982;20:93–117.

Krattinger SG, Lagudah ES, Spielmeyer W, Singh RP, Huerta-Espino J, McFadden H, Bossolini E, Selter LL, Keller B. A putative ABC transporter confers durable resistance to multiple fungal pathogens in wheat. Science. 2009;323:1360–3.

Leonards-Schippers C, Gieffers W, Schafer-Pregl R, Ritter E, Knapp SJ, Salamini F, Gebhardt C. Quantitative resistance to *Phytophthora infestans* in potato: a case study for QTL mapping in an allogamous plant species. Genetics. 1994;137:68–77.

Manosalva PM, Davidson RM, Liu B, Zhu X, Hulbert SH, Leung H, Leach JE. A germin-like protein gene family functions as a complex quantitative trait locus conferring broad-spectrum disease resistance in rice. Plant Physiol. 2009;149:286–96.

Marcel TC, Varshney RK, Barbieri M, Jafary H, de Kock MJD, Graner A, Niks RE. A high density consensus map of barley to compare the distribution of QTLs for partial resistance to *Puccinia hordei* and of defence gene homologues. Theor Appl Genet. 2007;114:487–500.

Niks RE. Early abortion of colonies of leaf rust, *Puccinia hordei*, in partially resistant barley seedlings. Can J Bot. 1982;60:714–23.

Niks RE. Failure of haustorial development as a factor in slow growth and development of *Puccinia hordei* in partially resistant barley seedlings. Physiol Mol Plant Pathol. 1986;28:309–22.

O'Connell RJ, Panstruga R. Tête à tête inside a plant cell: establishing compatibility between plants and biotrophic fungi and oomycetes. New Phytol. 2006;171:699–718.

Ongena M, Duby F, Rossignol F, Fauconnier ML, Dommes J, Thonart P. Stimulation of the lipoxygenase pathway is associated with systemic resistance induced in bean by a nonpathogenic *Pseudomonas strain*. Mol Plant-Microbe Interact. 2004;17:1009–10081.

Ou SH, Nuque FL, Bandong JM. Relation between qualitative and quantitative resistance to rice blast. Phytopathology. 1975;65:1315–6.

Parker D, Beckmann M, Zubair H. Metabolomic analysis reveals a common pattern of metabolic re-programming during invasion of three host plant species by *Magnaporthe grisea*. Plant J. 2009;59:723–37.

Parlevliet JE. Components of resistance that reduce the rate of epidemic development. Annu Rev Phytopathol. 1979;17:203–22.

Pramila T, Dubey NK. Exploitation of natural products as an alternative strategy to control post harvest fungal rotting of fruit and vegetables. Postharvest Biol Technol. 2004;32:235–45.

Qi X, Niks RE, Stam P, Lindhout P. Identification of QTLs for partial resistance to leaf rust (*Puccinia hordei*) in barley. Theor Appl Genet. 1998;96:1205–15.

Ribeiro do Vale FX, Parlevliet JE, Zambolim L. Concepts in plant disease resistance. Fitopatologia Bras. 2001;26:577–589.

Solomon PS, Tan K-C, Oliver RP. The nutrient supply of pathogenic fungi; a fertile field for study. Mol Plant Pathol. 2003;4:203–10.

Swarbrick P, Schulze-Lefert P, Scholes J. Metabolic consequences of susceptibility and resistance (race-specific and broad spectrum) in barley leaves challenged with powdery mildew. Plant Cell Environ. 2006;29:1061–76.

Thomma BP, Penninckx IA, Broekaert WF, Cammue BP. The complexity of disease signaling in *Arabidopsis*. Curr Opin Immunol. 2001;13:63–8.

Thomma B, Eggermont K, Penninckx I. Separate jasmonate-dependent and salicylate-dependent defense-response pathways in *Arabidopsis* are essential for resistance to distinct microbial pathogens. Proc Natl Acad Sci USA. 1998;95:15107–11.

Van Berloo R, Aalbers H, Werkman A, Niks RE. Resistance QTL confirmed through development of QTL-NILs for barley leaf rust resistance. Mol Breed. 2001;8:187–95.

Van Loon LC, Van Strien EA. The families of pathogenesis-related proteins, their activities, and comparative analysis of PR-1 type proteins. Physiol Mol Plant Pathol. 1999;55:85–97.

Xue HQ, Upchurch RG, Kwanyuen P. Ergosterol as a quantifiable biomass marker for *Diaporthephaseolorum* and *Cercospora kikuchii*. Plant Dis. 2006;90:1395–8.